Pulse Width Modulation for Control and Automation

Pulse Width Modulation for Control and Automation

Ashok Kumar Loganathan
Albert Alexander Stonier

WILEY

Published by John Wiley & Sons, Inc., Hoboken, New Jersey.

Library of Congress Cataloging-in-Publication Data applied for:

Hardback ISBN: 9781394177332

Cover Design: Wiley
Cover Image: © Pavel Ignatov/Shutterstock

Set in 9.5/12.5pt STIXTwoText by Straive, Chennai, India

Contents

About the Authors

Dr. Ashok Kumar Loganathan was a Postdoctoral Research Fellow from San Diego State University, California. He is a recipient of the BHAVAN fellowship from the Indo-US Science and Technology Forum and the SYST Fellowship from DST, Govt. of India. His current research focuses on the integration of Renewable Energy Systems in the Smart Grid and Wearable Electronics. He has 3 years of industrial experience and 19 years of academic and research experience. He has published 167 technical papers in International and National journals and presented 157 papers in National and International Conferences. He has completed 26 Government of India-funded projects, and currently, 7 projects are in progress. His PhD work on wearable electronics earned him a National Award from ISTE, and he has received 24 awards at the national level. Ashok Kumar has seven patents to his credit. He has guided 127 graduate and postgraduate projects, and he has also produced 4 PhD scholars, and 12 scholars are pursuing their PhD work. He is a member and holds prestigious positions in various national forums. He has visited many countries for institute-industry collaboration and as a keynote speaker. He has been an invited speaker in 178 programs. Also, he has organized 72 events, including conferences, workshops, and seminars. He completed his graduate program in Electrical and Electronics Engineering from the University of Madras and his postgraduate studies at PSG College of Technology, India, and a Master's in Business Administration from IGNOU, New Delhi. After completion of his graduate degree, he joined as a project engineer for Serval Paper Boards Ltd., Coimbatore (now ITC Unit, Kovai). Presently, he is working as a Principal at Thiagarajar College of Engineering, Madurai, and is also conducting research work in Wearable Electronics, Smart Grid, solar PV, and Wind Energy Systems. He is also a Certified Chartered Engineer and BSI-Certified ISO 500001 2008 Lead Auditor. He has authored the following books in his areas of interest: (i) *Computational Intelligence Paradigms for Optimization Problems Using MATLAB®/SIMULINK®*, CRC Press, (ii) *Solar PV and Wind*

Energy Conversion Systems—An Introduction to Theory, Modeling with MATLAB/SIMULINK, and the Role of Soft Computing Techniques—Green Energy and Technology, Springer, USA, (iii) *Electronics in Textiles and Clothing: Design, Products and Applications,* CRC Press, (iv) *Power Electronics with MATLAB,* Cambridge University Press, London, (v) *Automation in Textile Machinery: Instrumentation and Control System Design Principles*—CRC Press, Taylor & Francis Group, USA, ISBN 9781498781930, April 2018, (vi) *Proceedings of International Conference on Artificial Intelligence, Smart Grid and Smart City Applications*, Springer International Publishing, Springer, (vii) *Deep Learning Using Python, Wiley India Publications, India,* (viii) *Monograph on Smart Textiles,* (ix) *Monograph on Information Technology for Textiles, and* (x) *Monograph on Instrumentation & Textile Control Engineering*. He is a Senior Member in IEEE and a Fellow Member in IE (India), IETE, and IET (United Kingdom).

Dr. Albert Alexander Stonier is the recipient of the prestigious Raman Research Fellowship from the University Grants Commission (Government of India). He is a Postdoctoral Research Fellow from Northeastern University, Boston, Massachusetts, United States. His current research focuses on fault diagnostic systems for solar energy conversion systems and smart grids. He has 19 years of academic and research experience. He has published 129 technical papers in International and National journals (including IEEE Transactions, IET, Elsevier, Taylor & Francis, Springer, and Wiley) and presented 96 papers at national and international conferences.

He has completed 10 Government of India-funded projects, and 3 projects are ongoing with the overall grant amount of Rs. 5.03 crore. His PhD work on power quality earned him a National Award from ISTE, and he has received 30 awards for his meritorious academic and research career (such as Young Engineers Award from IE(I), Young Scientist Award from SPRERI, Gujarat, and Best researcher award from IEEE Madras section). He has also received the "National Teaching Innovator Award" from MHRD (Government of India). He is an approved "Margadarshak" from AICTE (Government of India). He is the approved "Mentor for Change" under Atal Innovation Mission. He is an AICTE-approved translator under the National Education Policy to translate engineering books into regional languages. He has guided 32 graduate and postgraduate projects. Six research scholars completed their PhDs, and five students are currently pursuing their PhDs under his guidance. He is a member in prestigious positions in various national and international forums (such as Senior Member, IEEE, and Vice President for Energy Conservation Society, India). He is the Chairman of the IEEE Power Electronics Society, IEEE Madras Section. He has been an invited speaker in 350 programs covering nine Indian states and also in the United States. He has organized 29 events, including faculty development programs, conferences, workshops, and seminars.

He completed his graduate program in Electrical and Electronics Engineering from Bharathiar University and his postgraduate and PhD programs from Anna University, India. Formerly, he worked as a Professor in the Department of Electrical and Electronics Engineering, Kongu Engineering College. Presently, he is working as a Associate Professor Senior in the School of Electrical Engineering, Vellore Institute of Technology, Vellore, and is also conducting research work in Artificial Intelligent-based controllers for power conversion, smart health care systems, smart grids, solar PV, and power quality improvement techniques.

He has authored the following books in his areas of interest: (i) *Development of 6G Networks and Technology,* Wiley Publishers, (ii) *Algorithm and Design Complexity,* CRC Press (Taylor and Francis), (iii) *Basic Electrical Engineering", Khanna Publishers (Under AICTE, Government of India),* (iv) *Design and Simulation of Electrical Machines with MATLAB,* Nova Publishers, (v) *Power Electronic Converters for Electrical Vehicles*, CRC Press, (vi) *Power Electronic Converters for Solar PV Systems*, Elsevier, (vii) *Advanced Deep Learning for Engineers and Scientists*, Springer, (viii) *Computational Paradigm Techniques for Enhancing Electric Power Quality,* CRC Press, (ix) *Basic Electrical, Electronics and Measurement Engineering*, Anuradha Publishers, and (x) *Special Electrical Machines*, Anuradha Publishers.

Preface

The concept of pulse generation is a fundamental aspect in control, automation, and monitoring in all sectors. As the world is moving toward a smart environment, control and automation play pivotal roles in meeting the intended expectations. This book will serve as a reference tool to understand the nuances in real-time implementation of pulse generation intended for the electrical and electronics domain with respect to detailed design in improving the overall performance of the system. The readers can experience the current research focus areas in pulse generation schemes under the categories of review, simulation, analysis, algorithmic procedures, and qualitative assessments.

Ashok Kumar Loganathan
Albert Alexander Stonier

Acknowledgments

The authors are always thankful to the Almighty for their perseverance and achievements. The authors owe their gratitude to Shri K. Hari Thiagarajan, Chairman and Correspondent of Thiagarajar College of Engineering, Madurai, and Dr. G. Viswanathan, Chancellor of Vellore Institute of Technology. The authors also owe their gratitude to all the trust members, Thiagarajar College of Engineering, Madurai, and Dr. V. S. Kanchana Bhaaskaran, Vice Chancellor, and Dr. T. Jayabarathi, Registrar, Vellore Institute of Technology, for their wholehearted cooperation and great encouragement in this successful endeavor.

Dr. Ashok Kumar Loganathan – I would like to take this opportunity to acknowledge those people who helped me in completing this book. I am thankful to all my research scholars and students who are doing their projects and research work with me. But the writing of this book is possible mainly because of the support of my family members, parents, and sisters. Most importantly, I am very grateful to my wife, Y. Uma Maheswari, for her constant support during writing. Without her, all these things would not be possible. I would like to express my special gratitude to my daughter, A. K. Sangamithra, for her smiling face and support; it helped a lot in completing this work. I would like to dedicate this work to her.

Dr. Albert Alexander Stonier – I would like to take this opportunity to acknowledge those people who helped me in completing this book. I am thankful to all my research scholars and students who are doing their projects and research work with me. But the writing of this book is possible mainly because of the support of my family members, parents, and brothers. Most importantly, I am very grateful to my wife, A. Lincy Annet, for her constant support during writing. Without her, all these things would not be possible. I would like to express my special gratitude to my son, A. Albin Emmanuel, for his smiling face and support; it helped a lot in completing this work.

About the Companion Website

This book is accompanied by a companion website:

www.wiley.com/go/loganathan/pulsewidth

To access the website materials, click on the link under Related Resources on the book's page on wiley.com. You can also find this by searching www.wiley.com for the ISBN 9781394177332.

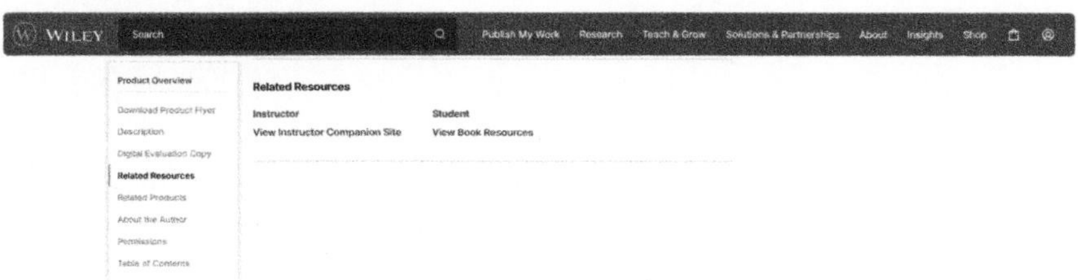

This website includes:

- Objectives
- Summary
- Tutorials
- Simulation exercises

1

Introduction to Pulse Width Modulation

1.1 Introduction and Outlook

The grid system basically comprises sources connecting to the common bus or load. Pulse width modulation (PWM) is a technique used to modulate the power flow from source to load. Effective voltage regulators also use the PWM. The main advantage of PWM lies in its very low power loss within the switching devices. When the switch is off, almost no current passes through it, and when it is on, the voltage drop across it is minimal while transferring power to the load. One technique for regulating the power delivered to a load without wasting energy is PWM. A method for using a microprocessor's digital outputs to drive analog circuitry is called PWM. The output will resemble a voltage at the required level by applying a voltage switch to the load with the proper duty cycle. The inductor and capacitor are typically used to filter the switching noise. This PWM method is also useful for analyzing different performance metrics. It is divided into several categories according to its use. Applications for PWM include power conversion and control as well as communications. Thus, the hardware and software techniques will be utilized to produce the PWM signal.

1.1.1 Definition of Pulse Width Modulation

PWM is an approach used to regulate the average voltage or power of electrical signals, whether AC or DC, by frequently switching them on and off and changing the width (or duration) of the ON period according to the specific application. PWM is a current regulation method that facilitates the control of motor speed, heater output, and several other applications in an energy-efficient and typically quieter manner.

Current applications for PWM encompass, but are not constrained to

- Variable Refrigerant Flow

PWM has transformed the world by greatly reducing the power use of motor-driven appliances, including inverter air conditioners, refrigerators, and washing machines. For example, inverter air conditioners will use less than half the energy of their non-inverter counterparts in some cases. The microprocessors utilize PWM to control digitally analog circuits. PWM has many uses, such as controlling power, converting signals, measuring, and communication.

1.2 Analog Signal

An analog signal is characterized by a continuously fluctuating value with theoretically infinite resolution in both time and amplitude. For example, a 9-volt battery is an analog device: its output voltage is rarely 9 V, it fluctuates over time, and it can take on any real-number value within its operating range. Similarly, the current drawn from the battery is not restricted to a finite set of possibilities.

In contrast, digital signals select values only from a finite set of specified possibilities, such as {0 V and 5 V}. Analog electronics will also be vulnerable to noise. Because analog signals have infinite resolution, every perturbation or noise alters the present value.

Analog voltages and currents can directly control objects, such as the volume of a radio. In a simple analog radio, a volume knob connects to a variable resistor (potentiometer). Turning the knob changes resistance, which in turn alters the current driving the speakers, thereby adjusting the volume. Analog circuits like this typically have outputs that are linearly proportional to their inputs.

Despite its simplicity, analog control isn't always practical or cost-effective.

Drift and tuning: Analog circuits are challenging to tune because their values tend to drift over time.

Size and cost: Achieving precision requires larger, heavier, and more expensive components (e.g., older home stereo equipment).

Heat generation: Analog circuits generate substantial heat. The power dissipated, which is often lost as heat, is proportional to the voltage across the active components multiplied by the current passing through them.

Figure 1.1 illustrates three distinct PWM signals. A PWM output operating at 10% duty cycle is displayed in Figure 1.1a. In other words, the signal is off for 90% of the time and on for 10%. PWM outputs are displayed at 50% and 90% duty cycles in Figure 1.1b and c, respectively. Three distinct analog signal values have been stored by these three PWM outputs at 10%, 50%, and 90% of their maximum power. For instance, a 0.9 V analog signal is produced when the duty cycle is 10%, and the supply is 9 V.

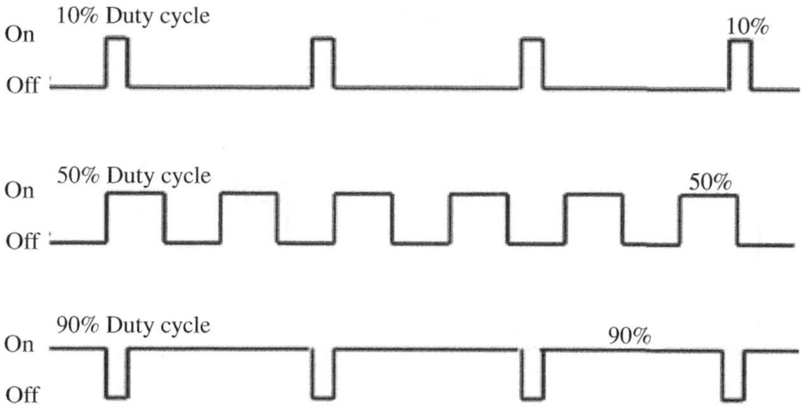

Figure 1.1 (a–c) PWM signals of varying duty cycles.

1.3 Digital Control

PWM is an approach for digitally encoding analog signal levels. By controlling analog circuits with a digital signal, system costs and power consumption are reduced significantly. Implementation is often straightforward because many microcontrollers and DSPs already feature on-chip PWM controllers. PWM utilizes high-resolution counters to modulate the duty cycle of a square wave, which encodes a specific analog value. The resulting PWM signal is fundamentally digital: At any given instant, the DC supply is either fully on or fully off.

The analog load receives power through a repeating series of on and off pulses:

- The on-time is the duration the DC supply is applied to the load.
- The off-time is the duration the supply is switched off.

By varying the ratio of on-time to the total period (the duty cycle), any desired analog value can be effectively encoded, provided the receiving system has sufficient bandwidth.

Figure 1.2 illustrates a circuit that is driven using PWM. Here, a 9 V battery is used.

Most loads, including inductive and capacitive types, require a much higher modulating frequency than 10 Hz. To illustrate, if a lamp were switched on for five seconds and off for five seconds, the duty cycle would still be 50%. However, the bulb would simply appear brightly lit for the first five seconds and then completely off for the next five. For the bulb to perceive an effective voltage of 4.5 V, the 50% equivalent of 9 V, the cycle period must be significantly shorter than the load's response time to the switch state change. In other words, to create the desired

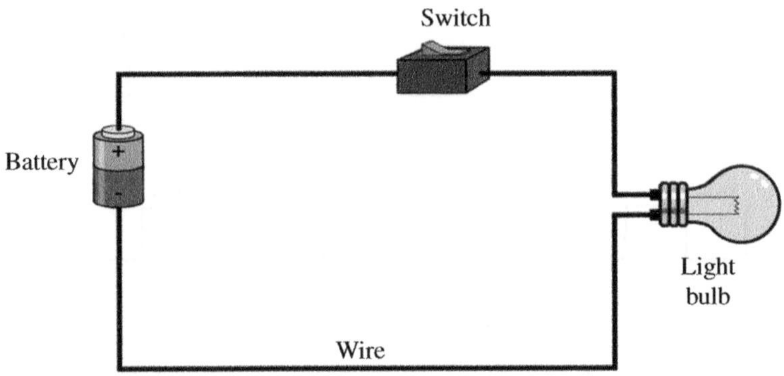

Figure 1.2 A simple circuit.

effect of a lamp that is always lit but dimmed, the modulating frequency must be increased.

This principle applies across all PWM applications. Typical modulating frequencies generally range from 1 to 200 kHz.

1.3.1 Hardware Controllers

For example, Microchip PIC16C67 has two such controllers, each with a customizable on-time and period.

The core relationships for the PWM signal are:

- Duty cycle is calculated by dividing the on-time by the period.
- The modulating frequency is the inverse of the period.

To begin generating the PWM signal, you need to set up the timer and the PWM module:

1) **Timer setup:** Program the timer/counter's period to determine the modulation frequency.
2) **PWM setup:** Define the desired duty cycle by setting the on-time value in the PWM register.
3) **I/O pin configuration:** Ensure the corresponding output pin is configured as a PWM output (setting its direction).
4) **Module activation:** Start the timer and then enable the PWM controller itself.

The basic concept of PWM remains consistent across various controller designs, primarily serving to bridge digital control with analog systems. A major benefit of PWM is maintaining the signal in the digital domain from the processor to the controlled system, eliminating the need for complex and costly digital-to-analog converters. This digital transmission significantly reduces noise susceptibility, as ambient noise must be powerful enough to flip a logic state (0–1 or vice versa). This improved noise immunity makes PWM an excellent choice for widening the communication channel length over what analog signals could reliably achieve. At the receiving end, simple RC or LC filter networks are used to smooth the high-frequency square wave and recover the desired analog equivalent signal.

PWM has diverse practical applications, such as controlling brakes or managing solar battery charging. In a PWM-controlled brake, the amount of clamping pressure (stopping power) is traditionally determined by an analog voltage or current input. By integrating PWM, a digital output is connected to a switch that powers the brake. To increase braking pressure, the software simply raises the PWM duty cycle. Achieving precise, repeatable pressure (e.g., 100 psi) requires the software to use a lookup table or formula, which may include adjustments for temperature or wear to determine the required duty cycle. For maximum accuracy, a system sensor can provide closed-loop feedback, allowing the software to continuously adjust the duty cycle until the exact pressure is reached.

In addition to braking and solar power management, PWM is widely used for controlling the speed of DC motors by effectively simulating a variable voltage between 0 V and the full supply voltage V_{CC}. This is achieved by rapidly switching V_{CC} on and off. Other common uses include fading LEDs to control brightness, generating different buzzer tones, and precisely adjusting the degree of spin in servo motors. Overall, PWM is a highly effective, inexpensive, space-saving, and noise-immune technique for converting digital instructions into proportional analog control (Figure 1.3).

Digital devices, such as microcontrollers, naturally operate with only two states: on and off. This makes them ideal for controlling simple binary functions, like setting an LED's status or toggling an electrical device using appropriate drivers (e.g., transistors, TRIACs, or relays).

However, many applications require proportional control beyond simple on/off switching such as adjusting an LEDs brightness or regulating a DC motor speed. PWM is a highly effective technology used to address this limitation.

A PWM signal is a square wave that remains "ON" (high) for a duration and "OFF" (low) for the remaining time within a cycle.

- **T_{on} (On-time):** The time the output remains high (e.g., 5 V).
- **T_{off} (Off-time):** The time the output remains low (0 V).
- **T (Time period):** The total duration of one cycle: $T = T_{on} + T_{off}$.

By rapidly switching between 0 V and the full supply voltage (e.g., 5 V), PWM creates an average voltage that can be used to control analog devices proportionally.

This method is preferred over practical analog control because it significantly reduces system cost and power use, streamlines design (often with on-chip

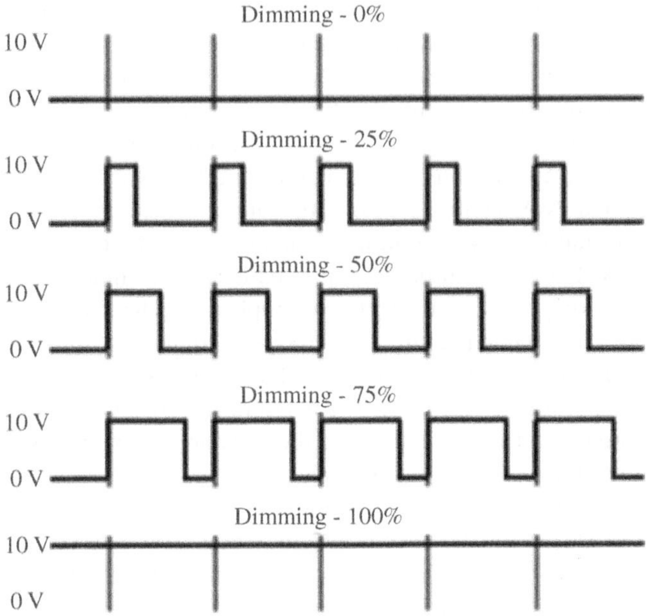

Figure 1.3 PWM application for light dimming.

controllers), and provides inherent noise immunity since the signal stays digital until it reaches the load. PWM is therefore essential for applications that need proportional control such as dimming LEDs, controlling DC motor speed, and managing brake pressure.

1.3.1.1 History of PWM

The military implemented variable (switching) power supplies in large quantities, which piqued the interest of commercial product designers.

In the 1960s, the Sinclair X10, an audio amplifier, was one of the first devices to use PWM.

The foundation for modern PWM control was established in 1976 when Bob Mammano of Silicon General designed the SG1524 integrated circuit. This development prompted other companies, leading to the widespread adoption and advancement of PWM techniques.

> Switching regulators are in the process of revolutionizing the power supply industry because of their low internal losses, small size and weight, and costs competitive with conventional series-pass or linear power supplies.
>
> Abraham Pressman

1.3.2 Characteristics of PWM

The primary purpose of PWM is to manage power delivery, especially to inertial electrical devices (like motors or lamps), by changing the average power of the signal. The output signal rapidly switches between on and off within a specified period.

The two basic factors defining a PWM signal are:

1) **Switching frequency:** The rate at which the signal cycle repeats per second.
2) **Duty cycle:** The ratio of the on-time T_{on} to the total period, expressed as a percentage.

PWM operates like a regulated low-loss switch. When the switch is on, power is delivered to the load with essentially no voltage drop. When the switch is off, no current flows. This pulse-based method is highly efficient because it reduces power losses compared to continuous (analog) methods.

The duty cycle indicates the fraction of time the signal is in the on position. For example, a 50% duty cycle means the signal is on for exactly half of the total period and off for the other half (Figure 1.4).

The PWM regulates the signals duty cycle, which in turn regulates the motor's speed. A duty cycle frequency of 50 Hz indicates that 50 duty cycles are executed every second, indicating the ratios ability to quickly adjust between high and low levels.

1.3.2.1 Duty Cycle

The duty cycle is a measure of the time the modulated signal is in its "high" state. It is generally recorded as the percentage of the signal period during which the

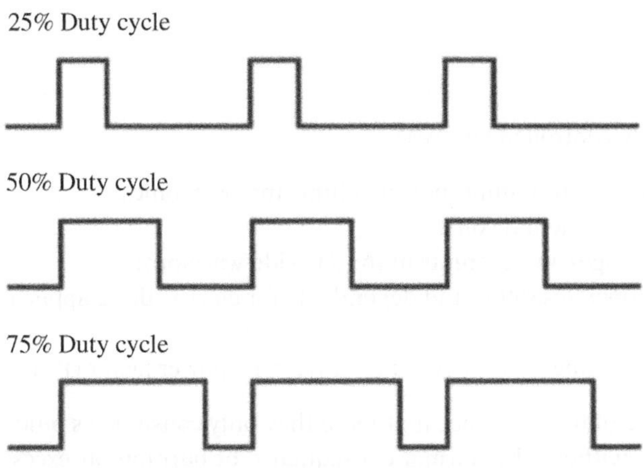

Figure 1.4 Duty cycle in terms of percentage.

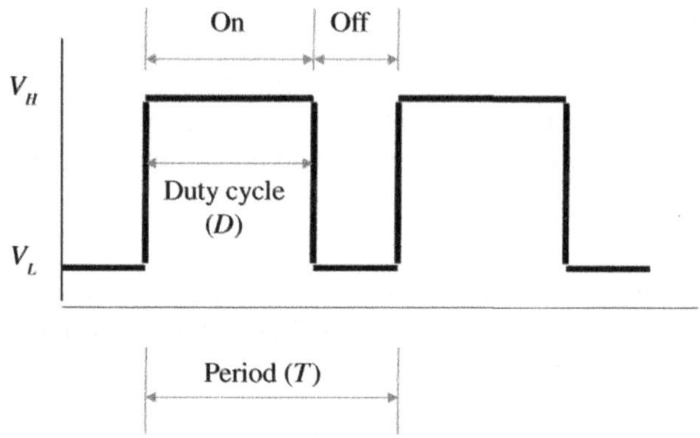

Figure 1.5 Relation between duty cycle and period.

signal is considered on. Duty cycle is determined by:

$$\text{Duty cycle} = \frac{\text{On Time}}{\text{Period}} \times 100\%$$

The average rate of a signal will be found as:

$$\bar{y} = \frac{1}{T}\int_0^T f(t)\text{dt}$$

In the overall analysis, V_L is set to zero volts for easier calculation.

$$V_{\text{avg}} = D \cdot \text{V}_H + (1 - D) \cdot V_L$$

Figure 1.5 shows the relationship between duty cycle and period for a simple digital ON/OFF waveform.

1.3.3 Need for PWM

Devices will be effectively controlled by PWM.

This approach reduces losses from improper switching and harmonics.
It controls the AC-side fundamental voltage.
It mitigates harmonic voltages and currents in the AC-side waveform.
It ensures the converter operates safely and dependably for electric drive applications.
It contributes to the improvement of the AC side converters' power factor (PF).

PWM adjusts the speed of the appliance motors so they only consume as much power as they need but without the normal consequence of burning off excess current as heat.

Lower limit	Upper limit
Motor operation is pulsed (not continuous) as inductance can't maintain current. The inverse of the period ($1/f$) is too long relative to the load's time constant.	The motor's inductance can cause the current drawn to become unstable.
High ripple voltages and currents lead to higher error. Frequency must be at least 10 times higher than the control system's update rate.	Limited by the resolution and maximum clock speed of the PWM controller hardware.
Low efficiency is caused by poor control and high ripple.	MOSFETs generate heat during rapid switching. Heat losses in electromagnetic materials are proportional to frequency squared ($\propto f^2$), leading to significant waste.
Causes audible sound disturbances (due to magnetostriction) if below 20 kHz, which is the range of human hearing.	High-frequency eddy currents are generated in coils, which cause adverse heating.

1.3.4 PWM Working

PWM is digital, which means that it has two states: on and off (which correspond to 1 and 0 in the binary context, which will become more relevant to you if using microcontrollers). The fundamental principle of PWM dimming depends on the relationship between the pulse on-time and the observed brightness. A longer pulse results in a brighter LED. The switching frequency is deliberately set very high, thousands of times per second, ensuring the LED never fully turns off and eliminating flicker. This control mechanism forms the basis of a PWM LED dimmer circuit. For advanced applications, designers can select from various PWM schemes such as square-wave, sinusoidal, regular-sampled, and specialized techniques like harmonic elimination PWM to best meet the desired performance characteristics (Figure 1.6).

PWM techniques are fundamentally used in inverters to eliminate or significantly reduce harmonics at the output while simultaneously providing precise control over the output voltage magnitude. There is significant focus on developing new PWM methods that are easily open to online, microprocessor-based implementation. This approach controls the advantages of microprocessors such as flexibility and precise digital control to achieve superior inverter performance.

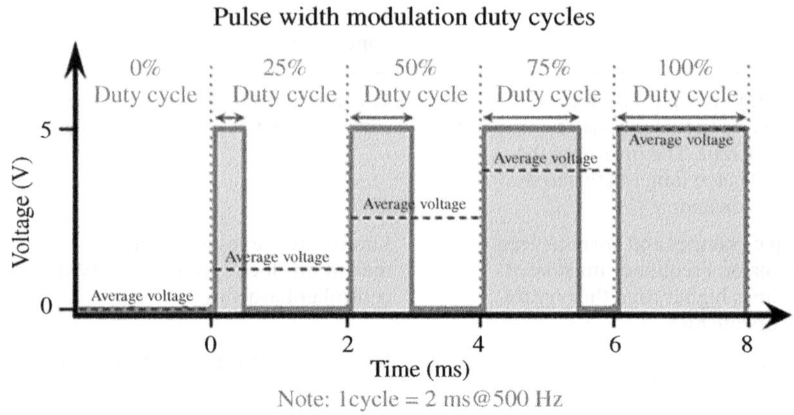

Figure 1.6 PWM duty cycles.

1.3.5 Types of PWM

PWM is extensively used in power electronic applications. The brightness of the LED circuit will also control the speed of the DC motors. Two factors are always associated with it: one is the switching frequency and the second is the duty cycle (Figure 1.7 and Table 1.1).

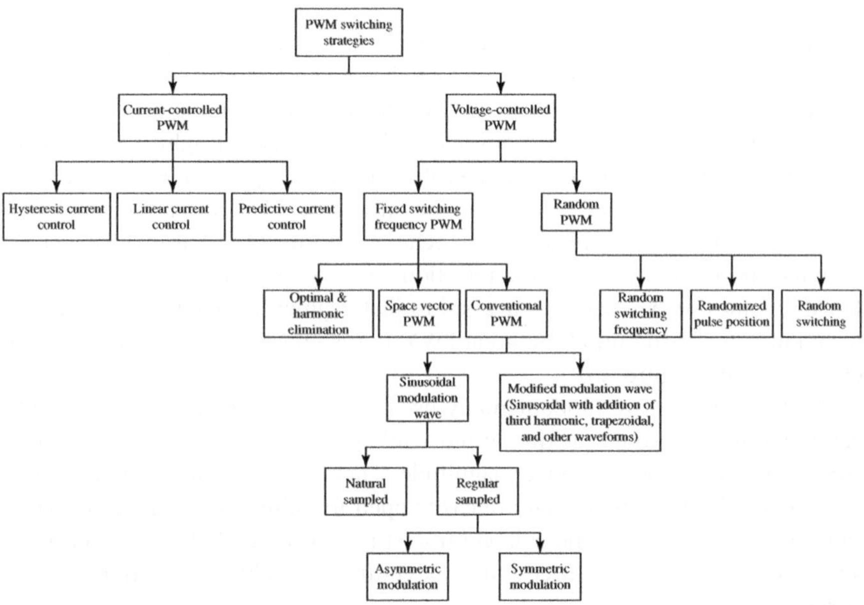

Figure 1.7 PWM switching strategies.

Table 1.1 Comparison of different types of PWM.

Parameter	SWPWM	NSPWM	SRSPWM	ARSPWM	HEPWM	RSHEPWM
Pulse width equation	Linear Algebraic	Transcendental nonlinear	Linear algebraic	Linear algebraic	Nonlinear equations	Linear algebraic
Number of calculations	one multiplication	Offline calculations	One addition and multiplication	Same as SRSPWM	Offline calculations	Same as HEPWM
Dominant harmonics	All odd types	$2p$, $2p \pm 1$, $4p \pm 1$, $p \pm 2$, $p \pm 4$, etc.	Same as NSPWM with increased amplitudes of the suppressed harmonics	Same as SRSPWM with lesser amplitudes	Eliminates q harmonics for $M = q + 1$	Like HEPWM
Remarks	Rarely used	Analog implementation preferred	Online implementation of CBPWM	Superior spectrum to SRSPWM	No defined modulation process	Modulation process for HEPWM

1.3.6 Types of PWM

1) Carrier-based PWM
2) Harmonic elimination PWM
3) Square wave PWM
4) Sinusoidal PWM
   a) Natural sampled PWM
   b) Regular sampled PWM
      i) Symmetric RSPWM
      ii) Asymmetric RSPWM
5) Harmonic elimination PWM
6) Regular sampled harmonic elimination PWM

Any periodic signal, and this periodic signal may have any shape; it does not matter, but it will be expressed as a sum of several sinusoids. Let us define the sinusoids a little more closely; these sinusoids may have certain frequencies. First, it could be 0 frequency, or there will be DC now. Next, it will have what is called the fundamental frequency, which is the same periodicity as the actual waveform, and it will also have frequencies of harmonic contents, which are the integral multiples of the basic frequency.

A periodic signal will be decomposed into a DC component, a fundamental component, and a set of harmonic components, as indicated here now. So, here it is given as (ωt). It has been variable instead of time and taken as an angle at the fundamental frequency. Here $\omega = 2\pi T$, where T is the time period of the waveform. The fundamental frequency or what we will call the fundamental angle. Now, this (ωt) could have a DC component, which is given by α02, and it could have several other frequency components; one of the frequency components could be "ω" itself.

If the frequency 0 is one that is DC, "ω" is another, one, and 2ω, 3ω, 4ω, and so on. This will have all these various frequencies now. So, the Fourier series helps us expand any given (ωt) into such a series. It helps us to compute the coefficients of numerous terms. Once we find the coefficients of various terms, we will know how much harmonic is there. For example, in this term, if we calculate $a0$, we know what the DC value is, which is $a02$, and if we calculate $a1$ and $b1$, we know what the sinusoidal component is. It will be $\sqrt{a12 + b12}$, and the root would be the scale of the primary component here. If you want the nth harmonic component, we must try and get this an and bn.

1.3.7 Modulation Index

The modulation index (MI) in PWM-controlled inverters is the ratio of the desired output voltage's fundamental component amplitude to half the available DC bus

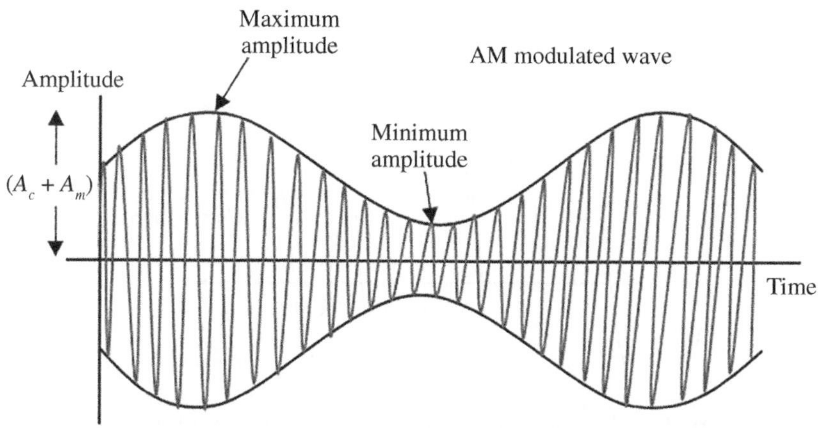

Figure 1.8 AM-modulated wave.

voltage. It effectively indicates how much of the system's total DC voltage is being utilized to generate the required AC output.

1.3.8 Types of Modulation Index

The MI of a modulation scheme describes how much the modulated variable of the carrier signal varies around its unmodulated level. It is defined differently in each modulation scheme.

1.3.8.1 Amplitude Modulation Index

The **MI** in amplitude modulation (AM) quantifies the degree to which the carrier signal's amplitude varies by the modulating signal. It measures the extent of amplitude change in the modulated carrier wave relative to its original, unmodulated amplitude level (Figure 1.8).

1.3.8.2 Frequency Modulation Index

Frequency modulation (FM) encodes information on a carrier wave by instantly varying its frequency. In analog FM (used for radio), the frequency deviation directly relates to the modulating signal's amplitude. For digital data, the approach is called frequency-shift keying (FSK), where discrete frequencies represent binary states (like "0" and "1"), commonly used in modems and low-frequency transmissions. The main benefit of FM is its higher signal-to-noise ratio compared to AM, making it highly resistant to interference, a key reason for its use in music broadcasting. FM and phase modulation (PM) are the two main forms of angle modulation, different from AM, where only the carrier's amplitude changes (Figure 1.9).

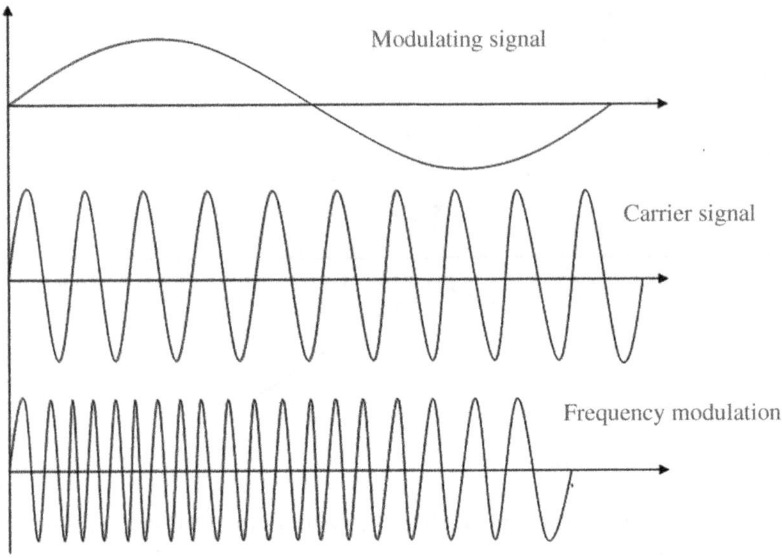

Figure 1.9 FM-modulated wave.

1.3.8.3 Phase Modulation Index

PM is an angle modulation technique with FM used to prepare communication signals for transmission. It works by encoding the message signal into a carrier wave by directly altering the carrier's instantaneous phase in proportion to the instantaneous amplitude of the baseband (message) signal. While the carrier amplitude and central frequency remain constant, these phase shifts transmit the information. PM is widely used in radio transmission and is a core component of many modern digital transmission schemes supporting technologies like Wi-Fi, GSM, and satellite television. Beyond communications, PM principles are also employed in digital synthesizers for sound and waveform generation (Figure 1.10).

1.3.9 Types of Modulation Index

MI is a fundamental metric in PWM used to quantify the relationship between the control signal and the output. It is formally defined as the ratio between the peak value of the modulating waveform and the peak value of the carrier waveform.

This index is essential because it directly establishes the connection between the available inverter's DC-link voltage and the resulting magnitude of the fundamental component of the pole voltage produced by the inverter's output. Higher modulation indices generally mean the inverter is utilizing more of the available DC voltage to generate the AC output.

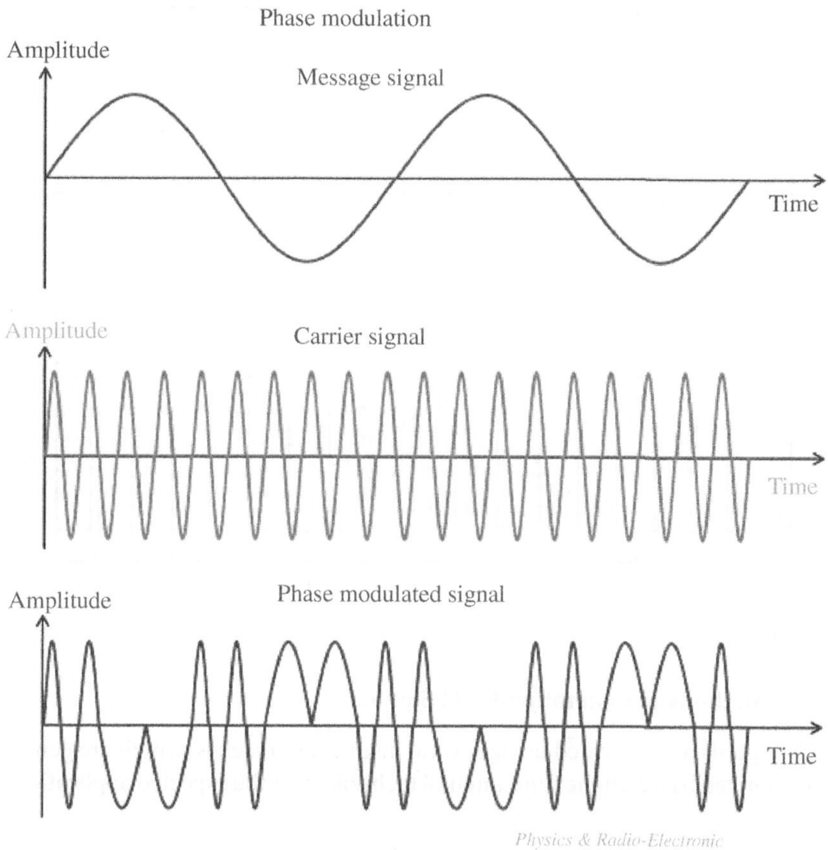

Figure 1.10 PM-modulated wave.

Various PWM techniques are employed in inverters and motor control systems, primarily categorized by how the pulse edges are aligned relative to the carrier signal:

1) **Lead edge modulation:** The leading edge of the output pulse is modulated (shifted in time), while the trailing edge remains fixed.
2) **Trail edge modulation:** The trailing edge of the output pulse is modulated, while the leading edge remains fixed.
3) **Pulse center two edge modulation (Phase Correct PWM):** Both the leading and trailing edges of the pulse are modulated symmetrically around the center of the carrier cycle. This is often preferred as it minimizes harmonic distortion (Figure 1.11).

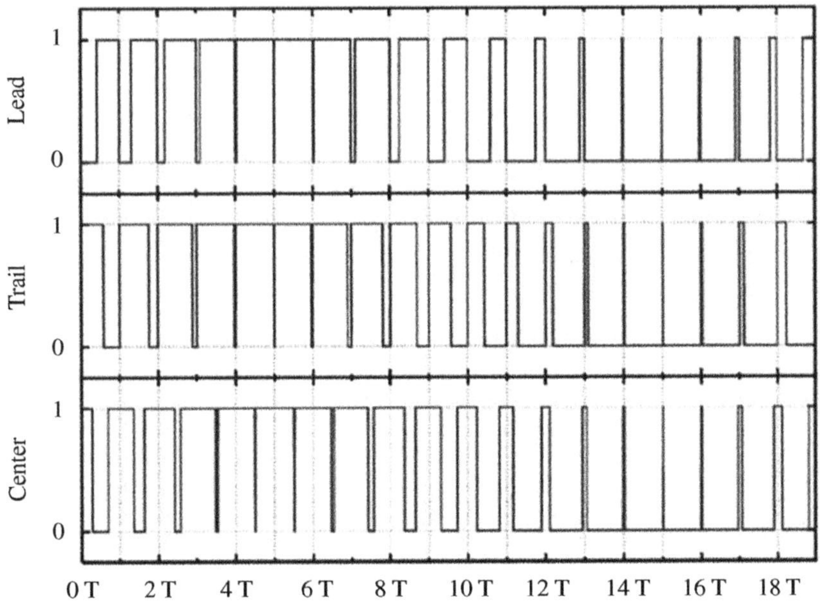

Figure 1.11 Phase correct PWM.

1.3.10 Methods of Generating PWM Signals

A PWM signal can be created using several distinct approaches, which are generally categorized based on the fundamental technology or the expected application:

1.3.11 Analog Generation Methods

The intersective method is an analog technique for generating the PWM signal. It works by comparing a reference sinusoid against a high-frequency triangular or sawtooth trigger signal. The duration of the resulting PWM pulses is entirely determined by where these two signals intersect. Specifically, the PWM output switches to the ON/high state when the reference sinusoid's amplitude is greater than the carrier signal's amplitude. Conversely, the PWM pulse switches to the OFF/low state when the sinusoidal amplitude is less than the carrier (Figure 1.12).

1.3.12 Digital Generation Methods

Digital PWM generation methods such as delta and delta-sigma modulation will depend on a control loop rather than on direct signal comparison. This technique generates the switching pattern using only the analog reference signal, the integrated output of the PWM signal, and a defined constant offset that establishes

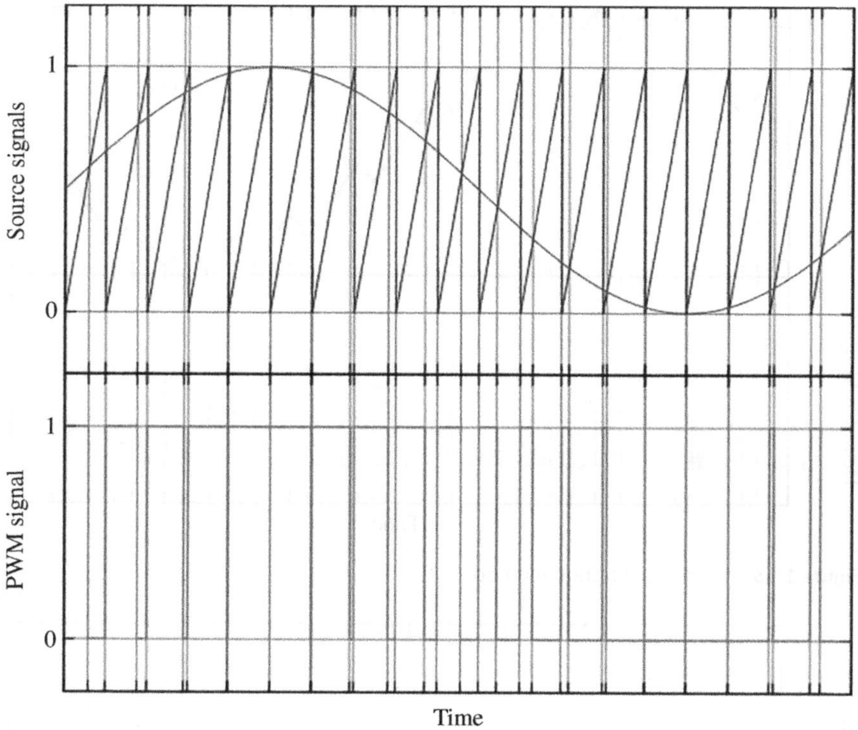

Figure 1.12 Analog selection method.

upper and lower limits around the reference. The method constantly integrates the PWM output and attempts to keep it adjusted toward the limits set around the reference. When the integrated output reaches one of these predefined boundaries, the PWM signal switches its state (from high to low or vice versa), thereby creating the necessary pulsing action (Figure 1.13).

1.3.13 Delta-Sigma Modulation

Like delta modulation, this method involves integration. However, an error signal is generated by subtracting the PWM signal from a reference sinusoid and then integrating it. Whenever this integrated error signal has reached predefined limits, the PWM signal changes its modes (Figure 1.14).

1.3.14 Generation of PWM Signal

A PWM signal is easily generated using a simple comparator circuit. This method takes two inputs: the modulating signal (representing the desired output level) and

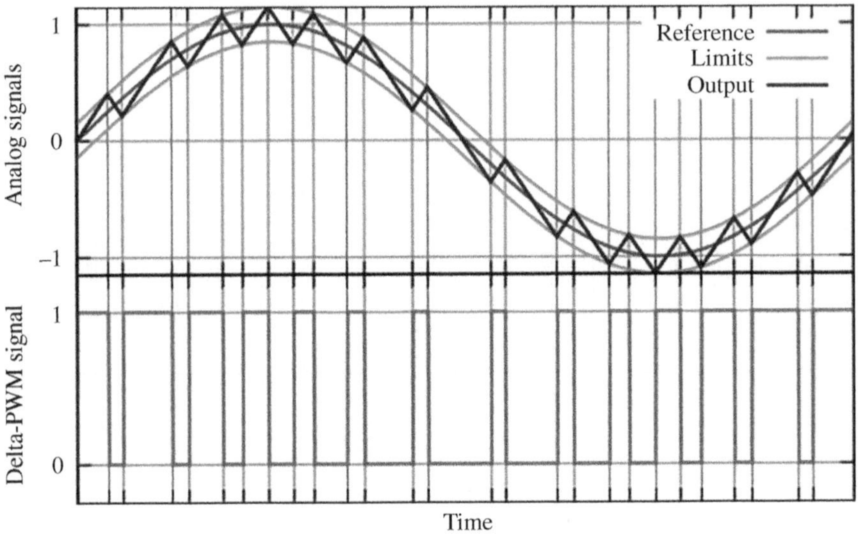

Figure 1.13 Digital generation method.

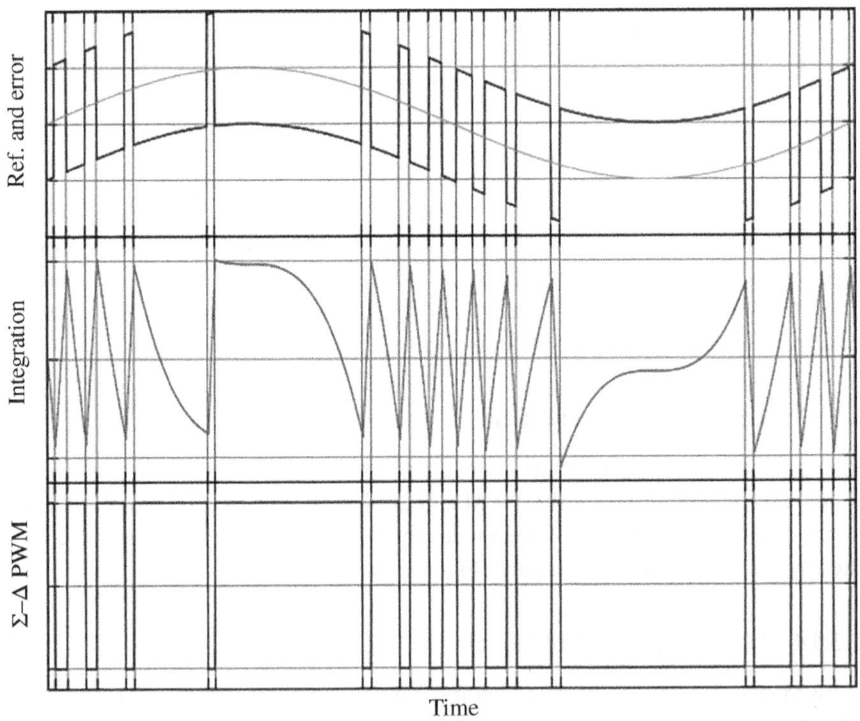

Figure 1.14 Delta-sigma modulation.

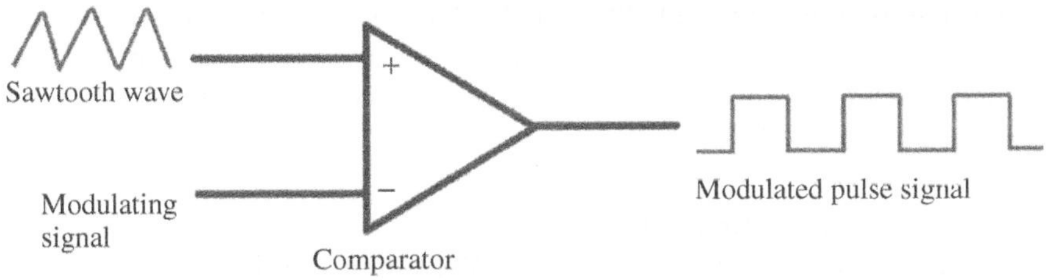

Figure 1.15 Generation of PWM signal.

a non-sinusoidal carrier wave (typically a sawtooth or triangular wave) operating at a high carrier frequency. The comparator continuously evaluates these two signals. If the carrier signal value is higher than the modulating signal value, the PWM output switches to a high state. Conversely, if the carrier signal is lower, the output remains in a low state. Therefore, the magnitude of the input modulating signal directly determines the comparator's output duty cycle, which defines the width of the pulse generated (Figure 1.15).

1.3.14.1 Single Pulse Width Modulation Technique

Single PWM is a technique applied to single-phase circuits to control the inverter's output voltage. It works by generating just one pulse per half-cycle and adjusting the width (on-time) of that pulse, which corresponds to the duration of the transistor's gate signal. Altering the pulse width directly modifies the transistor's conduction time, which in turn changes the magnitude of the inverter's output voltage. If the gate pulse width generated by the PWM comparator is d, the width of the resulting output voltage pulse will also be d.

1.3.14.2 Third Harmonic Injection Pulse Width Modulation

Third harmonic injection PWM is an advanced method used to increase the achievable output voltage from an inverter without causing distortion. This technique involves adding a third harmonic voltage component to the low-frequency sinusoidal reference signal. Because the third harmonic component is eliminated from the phase voltages and line-to-line voltages (assuming a three-phase load with an isolated neutral point), it does not distort the final output phase voltages. By incorporating this signal, the MI will be increased, allowing the inverter to utilize the DC-link voltage more effectively and achieve a higher fundamental output voltage amplitude.

1.3.14.3 Multiple Pulse Width Modulation

Multiple PWM features several pulses per half-cycle of the output voltage, with each pulse being modified by the carrier signal. The triangular wave frequency is higher than that used in single PWM and determines the number of gating signals per half-cycle. Operative multiple pulses reduce harmonic content, and the frequency determines the number of pulses per half-cycle. Modulation control controls the output voltage.

1.3.14.4 Sinusoidal Pulse Width Modulation

This common industrial method produces PWM output by using a sine wave as the modulating signal. The exact switching times are made by constantly comparing this sinusoidal reference wave against a high-frequency triangular carrier wave. By changing the MI, the effective RMS value of the output voltage will be precisely controlled. A key benefit of SPWM is its ability to improve the distortion factor and inherently eliminate all harmonics up to the $2p-1$ order, where p is the number of pulses, allowing remaining high-order harmonics to be easily removed with a simple filter.

1.3.14.5 Hysteresis Band Pulse Width Modulation

In this control strategy, the output (typically current or voltage) is allowed to fluctuate freely within a predefined error range called the hysteresis band. Unlike other methods, the switching frequency is not fixed, and it changes dynamically depending on the magnitude and frequency of the reference signal. The main advantage of this technique is its very quick response to sudden changes or disturbances in the load or input line voltage.

1.3.14.6 Space Vector Pulse Width Modulation

Space vector PWM (SVPWM) is an advanced algorithm-based PWM technique predominantly used to control 3-phase AC-powered motors by driving them at variable speeds from a DC source (often using multiple class-D amplifiers). SVPWM is considered an optimal modulation method because it significantly reduces the harmonic content present in the output, thereby providing a cleaner output current or voltage waveform to the motor.

1.3.15 Pulse Width Modulation Technique

PWM signals are pulse trains with fixed frequency and magnitude and variable pulse width. There is one pulse of fixed magnitude in every PWM period. The characteristic of a PWM signal is that the width of its pulses changes from one cycle to the next, following the modulating signal (the desired control signal). When this PWM signal drives the gate of a power transistor, it causes the transistors' on-time and off-time to vary precisely according to that same modulating signal.

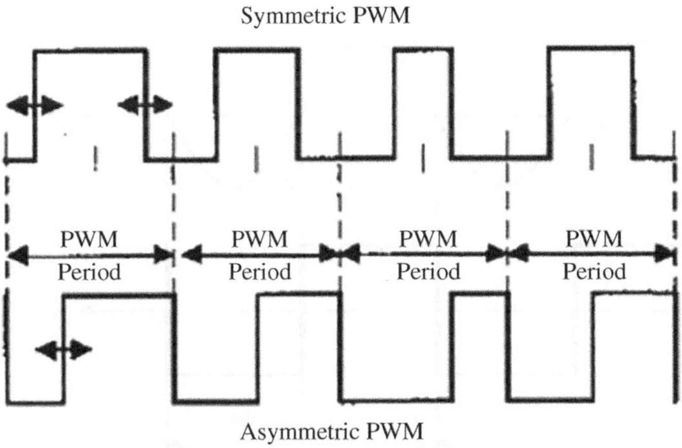

Figure 1.16 Symmetric and asymmetric PWM.

For this control to be effective, the frequency of the PWM signal (the carrier frequency) must be significantly higher than the frequency of the modulating signal (the fundamental frequency). This ensures that the total energy delivered to the load (like a motor) is predominantly controlled by the modulating signal, leading to smooth, proportional control rather than changing operation.

Figure 1.16 shows two different types of PWM signals, such as symmetric and asymmetric.

1.3.15.1 Single Pulse Width Modulation

In this approach, only one pulse per half-cycle is used, and the pulse width will change according to the controlled output voltage. Figure 1.17 shows the generation of gating signals of a single PWM. The gating signals are generated by:

Single PWM converts the reference signal (often a sine wave) into a single-pulse square-wave output. This conversion is achieved using a zero-crossing detection circuit that examines the reference signal. When the input signal is in its positive half (above zero), the output square wave is high (positive). Conversely, when the input signal is in its negative half (below zero), the output square wave is low (negative), as illustrated in Figure 1.17.

1.3.15.2 Multi-Pulse Width Modulation

Multi-PWM significantly reduces the harmonic content in the output voltage by generating several pulses during each half-cycle. The gating signals required to turn the inverter transistors on and off are produced by comparing the reference signal (modulating wave) with a high-frequency triangular carrier wave. The frequency of the reference signal sets the fundamental output frequency, while the

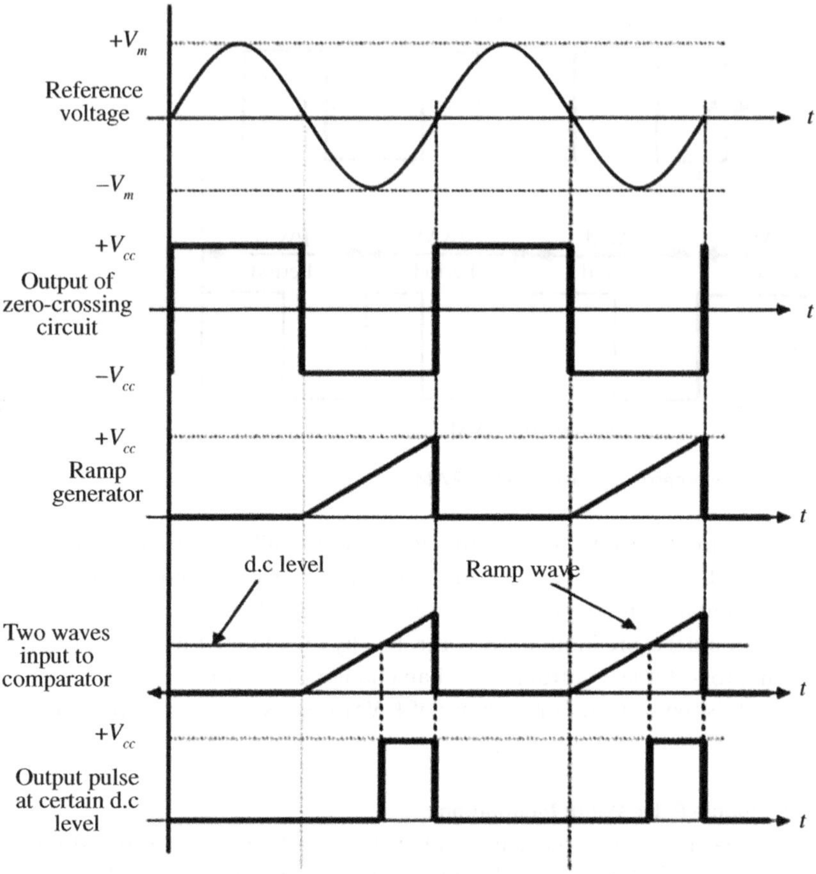

Figure 1.17 Single PWM.

carrier frequency determines the actual number of pulses generated per half-cycle (Figure 1.18).

$$p = \frac{f_c}{2f_0}$$

1.3.15.3 Carrier-Based PWM Technique

The carrier-based PWM technique is used to ensure the AC output voltage $V_0 = V_{an}$ of a VSI continuously follows a desired waveform (e.g., sinusoidal). This is achieved by properly controlling the ON and OFF states of the inverter's power switches.

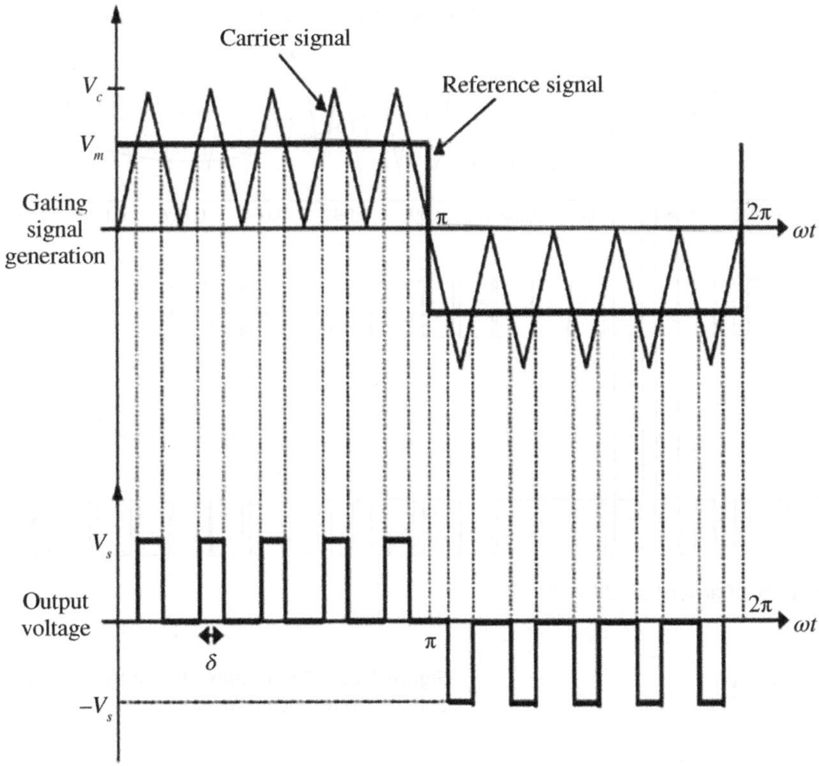

Figure 1.18 Multi-PWM.

The process defines the switching states for one leg of the VSI by comparing the modulating signal, which represents the desired AC output voltage, with a high-frequency triangular carrier signal Δ_V.

The switching rules are:

When the modulating signal is greater than the carrier signal V, the upper switch is turned ON, and the lower switch is turned OFF.

When the modulating signal is less than the carrier signal, the upper switch is turned OFF, and the lower switch is turned ON (Figure 1.19).

1.3.16 Space Vector Pulse Width Modulation Technique

This method is a more advanced and computationally demanding approach for producing sine waves, offering higher voltage with reduced total harmonic

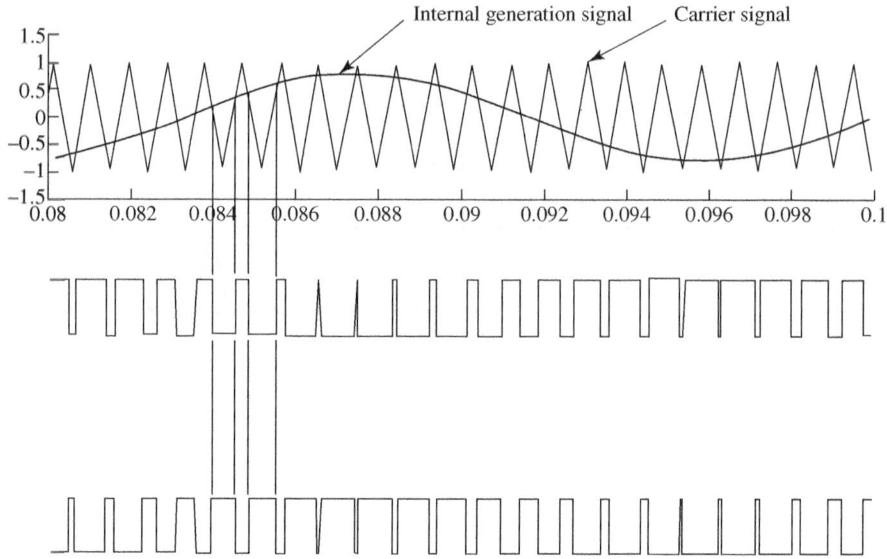

Figure 1.19 Carrier-based PWM.

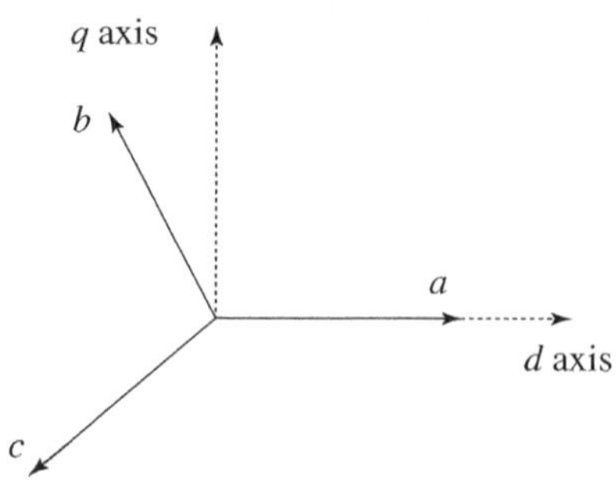

Figure 1.20 The relationship of the (*a*-*b*-*c*) and (*d*-*q*) reference frames.

distortion (THD) and is arguably the most effective among all PWM techniques. It restricts space vectors to be applied based on the region of the output voltage vector. SVPWM has become widely used due to its excellent performance characteristics, as its primary goal is to achieve a variable output voltage with the maximum fundamental component and minimal harmonics. SVPWM is favored over other modulation techniques because it generates less harmonic distortion in output voltages/currents and provides a more efficient utilization of the supply voltage (Figure 1.20).

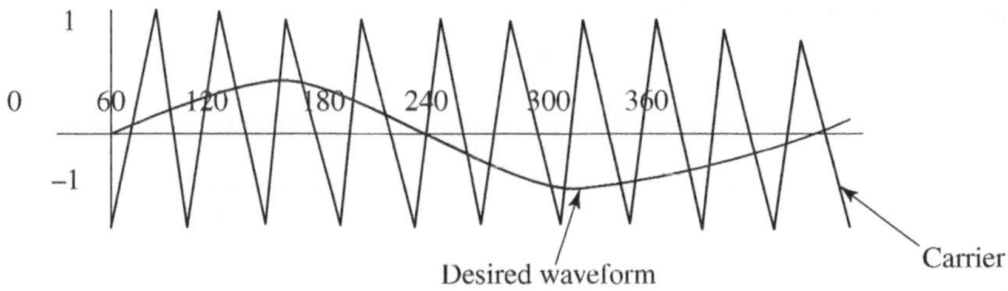

Figure 1.21 Sinusoidal PWM.

SVPWM is fundamentally a special switching sequence applied to the upper power switches of a three-phase power inverter.

- The inverter utilizes **six power switches,** S1–S6, which are controlled by switching variables (a, a', b, b', c, c').
- Since the upper and lower transistors in each leg operate in a complementary fashion (e.g., if the upper switch S1 is ON, the corresponding lower switch S4 or S5 is OFF), the output voltage is determined solely by the states of the **three upper switches S1, S3, and S5**.

To implement SVPWM, the voltage equations initially defined in the three-phase *abc* reference frame are mathematically transformed into the stationary d–q reference frame. This d–q frame consists of the horizontal d and vertical q axes and is used to define the voltage space vectors that guide the optimal switching strategy (Figure 1.21).

1.3.16.1 Modified Sinusoidal Waveform PWM

A modified sinusoidal PWM (SPWM) waveform is employed to manage power and enhance the PF by adjusting the current delay to align with the grid voltage, improving efficiency.

Multiple PWM involves generating several output pulses of varying values but constant duration, enabling inverters to operate at high voltages. A harmonic is a periodic waveform whose frequency is an integral multiple of the fundamental power frequency. PWM itself regulates the converter output voltage by adjusting the pulse width.

Advantages of PWM include low power consumption and high efficiency (up to 90%) by minimizing heat losses, being inexpensive to produce, offering high power handling, and providing robust noise immunity with reduced filter requirements. Additionally, the average output value is proportional to the duty cycle, and amplitude and frequency can be independently controlled. However,

Table 1.2 Comparison of PWM techniques.

Parameter	Peak fundamental (V_{dc})	THD (%)	Dominating harmonics
Square Wave	1.271	48.3	3rd, 5th, 7th
Multi PWM	1.23	38.5	3rd, 5th, 7th
Unipolar PWM	0.866	3.76	43rd and 45th
Trapezoidal PWM	1.05	38.5	3rd, 5th, 7th
Selective Harmonics	1.18	4.88	27th
Modified SPWM	0.866	3.76	43rd and 45th
AC PWM	1.10	22.35	19th and 21st

PWM drawbacks include the increased cost and complexity of integrated circuitry, limitations of RF and EMI, voltage spikes caused by rapid switching, and the necessity of using expensive fast-switching semiconductor devices. Furthermore, communication applications require a large bandwidth, and high switching frequencies lead to increased switching loss (Table 1.2).

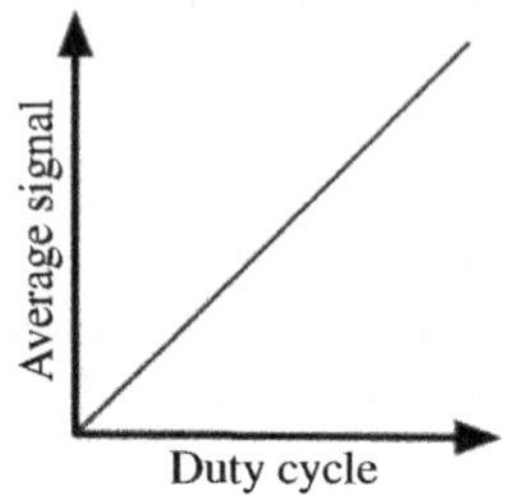

1.3.17 Use of PWM Techniques for Power Quality Improvement

1.3.17.1 Advantages of PWM for Power Quality Improvement

This control system is valued for its easier implementation and control and its compatibility with modern digital microprocessors. It eliminates the issues of drifting or degradation in linearity caused by temperature variations and component aging, ensuring stable long-term performance. Furthermore, it offers lower power dissipation and enables linear amplitude control over the output voltage or current, a functionality that was often absent in previous methods.

1.3.17.2 PWM Rectifier

To maintain some kind of minimum MI, to maintain this thing, it should not touch you when the input voltage is at its peak, and we require some mechanism that has been listed here, which is to avoid distortion near zero crossing. The converter should be capable of producing, actually, this MI approaching infinity. So it is not possible, but we should have. It means that actually the switch should be continuously on close to the zero crossings. The above expression also neglects the converter dynamics, so you will not have this kind of crossing because you require short-circuit protection. Short-circuit protection will come into the picture, and ultimately, it will shut it down. What happens is that boost, buck-boost, Cuk, SEPIC, and other DC-to-DC converters with similar conversion ratios are also suitable. These are essentially non-isolated DC-to-DC converters. We also have another type of DC-to-DC converter. We will be discussing the DC-to-DC converter later, so these are called isolated DC-to-DC converters that come with a high-frequency transformer incorporated inside, and the boost converter exhibits lower transistor stresses with a switch.

1.3.17.3 Disadvantages of PWM

- There is a noticeable reduction in the amplitude of the desired fundamental component of the PWM output waveform, specifically decreasing the voltage from 1 to 0.866 pu.
- Drastic increase in switching frequencies. This higher frequency imposes greater stress on the associated power switching stress (like transistors), often necessitating the derating (reducing the operational limits) of those components to prevent failure and
- Generation of high-frequency harmonic components.

1.3.17.4 The Objectives of Power Quality Improvement Using PWM

The objective of reporting the THD of the unfiltered inverter output line voltages is to quantitatively compare the quality of power generated by three widely used multi-carrier PWM techniques like alternative phase opposition disposition (APOD), in-phase disposition (PD), and phase opposition disposition (POD). These techniques are critical for motor control and power systems, as the primary goal of any PWM scheme is to maximize the fundamental voltage component while minimizing unwanted harmonics, which directly translates to a lower THD value. Reporting the unfiltered THD is essential for assessing the inherent performance of each technique, as a lower THD indicates a cleaner waveform that requires smaller, less expensive output filters and ultimately leads to more efficient power delivery to the load.

1.3.18 Applications

PWM is a highly adaptable technique applied across various engineering domains. In servo control and telecommunications, PWM pulses encode data where the pulse width corresponds to a specific data value. In fields like telecommunications, the leading edge of the data pulse can serve as a clock reference. PWM efficiently creates an approximate desired DC output voltage by switching the input to the load with the correct duty cycle, with an inductor and a capacitor typically used to filter the switching noise. For example, a basic regulator constantly checks the output voltage and adjusts the switch state (ON when low, OFF when high) to maintain the target level.

PWM is widely utilized in power delivery to regulate the energy supplied to a load without the high losses associated with resistive, linear methods. High-frequency switching using modern semiconductor devices like MOSFETs or IGBTs is key, as these switches dissipate almost no power in their fully ON or fully OFF states. Power loss only occurs briefly during the transitions. This results in high efficiencies, often exceeding 98% in frequency converters for AC motors or 70–80% in switching power supplies. Applications include variable-speed computer fan controllers and home light dimmers (though the latter often use simple phase-cut PWM on the AC cycle). For rapid-response loads like LEDs, a high PWM frequency (faster than the flicker fusion threshold) is essential to prevent perceptible flickering.

In audio systems, PWM forms the basis of highly efficient class-D amplifiers, which convert an analog input into a PWM-equivalent signal and feed it through a filter to the loudspeaker, achieving efficiencies of 90% or more with a compact size. In music synthesis, varying the pulse waveforms' duty cycle creates unique variations. The 50% duty cycle (square wave) is distinctive, as it eliminates even-numbered harmonics. Furthermore, the DSD sound encoding method uses a generalized form of PWM called pulse density modulation (PDM) at extremely high sampling rates to encode audio with high fidelity, paralleling the principles used in class-D amplification.

1.3.18.1 Electrical

Sine-triangle PWM (SPWM) is important in microinverter design for applications like solar and wind power. The efficiency of these devices depends on the harmonic content of the SPWM switching signals fed to the FETs, prompting research into using modified carrier signals (instead of the classic sawtooth) to reduce power losses and enhance fundamental strength. SPWM is also commonly applied in robotics to control motor speeds.

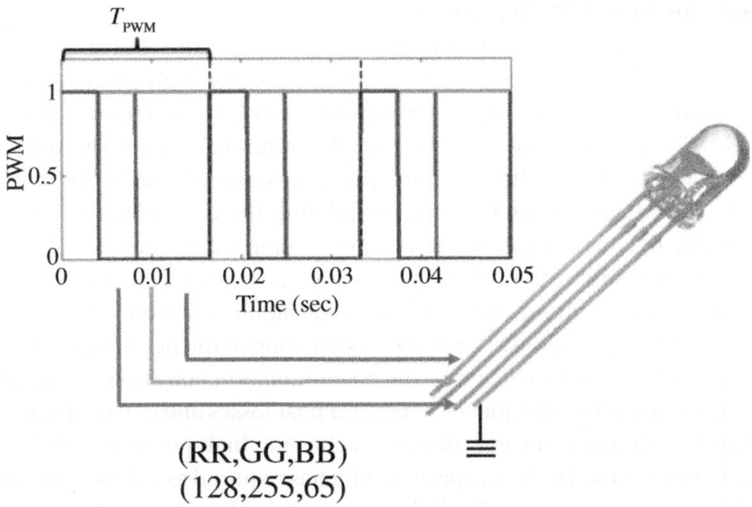

Use of PWM Techniques for Power Quality

1.3.19 Analysis of PWM with Performance Parameters

1.3.19.1 Analysis of Line Ripple Current

Switching energy converges to the harmonics of switching frequency; the electromagnetic interference (EMI) noise peak values will also be high around these frequencies, making the EMI problem serious. Beyond the common constant switching frequency PWM (CSFPWM), VSFPWM is being developed to take advantage of the flexibility offered by a non-fixed frequency. The core of VSFPWM involves dynamically adjusting the switching frequency based on the predicted ripple current distribution. For a single-phase inverter, the ripple current in any switching cycle can be accurately predicted using the duty cycle (d), the switching period (T_s), the output inductance (L), and the DC voltage (V_{dc}).

By controlling the frequency based on this prediction, the current ripple can be significantly reduced to meet specific requirements (e.g., peak current ripple or RMS current ripple). This method allows for a reduction in switching losses while still satisfying ripple requirements. It is important to note that extending this technique to a three-phase converter is considerably more complex, as the ripple current in any single phase is subject to the switching actions occurring across all three phases.

1.3.19.2 Analysis of DC Link Current

PWM is an essential technique for achieving variable control over analog loads using digital microprocessors, offering advantages like high efficiency (>90%), low power dissipation, and superior noise immunity compared to resistive analog methods. The core principle involves rapidly switching the full DC voltage ON and OFF, where the resulting average power is proportional to the duty cycle (T_{on}/T). The MI quantifies this control, relating the desired output voltage to the available DC bus voltage. Numerous modulation techniques exist, from basic SPWM, which is simple to implement, to advanced methods like SVPWM, which maximizes voltage output and minimizes harmonic distortion, a critical goal across all PWM applications. The required switching frequency must be high (e.g., >20 kHz) to eliminate audible noise and ensure smooth, non-flickering operation, though excessively high frequencies increase heat losses and device stress.

A major challenge in inverter design is achieving high power density by minimizing the size of the DC-link capacitor, which is required to stabilize and smooth the DC-link voltage. Since the DC-link voltage ripple plays an important role in the sizing criteria, analytical methods are used to derive simple and effective design guidelines, considering both high-frequency switching ripple and low-frequency components (100 Hz). The design is further complicated by the choice of modulation strategy (like SPWM vs. continuous symmetric centered PWM) and the output phase angle.

For multiphase inverters, an important finding regarding the DC-link capacitor size is its relationship to the number of phases. It is established that for the same total output current, the required DC-link capacitor size is reduced by increasing the number of phases. This is a significant factor in achieving a good compromise between capacitor size and voltage ripple. However, this benefit does not continue indefinitely. There are no further capacitor sizing benefits to increasing the number of phases beyond a certain point. This analysis provides practical equations based on maximum peak-to-peak voltage ripple to guide the design of reliable and compact multiphase inverter systems.

1.3.19.3 Analysis of Torque Ripple

Torque ripple is a significant disadvantage in motor drives, especially at low speeds, arising from phase current commutation discrepancies caused by phase inductance and limited inverter voltages. This ripple generates noise and impairs speed control. To minimize these pulsations, two main strategies are employed: enhanced motor designs (using skewing, fractional-slot windings, etc.) and improved control schemes (such as adaptive control, preprogrammed current waveforms, and selective harmonic injection). A key control advancement involves applying a modified PWM signal during the start-up of the commutation sequence to allow the incoming phase current to build up gradually, thereby minimizing ripples in BLDC drives without needing additional hardware.

1.3.19.4 Effect of Over-Modulation

The over-modulation principle is a technique used in inverters to achieve a low THD while simultaneously increasing the first voltage harmonic and the RMS output voltage. This phenomenon effectively leads to a reduction in the DC converter input voltage, which is highly desirable for systems using low-voltage sources (like solar panels or fuel cells). By reducing the required input DC voltage while maintaining output quality, over-modulation improves system efficiency.

For general AC–DC converters, a common issue is that the PF decreases as the output voltage is reduced (by increasing the firing angle). Three schemes are used for PF improvement (PFI): extinction angle control, symmetrical angle control, and PWM control. Unlike single-pulse methods, where the lowest-order harmonic (third) is difficult to filter, PWM control turns switches on and off multiple times per half-cycle, allowing better control over harmonics. Modern PWM rectifiers often use a double closed-loop PI control system: the inner loop controls the input current to maintain a near-unity PF, and the outer loop controls the DC output voltage. The process requires a precise reference to the current generation based on the required output power, with an inductor playing a major role in achieving the unity PF. The required pulse generation for these controllers can be implemented using integrated circuits, OP-AMPs, or timers, though microprocessor-based systems are preferred for the complex logic.

1.3.20 Software Package

There are different software packages used to generate pulses.

MATLAB

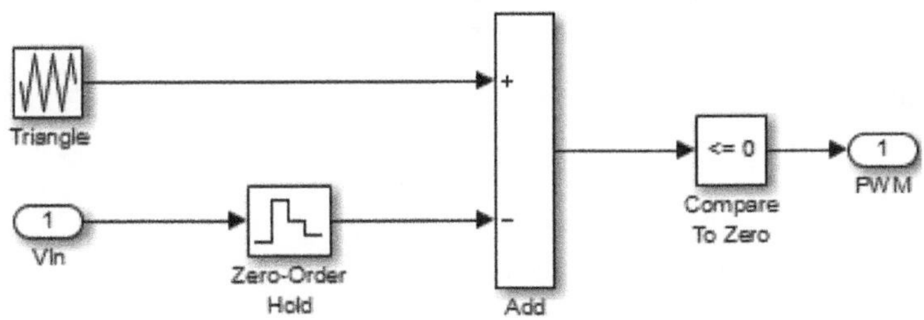

LABVIEW

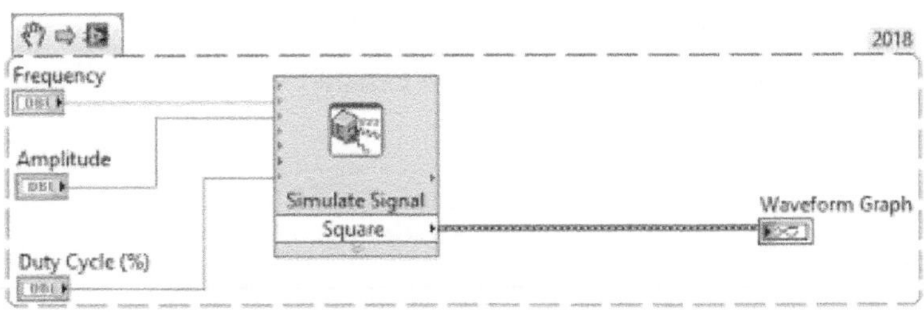

PROTEUS

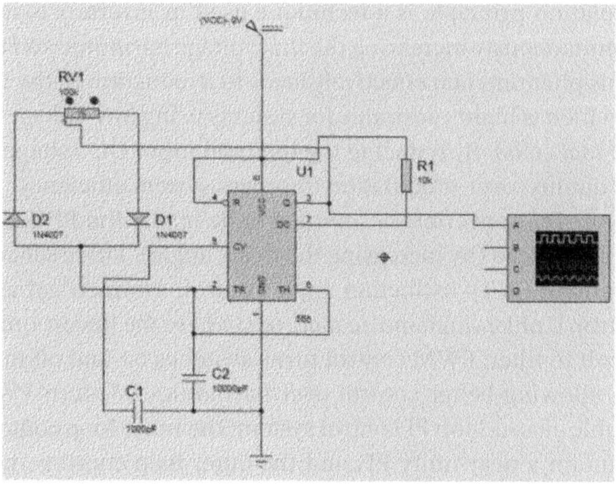

PSPICE

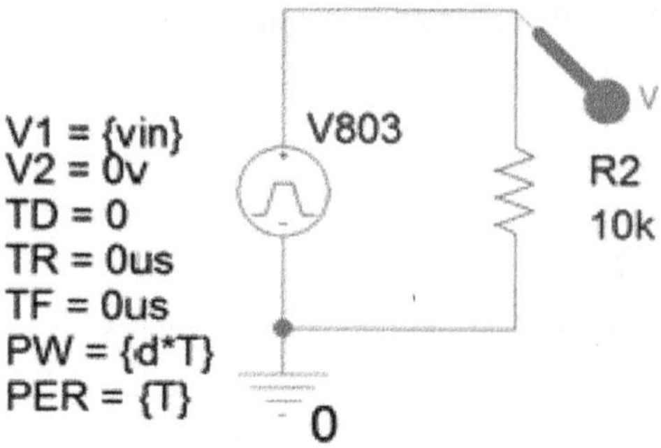

MULTISIM

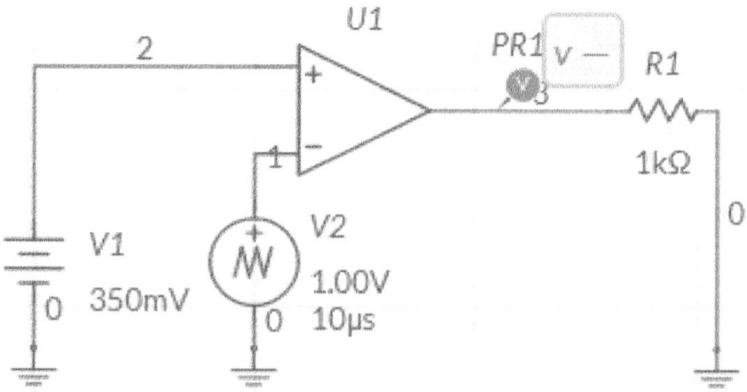

Various software tools are utilized to simulate the pulses necessary for the converter. These software packages include MATLAB, LabVIEW, PROTEUS, PSPICE, and MULTISIM.

1.3.21 Hardware Interface Package

PWM is a technique that uses a digital signal to effectively generate a variable analog output. It achieves this by quickly switching the digital signal between its high (1) and low (0) states, thereby controlling the signal's average value. PWM signals are commonly generated using modern digital platforms, including microcontrollers, Arduino, FPGAs, Raspberry Pi, and digital controllers. While the PWM interface provides necessary access to a specific PWM channel on a device, it often grants privileged access to the underlying hardware, which will force certain limitations.

1.4 Summary

In conclusion, PWM signals are employed to manage the flow of power from the source to the load without incurring losses. The choice of switching devices for semiconductor applications is important, depending on the applications of PWM signals in various converters. PWM signals are generated using both analog and digital circuits. Different types of PWM signals are used in the converter, and based on the type, performance parameters like ripple factor and harmonics are evaluated. Converters serve as the system's core, while PWM signals function as the system's intelligence, facilitating lossless power flow. The use of PWM signals in converters significantly enhances the system's power quality. PWM signals are analyzed using software packages such as MATLAB, PROTEUS, PSPICE, LabVIEW, MULTISIM, and MODLICA, alongside hardware techniques employing advanced processors and controllers.

References

http://en.wikipedia.org/wiki/Pulse-width_modulation. (accessed on October 2, 2025).

http://tutorial.cytron.com.my/2012/01/14/basic-pulse-width-modulation-pwm/. (accessed on October 2, 2025).

http://www.societyofrobots.com/member_tutorials/book/export/html/228. (accessed on October 2, 2025).

http://powerelectronics.com/power-management/pwm-single-chip-giant-industry.

http://www.freescale.com/files/microcontrollers/doc/data_sheet/MC9S12C128V1.pdf. (accessed on October 2, 2025).
http://en.wikipedia.org/wiki/Pulse-width_modulation#Telecommunications. (accessed on October 2, 2025).
http://www.analog.com/en/content/ta_fundamentals_of_voltage_regulators/fca.html. (accessed on October 2, 2025).
http://www.mathworks.com/help/matlab/ref/imread.html. (accessed on October 2, 2025).
www.rapidtables.com/web/color/RGB_Color.htm. (accessed on October 2, 2025).
http://www.monkeylectric.com. (accessed on October 2, 2025).

2

Basic PWM Approaches

Learning Objectives

1) Understand the fundamentals of pulse width modulation (PWM):
   Define PWM and its role in power electronics.
   Explain the basic principles of how PWM controls power delivery.
2) Identify different PWM techniques:
   Describe the operational principles of triangle comparison-based PWM.
   Explain the mechanism and benefits of bus clamping PWM.
   Differentiate between unipolar PWM and bipolar PWM.
   Understand how voltage-controlled PWM and current-controlled PWM regulate power.
   Describe the compensation strategies used in compensated PWM.
3) Analyze advanced PWM methods:
   Understand the vector representation of voltages in space vector PWM.
   Explain how space vector PWM optimizes the output waveform and reduces harmonic distortion.
   Identify the scenarios where hybrid PWM is beneficial and understand its implementation.
4) Compare and contrast PWM techniques:
   Evaluate the advantages and disadvantages of each PWM method.
   Determine the appropriate PWM technique for specific power electronic applications based on performance criteria.
5) Implement PWM techniques in MATLAB:
   Develop MATLAB scripts to generate triangle comparison-based PWM, bus clamping PWM, unipolar PWM, and bipolar PWM signals.
   Simulate the behavior of voltage-controlled PWM and current-controlled PWM.
   Explore the implementation of compensated PWM in MATLAB.

Simulate space vector PWM and understand its output characteristics.
Design a hybrid PWM strategy using MATLAB.

By mastering these objectives, readers will gain a comprehensive understanding of various PWM techniques and their applications in power electronics, enabling them to design efficient and effective power control systems.

2.1 Introduction

In the field of electronics and signal processing, pulse width modulation (PWM) is an extensively used method due to its good control of power delivery, signal generation, and other applications. At its core, PWM is interested in controlling the pulse width in a pulse train to code information or regulate the extent of power delivered to a load. This modulation is achieved by varying the duty cycle, which is the ratio of the pulse's on-time (T_{on}) to the total period (T) of the waveform.

2.2 Basics of PWM

2.2.1 Duty Cycle

The duty cycle of a PWM signal is defined as the percentage of one period during which a signal is active. It is given by the formula:

$$\textbf{Duty cycle} = \boldsymbol{T_{on/}T_{off}}$$

where T_{on} is the time in which the signal is high (on), and T_{off} is the time the signal is low (off). A 50% duty cycle means the signal is high for half the period and low for the other half.

2.2.2 Frequency

The frequency of a PWM signal is the number of times the pulse train replicates in one second. It is the inverse of the period (T), which is the sum of T_{on} and T_{off}.

$$\boldsymbol{F = 1/T}$$

where $T = T_{on} + T_{off}$

2.2.3 Amplitude

The amplitude of a PWM signal is the height of the pulse, which in general relates to the voltage level when the signal is extreme (Figure 2.1).

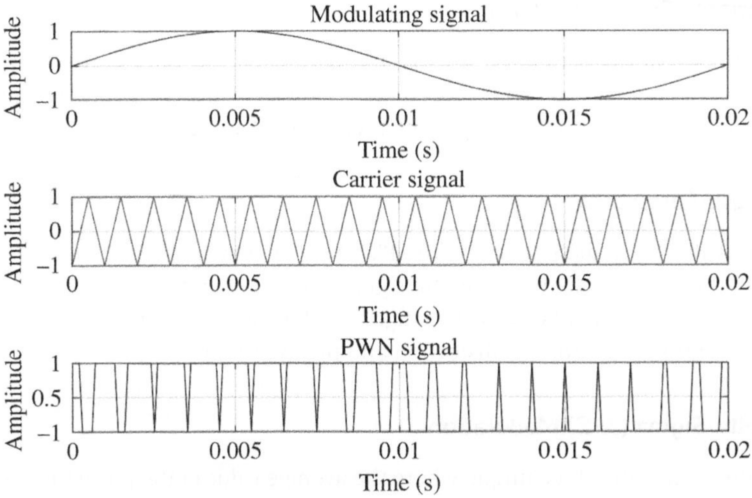

Figure 2.1 Generation of PWM signal.

2.3 Types of PWM Techniques

Numerous PWM techniques can be used based on application constraints. The most common are shown below (Figure 2.2):

2.3.1 Fixed-Frequency PWM

In this technique, the frequency of the PWM signal is kept constant while changing the duty cycle to regulate the output. It is extensively used in motor control and power supplies, where upholding a constant switching frequency is vital to avoid intrusion and ensure expected performance.

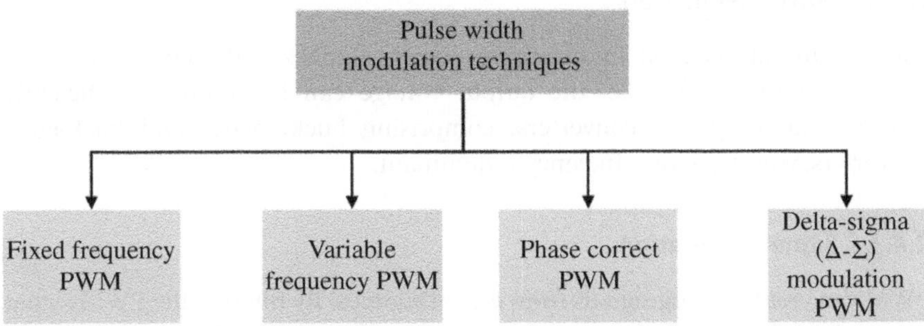

Figure 2.2 Types of PWM based on application constraints.

2.3.2 Variable-Frequency PWM

In this technique, it involves altering both the frequency and duty cycle of the PWM signal. It is used in wide applications like motor control systems and power inverters where variable speed and efficient energy usage are required.

2.3.3 Phase-Correct PWM

Phase-correct PWM is a method where the PWM signal is symmetrical about the midpoint of the period, which means the signal counts up and down to the top value. It is beneficial for applications like digital audio and servo control systems, which require minimal harmonic distortion and precise timing.

2.3.4 Delta-Sigma (Δ-Σ) Modulation

This is an innovative PWM technique where the average value of the pulse train is proportional to the input signal. Delta-sigma modulation is used in high-reliability digital-to-analog conversion, and in applications, it requires precise control over a widespread range of frequencies.

2.4 Applications of PWM

2.4.1 Motor Control

PWM is widely used to control the speed and torque of electric motors. The speed of the motor will get adjusted by controlling the amount of power delivered to the motor, and it can be possible by varying the duty cycle of the PWM signal applied to the motor's input. DC motor speed controllers and brushless motor drivers are widely used for these techniques (Figure 2.3).

2.4.2 Power Regulation

PWM technique is used in switching power supplies and voltage regulators. By altering the duty cycle, the output voltage can be controlled efficiently. This is vital in power converters, comprising buck, boost, and buck-boost converters, where energy efficiency is dominant.

2.4.3 Signal Generation

PWM can create analog signals from digital sources. By filtering the PWM signal, a smooth analog waveform can be formed. This method is used in audio synthesis, signal processing, and communication systems.

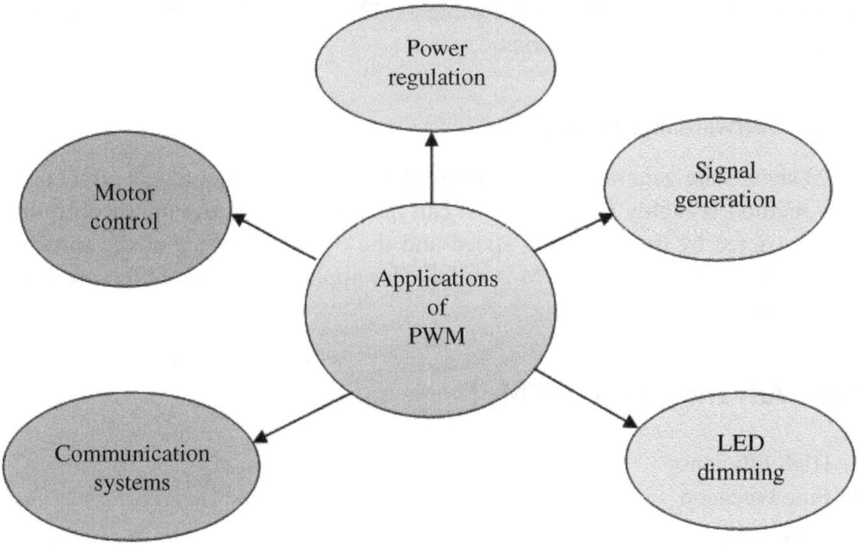

Figure 2.3 Applications of PWM.

2.4.4 LED Dimming

PWM is used in LED dimming applications to control the illumination of LEDs. By varying the duty cycle, the average current through the LED is changed, which alters its brightness devoid of affecting its color accuracy.

2.4.5 Communication Systems

PWM can encode information in digital communication systems. Pulse Position Modulation (PPM), a derivative of PWM, is used in some RF communication systems to transmit data by varying the position of the pulses.

2.5 PWM Implementation

PWM can be implemented using both hardware and software methods:

2.5.1 Hardware Implementation

PWM implementation in hardware is usually produced using dedicated PWM controllers/timers in microcontrollers. High-frequency PWM signals are produced by these controllers with precise scheduling, constituting them a suitable choice for high-speed functions. Examples: PWM modules in microcontrollers like AVR, Arduino, and PIC series.

2.5.2 Software Implementation

PWM signals are generated by utilizing software routines that toggle GPIO pins. This method is highly flexible, and it can be employed on any microcontroller; it is restricted by its processing speed and the timing accuracy of the software implementation. Software PWM is suitable for applications where high precision is not important.

2.6 Advantages of PWM

1) High Efficiency
2) Fine Precision
3) Simplicity
4) Flexibility

2.7 Low Switching Frequency Operation

In order to decrease the number of switch transitions per unit time, low switching frequency operation in PWM technique is employed. This method can lead to major benefits in several applications, predominantly in high-power and high-efficiency systems. Here, the frequency of the carrier signal is reduced. This consequences to fewer switching events per second. For example, if the PWM switching frequency is 20 kHz, a low switching frequency may be 2 kHz or lower (Figure 2.4).

2.7.1 Advantages of Low Switching Frequency

2.7.1.1 Reduced Switching Losses Sensitive to Small Changes in the Reference Signal or Carrier Waveform

Switching losses will occur in power electronic devices like transistors due to the shift between the on and off states. These losses are proportionate to the switching frequency. By lowering the switching frequency, the total energy lost through switching transitions is reduced.

2.7.1.2 Improved Efficiency

Whenever the switching losses are low, the overall efficiency is also increased. This is predominantly important in high-power applications where even a small

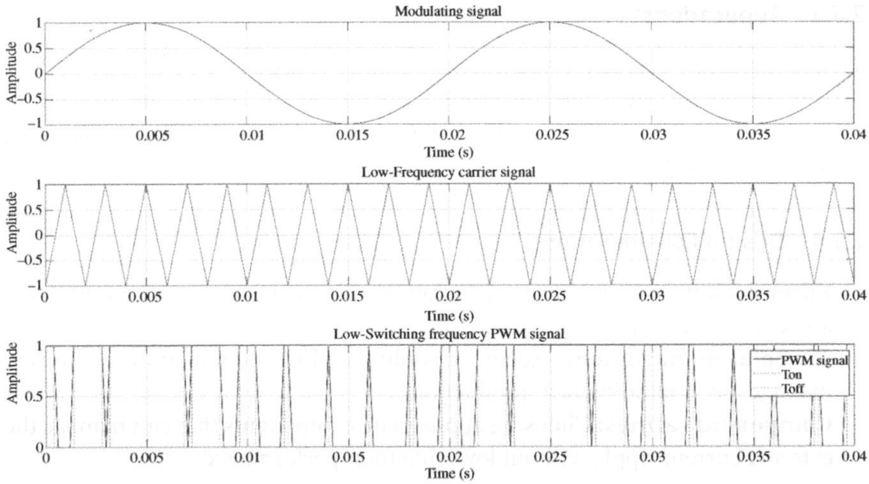

Figure 2.4 Generation of a low switching frequency signal.

increment in efficiency can lead to considerable energy savings and lowered heat dissipation.

2.7.1.3 Thermal Management

If the switching frequencies are low, it results in less heat production inside the switching devices. This can increase the thermal management constraints, decrease the requirement for huge cooling systems, and it improves the reliability and lifetime of the devices.

2.7.2 Disadvantages of Low Switching Frequency

2.7.2.1 Increased Output Ripple

Due to lower switching frequency, ripples will be high in the signal output. These ripples are produced due to the alteration of the output from the desired average value and are inversely proportional to the switching frequency. Due to these ripples, additional filtering is needed to smooth out the signal, which leads to an increase in the system's complexity and cost.

2.7.2.2 Slower Response Time

The system's ability to react quickly to variations in the modulating signal is decreased, and it will be a drawback in applications that require fast dynamic response.

2.7.3 Applications

a) Low switching frequency operation is commonly used in
b) High-power inverters and converters
c) Applications where efficiency is predominant, such as electric vehicles (EVs) and power supplies.

2.7.4 Design Considerations

a) **Filter design:** Appropriate design of output filters to smooth the output signal and decrease ripples.
b) **Thermal management:** Adequate cooling and heat dissipation methods to manage any remaining heat production.
c) **Component ratings:** Choosing appropriate components that can manage the extreme current ripple without lowering their performance.

2.8 Triangle Comparison-Based PWM

This method is widely used in power electronics to produce pulse-width modulated signals. It works by comparing a high-frequency triangular waveform, which is the carrier signal, with a lower-frequency reference signal; normally, a sinusoidal waveform is used. The points where the reference signal crosses the triangular waveform determine the switching instants of the PWM signal. The duty cycle of the PWM signal is determined by the ratio of on time to total period, which is determined by the characteristics of the reference signal and the carrier waveform. By changing the parameters like amplitude or frequency of the reference signal, the duty cycle, the average output voltage, or current can be controlled. Triangle comparison-based PWM is frequently employed in numerous power electronics applications related to the inverters for motor drives, audio amplifiers, and power converters. Advantages include ease of implementation, effective generation of sinusoidal waveforms, and versatility in controlling output voltage or current. However, it may suffer from limited accuracy when generating complex waveforms and can be sensitive to small changes in the reference signal or carrier waveform (Figure 2.5).

2.8.1 Bus Clamping PWM

The bus clamping PWM method is used in power electronics, especially in three, the PWM voltage is admitted switching to the opposite polarity or to vary according to the reference signal. This switching operation continues as per the PWM modulation.

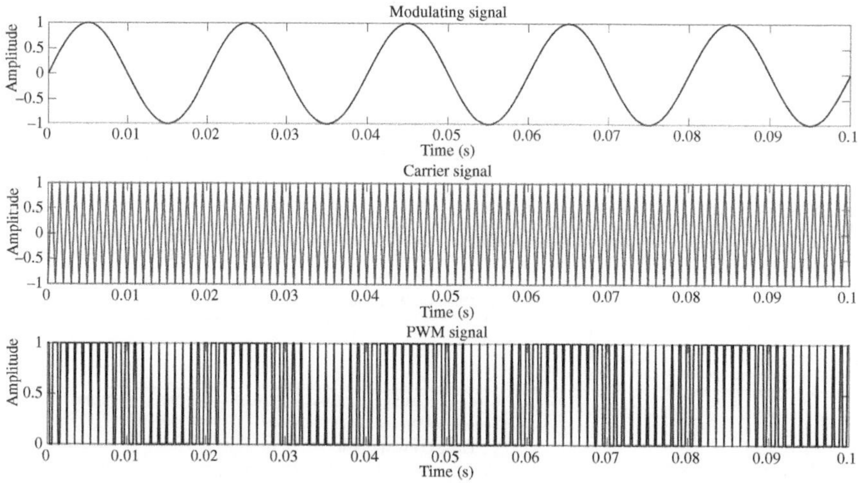

Figure 2.5 Triangle comparison-based PWM.

The bus clamping PWM method is used in power electronics, especially in three-phase inverters, to reduce switching losses and electromagnetic interference (EMI). It works by clamping the PWM output voltage to one of the DC bus voltage levels for a specific duration during each switching cycle.

1) **DC bus voltage:** In power electronic systems such as inverters, power is usually supplied from a DC source, well known as the DC bus. This bus voltage will act as the energy source for generating the output AC voltage.
2) **PWM generation:** PWM is used to manage the output voltage of the inverter. PWM signals are generated by comparing a reference signal with a high-frequency carrier waveform.
3) **Clamping operation:** In bus clamping PWM, additional circuitry is established to observe the PWM voltage. When the PWM voltage reaches one of the DC bus voltage levels, either positive or negative, the clamping circuitry activates.
4) **Clamping duration:** The clamping circuitry ensures that the PWM voltage remains clamped to the DC bus voltage level for a certain duration during each switching cycle. This duration is normally decided based on the system requirements and design considerations.
5) **Switching operation:** After the clamping duration, the PWM voltage is admitted, switching to the opposite polarity or varying according to the reference signal. This switching operation continues as per the PWM modulation (Figure 2.6).

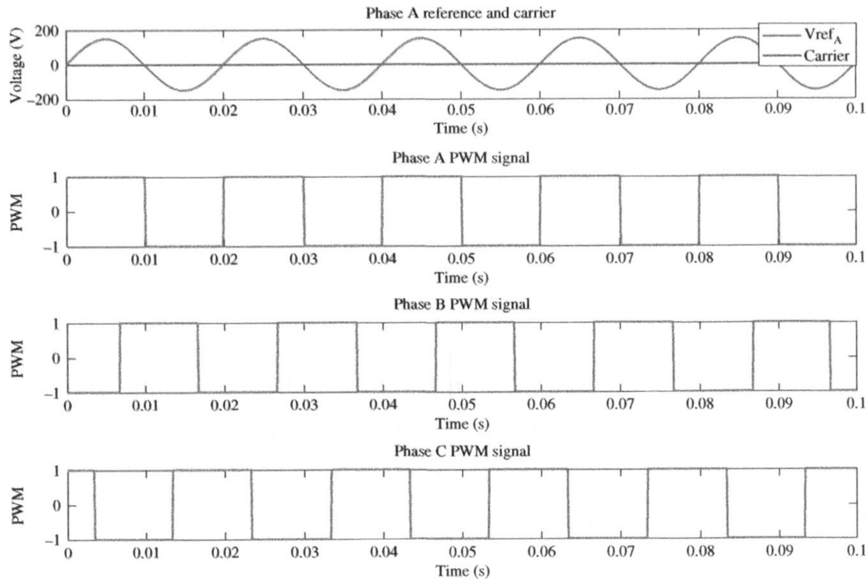

Figure 2.6 Bus clamping PWM.

2.8.2 Advantages

Reduced switching losses: By clamping the PWM voltage to the DC bus voltage level, the number of switching transitions in each cycle is reduced. This leads to lower switching losses in the power electronic devices like transistors.

Lower electromagnetic interference: Due to the reduction in switching transitions, low EMI and only fewer high-frequency harmonics are present in the waveform output.

Improved efficiency: By reducing switching losses and EMI, bus clamping PWM improves the overall system efficiency, specifically in high-power applications where the efficiency is most important.

2.8.3 Applications

Three-phase inverters: Bus clamping PWM is commonly used in three-phase inverters for motor drives, renewable energy systems, and industrial applications.

Electric vehicle drives: In EVs, where efficiency and EMI control are important, bus clamping PWM can help improve performance and reliability.

Renewable energy systems: In inverters for solar or wind power systems, bus clamping PWM can improve efficiency and reduce grid instability.

Bus clamping PWM puts forward significant improvements in terms of reduction in switching losses and EMI in power electronic systems, generating it a valuable technique for enhancing efficiency and system performance.

2.9 Unipolar PWM

To control the output voltage of inverters, converters, and motor drives, the unipolar PWM technique is used commonly in power electronics. In unipolar PWM, the output voltage swaps between zero and a positive or negative voltage level, rather than the voltage level existing between positive and negative voltages as in bipolar PWM.

If the output voltage needs to be positive, the upper switch of one half-bridge and the lower switch of the other half-bridge are switched on and off. If the output voltage is required to be negative, the switch condition will get inverted (Figures 2.7–2.10).

1) **Reference signal:** A lower-frequency sinusoidal signal will be operated as the desired output voltage or current. This reference signal defines the shape of the output waveform.

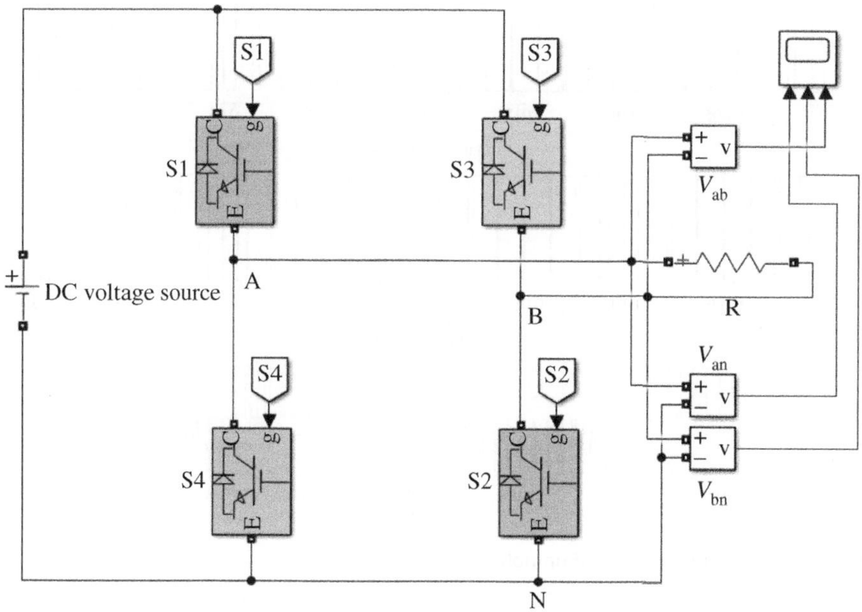

Figure 2.7 Single-phase H-bridge inverter.

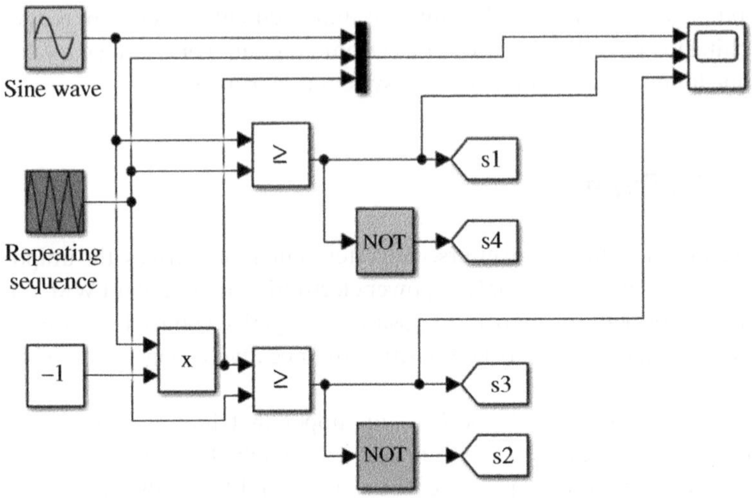

Figure 2.8 Simulink model for the generation of gate pulses.

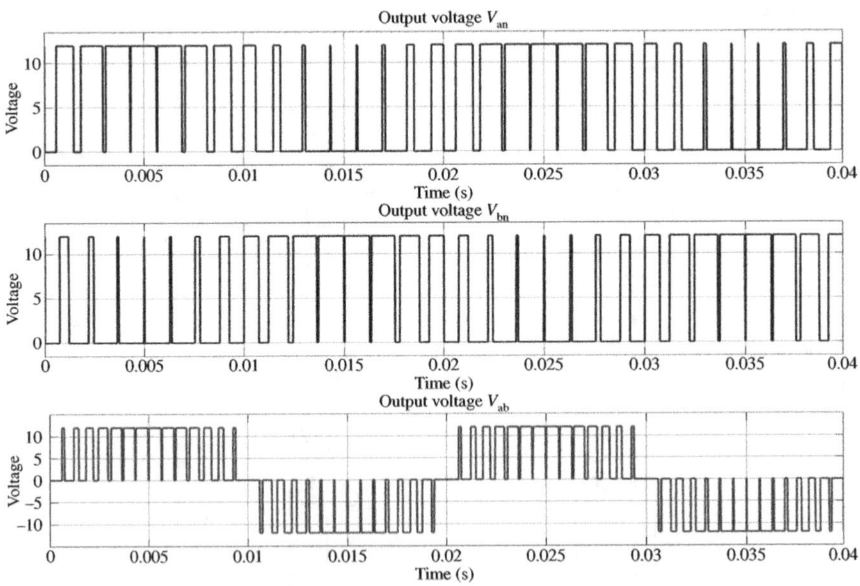

Figure 2.9 Output voltage of unipolar PWM.

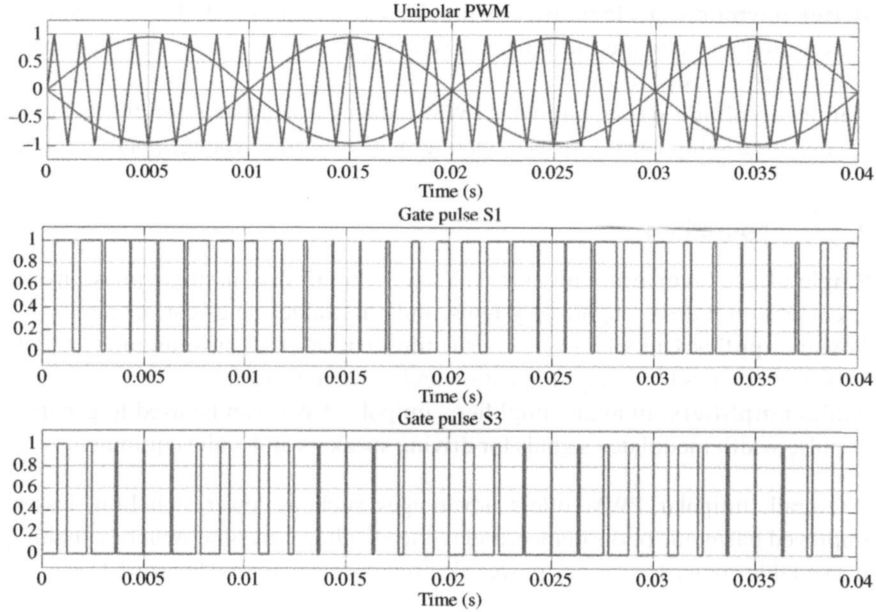

Figure 2.10 Generation of gate pulses.

2) **Carrier waveform:** A high-frequency triangular or sawtooth waveform, known as the carrier signal, is generated. This carrier signal acts as the timing reference for the PWM modulation process.
3) **Comparison:** The reference signal is compared with the carrier waveform. When the amplitude of the reference signal exceeds the amplitude of the carrier waveform, the PWM output is high; otherwise, it is low.
4) **Switching operation:** In unipolar PWM, the PWM output voltage switches between zero and a positive (or negative) voltage level. The duration of the on state is regulated by the amplitude of the reference signal compared to the carrier waveform.
5) **Output voltage generation:** By adjusting the frequency and amplitude of the reference signal, the average output voltage can be controlled. Unipolar PWM produces a series of pulses with varying widths, resulting in an output voltage that closely follows the shape of the reference signal.

2.9.1 Advantages

Lower switching losses: Compared to bipolar PWM, unipolar PWM typically results in lower switching losses since the output voltage only switches between zero and one polarity, reducing the number of switching transitions.

Better harmonic performance: Unipolar PWM can provide better harmonic performance compared to bipolar PWM, especially in applications requiring low harmonic distortion.

Reduced voltage stress: Unipolar PWM reduces voltage stress on the switches compared to bipolar PWM, as it operates within a single voltage polarity range.

2.9.2 Applications

Single-phase inverters: Unipolar PWM is commonly used in single-phase inverters for motor drives, lighting systems, and renewable energy converters.

Power supplies: Unipolar PWM is also utilized in DC–DC converters and AC–DC power supplies for voltage regulation and power conversion.

Audio amplifiers: In audio amplifiers, unipolar PWM can be used to generate pulse-width modulated signals for driving speakers and audio equipment.

Overall, unipolar PWM offers advantages such as lower switching losses, improved harmonic performance, and reduced voltage stress on switches, making it a suitable choice for various power electronics applications (Figure 2.11).

2.9.3 Bipolar PWM

In bipolar PWM, the voltage applied to the load alternates between the positive and negative supply voltages. This switching method results in the load experiencing a full voltage swing from positive to negative during each PWM cycle. This

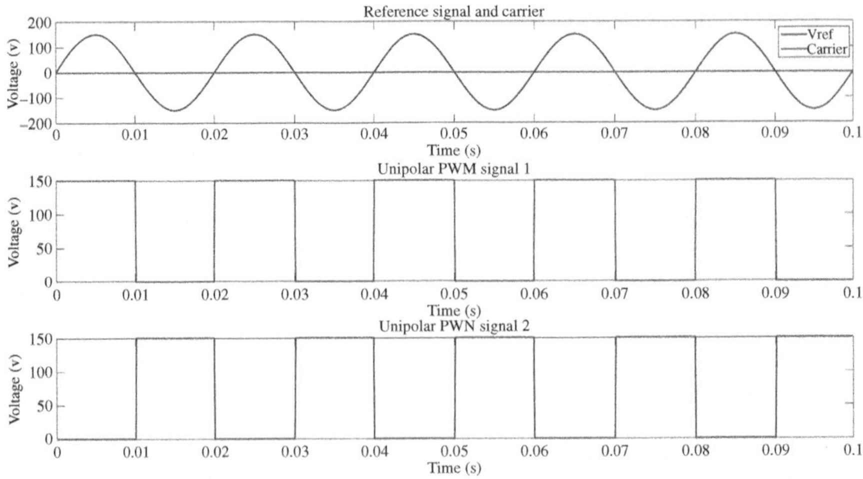

Figure 2.11 Generation of unipolar PWM signals for reference sinusoidal waveform.

modulation technique switches the output voltage between the positive and negative supply voltages, generating a square wave so that it can be filtered to generate a smooth AC waveform. These applications are used in electric drives.

1) The switches are controlled such that during one half of the PWM period, the load is connected to +V, and throughout the other half, it is connected to −V.
2) This results in the load voltage varying directly between the positive and negative supply levels (Figures 2.12 and 2.13).

In bipolar PWM, the output voltage oscillates between the positive and negative supply voltages, and it is typically represented as $+V_{dc}$ and $-V_{dc}$. This switching results in the load voltage oscillating between these two extremes during each PWM cycle. The duty cycle of the PWM signal is the ratio of the time the signal is in the high state ($+V_{dc}$) to the total period of the PWM cycle. By varying the duty cycle, the average voltage delivered to the load can be controlled. For example, a 50% duty cycle means the output voltage is $+V_{dc}$ for half the cycle and $-V_{dc}$ for the other half, resulting in an average voltage of 0 (Figures 2.14–2.16).

2.9.4 Voltage-Controlled PWM

Voltage-controlled PWM is a commonly used method in electronics and power systems for regulating the power supplied to a load by modifying the duty cycle

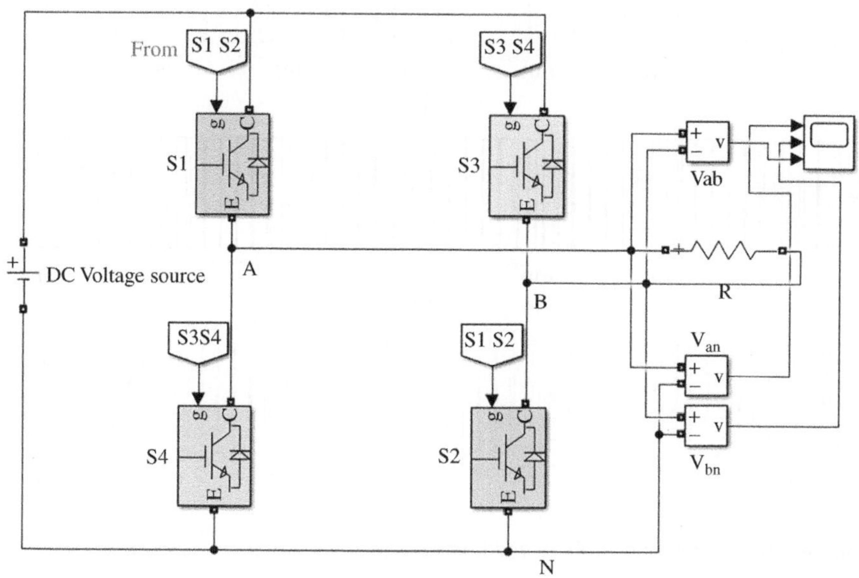

Figure 2.12 Single-phase H-bridge inverter.

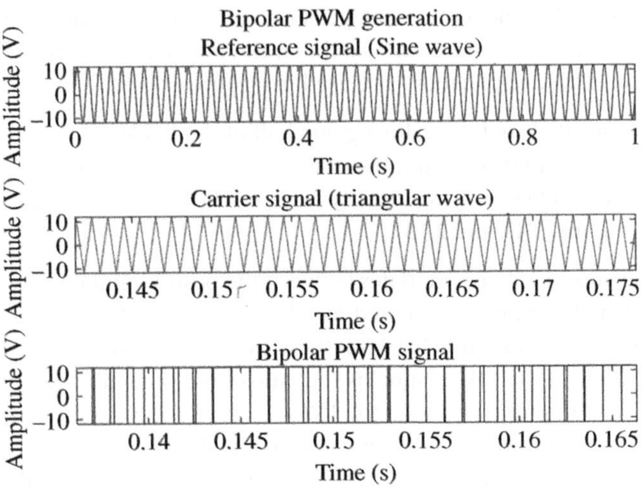

Figure 2.13 Generation of bipolar PWM.

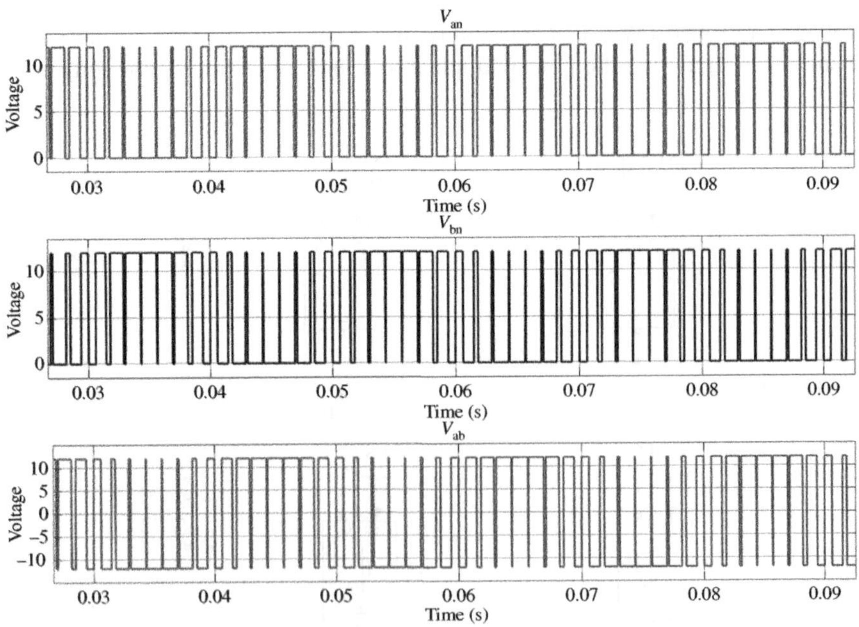

Figure 2.14 Generation of phase and line voltage.

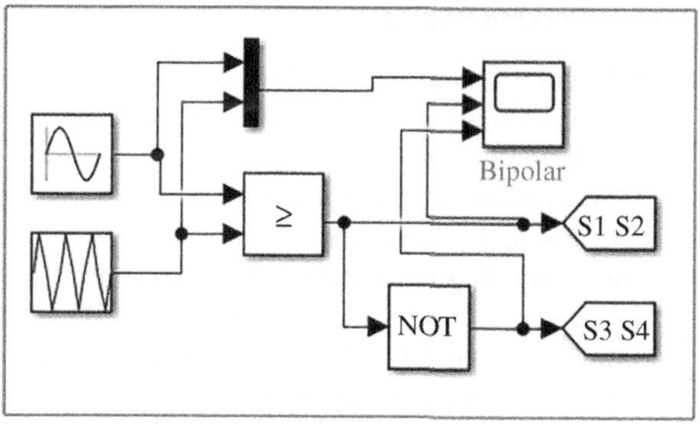

Figure 2.15 Simulink model for the generation of gate pulses.

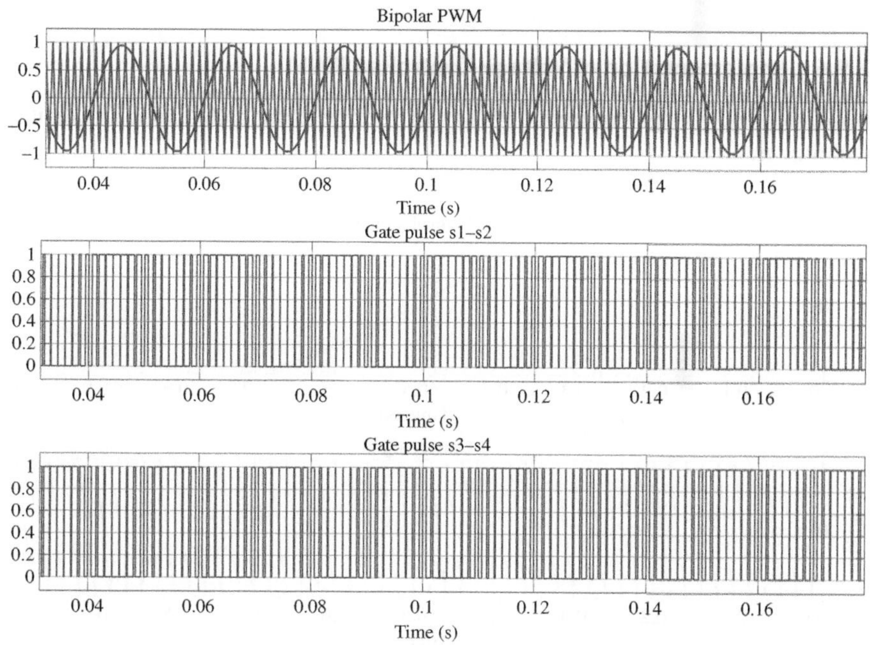

Figure 2.16 Generation of gate pulses.

of a digital signal. This technique is particularly beneficial in applications where precise control of power, speed, or voltage is needed, such as in light dimming, motor drives, and SMPS. The duty cycle of the PWM signal is mainly affected by the input control voltage. This permits flat and continuous control of the output, as opposed to discrete steps.

2.9.5 Working Principle

1) **Reference signal generation:** Normally, a periodic waveform, usually triangular or sawtooth, is produced. This waveform will act as a reference for modulation.
2) **Comparison:** The control voltage is continuously compared to the reference signal using a comparator.
3) **Duty cycle determination:** When the control voltage is greater than the reference signal, the output of the comparator is high, producing the "on" portion of the PWM signal. On the contrary, when the control voltage is low, the output is also low, representing the "off" state.
4) **PWM output:** The resultant signal is a PWM waveform where the duty cycle is proportionate to the control voltage (Figure 2.17).

2.9.6 Implementation of Voltage-Controlled PWM

Implementing voltage-controlled PWM can be done using both analog and digital techniques.

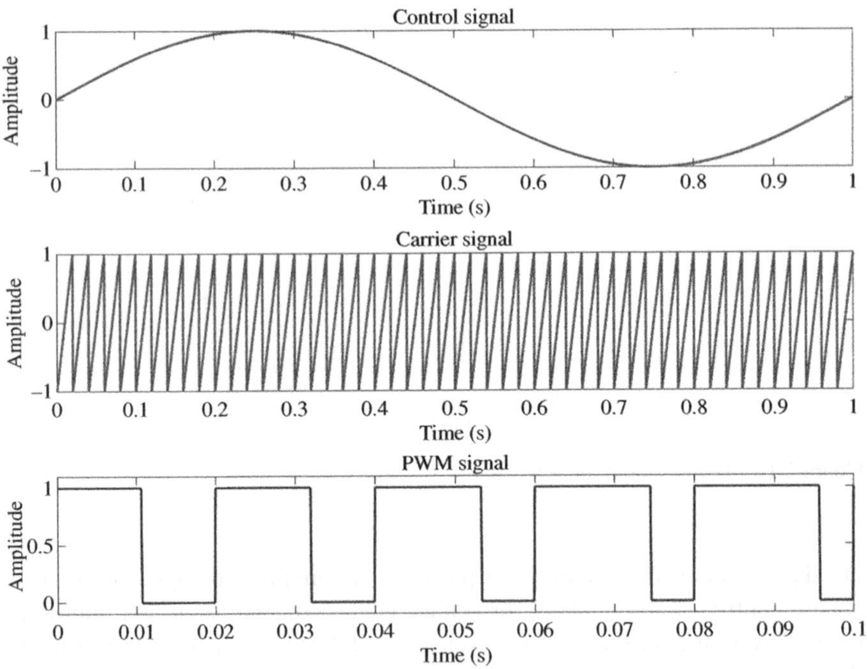

Figure 2.17 Generation of voltage-controlled PWM.

2.9.6.1 Analog Implementation

1) **Voltage-controlled oscillator (VCO):** A VCO produces a frequency-modulated signal whose frequency is controlled by the input voltage.
2) **Comparator circuit:** An operational amplifier designed as a comparator is operated to compare the control voltage with a reference waveform.
3) **Pulse generator:** The comparator's output leads a transistor or MOSFET to produce the PWM signal.
4) **Filter and amplifier:** An optional low-pass filter can be employed to smooth out the PWM signal if an analog output is essential.

2.9.6.2 Example Circuit

1) **Sawtooth generator:** An op-amp integrator circuit is used to generate the reference waveform.
2) **Comparator:** Another op-amp is utilized by comparing the control voltage to the sawtooth waveform.
3) **PWM output stage:** The output from the comparator can drive a transistor or MOSFET switch to change the power delivered to the load.

2.9.6.3 Digital Implementation

1) **Microcontroller/DSP:** A microcontroller or digital signal processor can be programmed to produce PWM signals based on an input control voltage.
2) **ADC:** Converts the control voltage to a digital value.
3) **PWM timer module:** Uses the digital value to adjust the duty cycle of the PWM signal.

2.9.6.4 Example Process

1) **Read control voltage:** The control voltage is inspected by the ADC.
2) **Calculate duty cycle:** The microcontroller analyzes the corresponding duty cycle constructed on the digital value.
3) **Generate PWM signal:** The PWM module within the microcontroller regulates the duty cycle and produces the PWM signal.

2.9.7 Applications of Voltage-Controlled PWM

1) **Motor control:** Voltage-controlled PWM is widely used to regulate the speed of DC motors. By adjusting the duty cycle, consequently, the speed of the motor and the average voltage can be controlled.
2) **Light dimming:** PWM is used in LED dimming to modify the brightness easily. The control voltage determines the duty cycle, which switches the average power supplied to the LEDs.

3) **Switch-mode power supplies:** PWM is crucial in controlling the output voltage of switch-mode power supplies, enhancing efficient power conversion and stable output.
4) **Audio amplifiers:** Class D audio amplifiers utilize PWM to convert the audio signal into a high-frequency PWM signal, and then it is filtered to construct the analog audio output.
5) **Signal generation:** PWM can be employed in signal generators to generate variable-duty-cycle signals for testing and measurement objectives.

2.9.8 Advantages of Voltage-Controlled PWM

1) **Efficiency:** PWM allows for effective power delivery with minimum losses, as the switching devices (transistors or MOSFETs) function either fully on or fully off.
2) **Precision control:** Voltage control supplies precise and continuous control over the output, making it ideal for purposes that require fine adjustments.
3) **Flexibility:** PWM can be used for a broad range of functions, from motor control to power supplies and in signal processing.

2.10 Current-Controlled PWM

Current-controlled PWM is a method used in numerous applications, particularly in motor control and power electronics, to control the current flowing through a load. Not like voltage-controlled PWM, which directly adjusts the output voltage, current-controlled PWM first adjusts the duty cycle of the PWM signal to attain the desired current.

1) **Current measurement:** The current flowing through the load is continuously determined using current sensors like shunt resistors, Hall effect sensors, or current transformers.
2) **Feedback loop:** The measured current is fed back to a controller, which compares it to a reference current.
3) **Error calculation:** The controller computes the error, which is the deviation between the measured current and the reference current.
4) **PWM adjustment:** Based on the error, the controller adjusts the duty cycle of the PWM signal. The main objective is to minimize the error by increasing or decreasing the PWM duty cycle.
5) **Output adjustment:** The adjusted PWM signal drives the power electronics (e.g., transistors, MOSFETs) that control the current through the load.

2.10.1 Applications of Current-Controlled PWM

1) **Motor control:** Used in brushless DC motors, stepper motors, and other kinds of motors to provide precise control over motor current, improving performance and efficiency.
2) **Power supplies:** In switched-mode power supplies, current-controlled PWM helps uphold a stable output current and guards against overcurrent conditions.
3) **LED drivers:** Used to control the current through LEDs to continue consistent brightness and protect against current surges.

2.10.2 Advantages of Current-Controlled PWM

1) **Improved performance:** Affords better control and stability in systems where current regulation is serious.
2) **Protection:** Helps to protect against overcurrent conditions, which can damage components.
3) **Efficiency:** Improves power delivery to the load, thereby enhancing overall system efficiency.

2.10.3 Example Implementation

A typical implementation of a current-controlled PWM system might involve:

1) **Current sensor:** A device to measure current, such as a shunt resistor.
2) **Controller:** A microcontroller or dedicated PWM controller IC that processes the current measurement and adjusts the PWM duty cycle.
3) **Power stage:** MOSFETs or IGBTs that switch the load current on and off according to the PWM signal.
4) **Feedback path:** Circuitry to feed the current measurement back to the controller.

Summary

Current-controlled PWM is a powerful technique for regulating current in various applications. By continuously adjusting the PWM duty cycle based on feedback from current measurements, it ensures precise control, enhancing performance, efficiency, and protection in electronic systems.

2.11 Compensated Pulse Width Modulation

Compensated PWM refers to the use of feedback mechanisms and control strategies to enhance the performance and stability of PWM-driven systems. This approach is vital in various applications, including motor control, power supplies, and audio amplifiers, where precise regulation of voltage, current, or speed is essential.

2.12 Fundamentals of PWM

PWM is a technique used to control the power supplied to electrical devices by varying the duty cycle of a digital signal. The duty cycle is the percentage of one period in which a signal is active. By adjusting the duty cycle, the effective voltage and, consequently, the power delivered to the load can be controlled.

2.13 Need for Compensation

While basic PWM can control power effectively, several factors necessitate the use of compensation:

1) **Nonlinearities:** Devices such as motors exhibit nonlinear behavior, requiring more sophisticated control to maintain performance.
2) **Disturbances:** External disturbances and load variations can affect system stability and performance.
3) **Component variations:** Differences in component values and environmental conditions can lead to discrepancies in system behavior.
4) **Dynamics:** Systems with dynamic responses, such as motors with varying loads, need adaptive control to maintain stability and performance.

2.13.1 Types of Controlling/Compensation Techniques

1) **PID controllers:** proportional-integral-derivative (PID) controllers are commonly used in compensated PWM systems. They adjust the duty cycle based on the error between the desired and actual output, incorporating proportional, integral, and derivative terms to enhance system stability and response (Figure 2.18).
2) **Feedforward control:** Involves predicting the required control action based on known system characteristics and disturbances. This technique is combined with feedback control to improve the system's transient response.

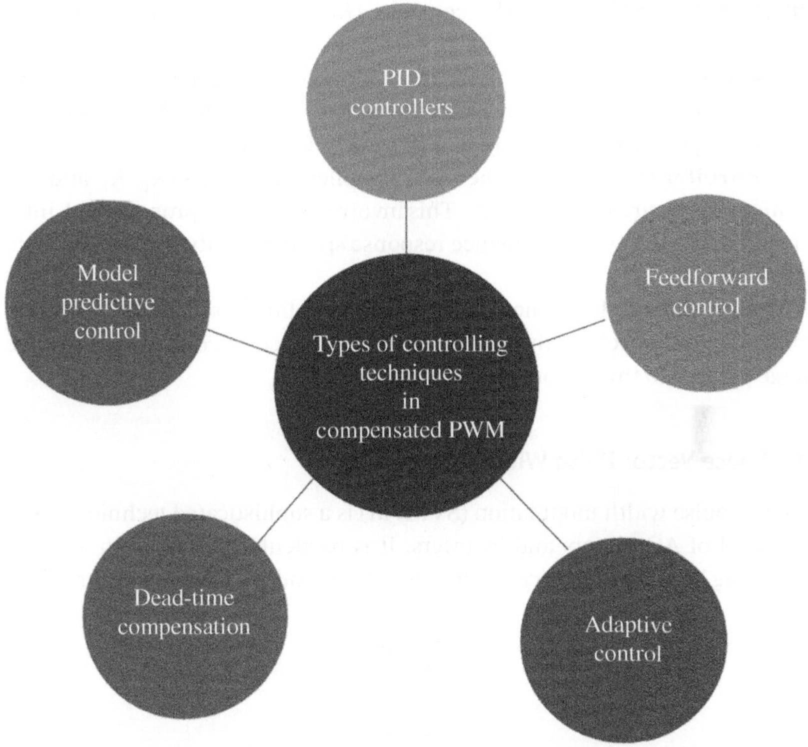

Figure 2.18 Types of controlling techniques in compensated PWM.

3) **Adaptive control:** Adjusts control parameters in real time based on system performance. This approach is suitable for systems with varying dynamics, such as variable load motors.
4) **Model Predictive Control (MPC):** Uses a model of the system to predict future behavior and optimize control actions over a defined horizon. MPC is effective for managing constraints and optimizing performance in complex systems.
5) **Dead-time compensation:** Addresses the delays inherent in switching devices, which can cause distortions and instabilities. Techniques such as predictive control or adding compensating delays help mitigate these effects.

2.13.2 Implementation of Compensated PWM

1) **System modeling:** Begin by modeling the motor and its dynamics. The motor's electrical characteristics can be represented by its resistance (R),

inductance (L), and back EMF constant (K_b). The mechanical dynamics include inertia (J) and friction (B).

2) **Feedback loop design:** Design a feedback loop to regulate the motor's current or speed. The feedback loop consists of a sensor to measure the desired quantity (current or speed), a controller (e.g., PID), and a PWM modulator.
3) **PID controller tuning:** Tune the PID controller parameters (K_p, K_i, and K_d) to achieve the desired performance. This involves setting the proportional, integral, and derivative gains to balance response speed, stability, and steady-state error.
4) **PWM modulation:** Implement the PWM modulation based on the controller output. The duty cycle of the PWM signal is adjusted to control the effective voltage applied to the motor (Figure 2.19).

2.13.3 Space Vector Pulse Width Modulation

Space vector pulse width modulation (SVPWM) is a sophisticated technique used in the control of AC motors and inverters. It is particularly popular in applications requiring high performance, such as EV drives, industrial motor control, and

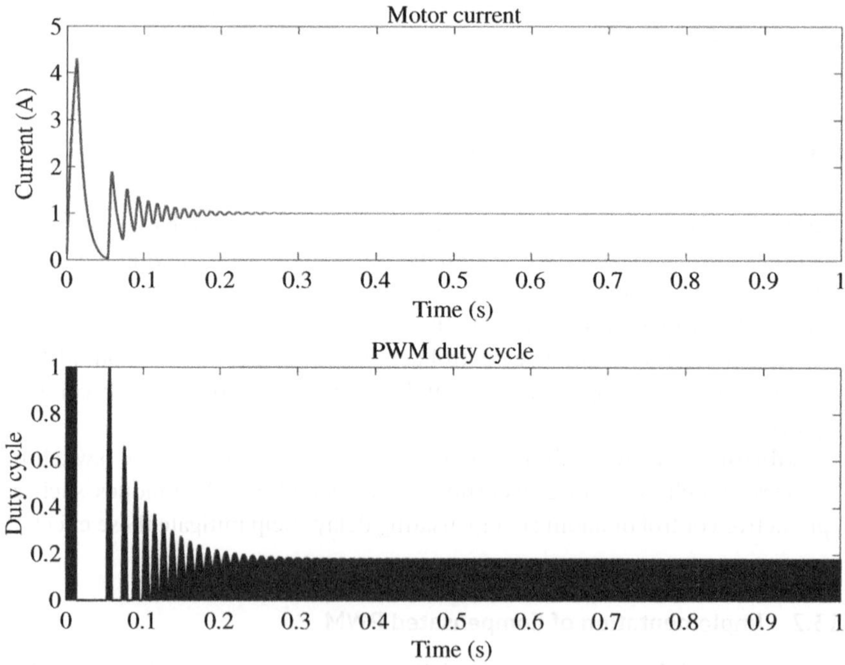

Figure 2.19 PWM control for DC motor.

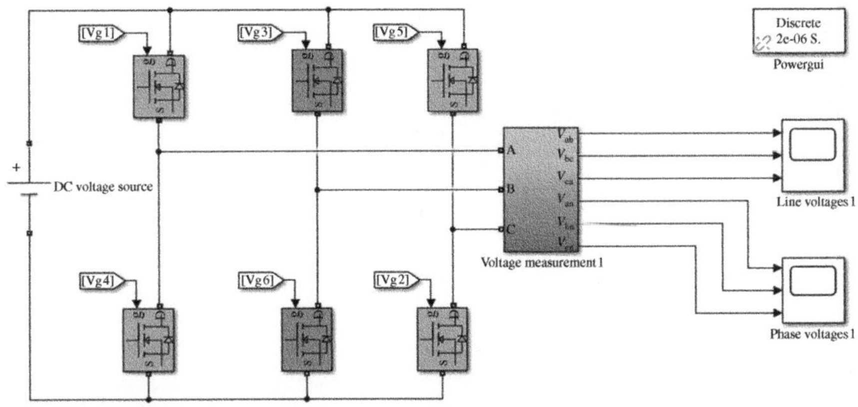

Figure 2.20 Simulation model of SVPWM in the H-bridge inverter.

renewable energy systems. SVPWM is favored for its efficiency, ability to generate sinusoidal output waveforms, and optimal use of the DC bus voltage (Figure 2.20).

2.13.4 Fundamentals of Space Vector PWM

SVPWM is based on the representation of the three-phase system in a two-dimensional space using space vectors. This technique transforms the three-phase signals (A, B, and C) into a two-dimensional plane (α-β) using the Clarke transformation, simplifying the control and modulation processes (Figure 2.21).

2.13.5 Clarke Transformation

The Clarke transformation converts the three-phase time-domain signals into two orthogonal components, α and β. This transformation effectively projects the three-phase system onto a two-dimensional plane, where a rotating vector (space vector) represents the instantaneous values of the three-phase signals (Figure 2.22).

2.13.6 Space Vector Concept

In SVPWM, the space vector represents the state of the inverter and is determined by the switching states of the inverter's switches. There are eight possible switching states in a three-phase inverter: six active states and two zero states. Each state corresponds to a unique space vector in the α-β plane. The active vectors form a hexagon, with each vector separated by 60 degrees. The zero vectors (all switches on or all switches off) are represented at the origin.

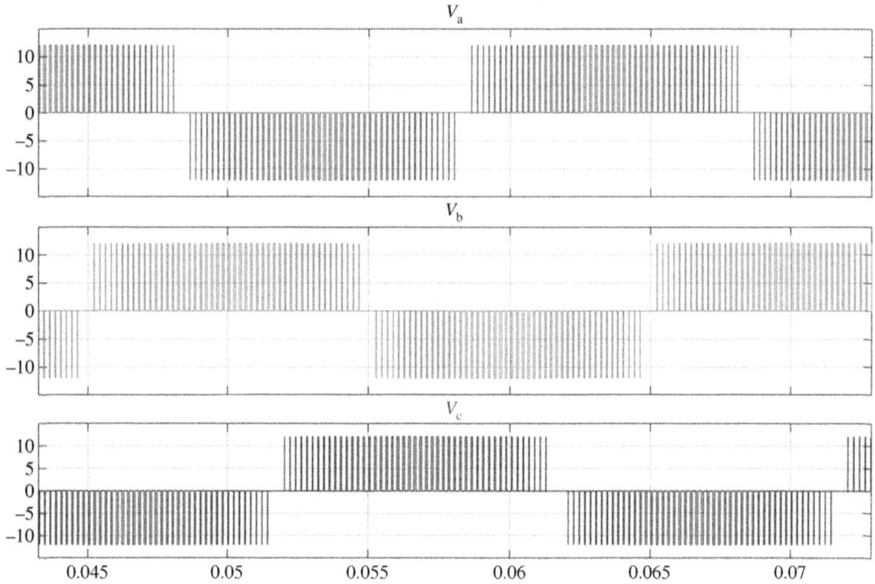

Figure 2.21 Generation of line voltages.

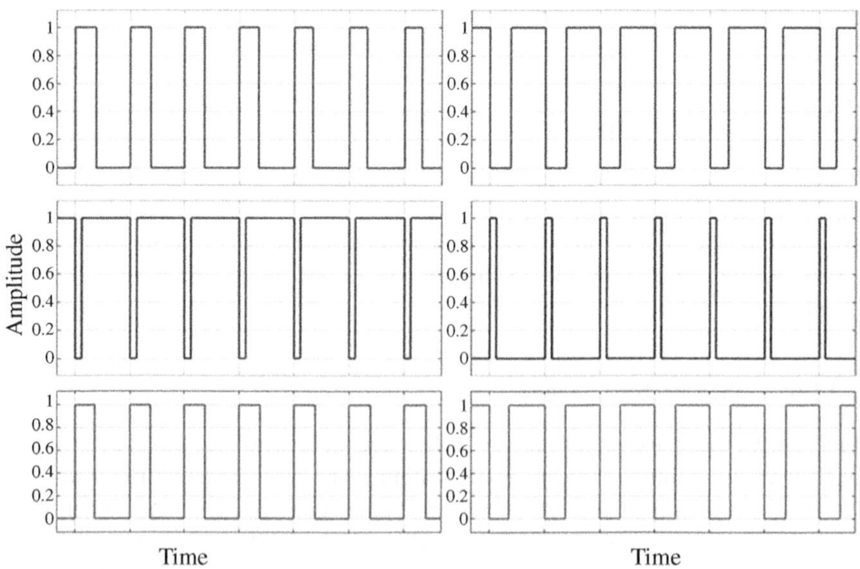

Figure 2.22 Generation of six clock pulses.

2.14 Modulation Process

1) **Reference vector generation:** The desired output voltage vector is calculated based on the desired voltage and frequency for the motor. This vector rotates in the α-β plane, representing the sinusoidal output.
2) **Sector identification:** The α-β plane is divided into six sectors, each corresponding to 60 degrees. The sector in which the reference vector lies is identified. This helps in determining the nearest two active vectors for modulation.
3) **Time calculation:** The time durations for which the inverter should apply the nearest active vectors and the zero vectors are calculated.
4) **Switching sequence:** The inverter switches are controlled to apply the calculated active and zero vectors for the respective time durations. The switching sequence is optimized to minimize switching losses and achieve a balanced operation.

2.14.1 Advantages of SVPWM

1) **Higher utilization of DC bus voltage:** SVPWM can achieve a higher output voltage (approximately 15% more) compared to traditional sinusoidal PWM (SPWM) by using the DC bus voltage more effectively.
2) **Reduced harmonics:** It generates lower total harmonic distortion (THD) in the output waveforms, improving the quality of the output voltage and current.
3) **Efficiency:** SVPWM reduces switching losses by optimizing the switching sequences, making it more efficient for high-power applications.
4) **Dynamic performance:** The technique offers better dynamic response, making it suitable for applications requiring precise control and fast response.

2.14.2 Applications

1) **Electric vehicles:** SVPWM is used in motor drives for EVs due to its high efficiency and precise control capabilities.
2) **Industrial motor control:** It provides smooth and efficient control of industrial motors, improving performance and energy efficiency.
3) **Renewable energy systems:** Inverters in solar and wind power systems use SVPWM for efficient power conversion and grid integration.
4) **HVAC systems:** Heating, ventilation, and air conditioning systems use SVPWM for efficient motor control, enhancing energy savings.

Space Vector PWM is a powerful modulation technique that optimizes the performance of AC motor drives and inverters. By effectively utilizing the DC bus voltage and reducing harmonic distortion, SVPWM enhances the efficiency and dynamic performance of controlled systems. Its application spans across various high-performance domains, making it a critical technology in modern electrical engineering.

2.15 Hybrid Pulse Width Modulation

Hybrid pulse width modulation (HPWM) is a sophisticated technique utilized in power electronics to improve the efficiency, performance, and functionality of inverters and other power conversion systems. This method combines the benefits of various PWM techniques to optimize the overall system performance. Here, we delve into the concept, types, applications, and advantages of HPWM.

2.15.1 Concept of Hybrid PWM

PWM is a method used to control the amount of power delivered to an electrical load by varying the width of the pulses in a pulse train. Traditional PWM methods include SPWM, SVPWM, and hysteresis PWM, each with distinct characteristics and advantages. Hybrid PWM aims to integrate the strengths of these techniques to mitigate their individual limitations.

2.15.2 Types of Hybrid PWM

1) SPWM and SVPWM Hybrid:
   SPWM: Generates a PWM signal by comparing a sinusoidal reference signal with a triangular carrier signal. It is simple and effective but can result in higher switching losses and lower efficiency.
   SVPWM: Utilizes space vector theory to generate PWM signals, improving the DC bus utilization and reducing harmonic distortion. However, it is more complex to implement.
   Hybrid approach: Combines SPWM for lower switching frequencies and SVPWM for higher frequencies. This hybrid technique optimizes efficiency and harmonic performance across a broader range of operating conditions (Figure 2.23).
2) Hysteresis and Fixed-Frequency PWM Hybrid
   Hysteresis PWM: Controls the inverter output by maintaining the current within a hysteresis band around the reference current. It offers fast dynamic response but can lead to variable switching frequency, causing EMI issues.
   Fixed-frequency PWM: Ensures a constant switching frequency, simplifying filter design and reducing EMI, but may suffer from slower dynamic response.
   Hybrid approach: Uses hysteresis control during rapid transients for quick response and switches to fixed-frequency PWM during steady-state operation to maintain EMI control and consistent performance.
3) Multi-carrier PWM Techniques
   Utilizes multiple carrier signals of different frequencies or phases to create a hybrid modulation scheme that can reduce harmonic distortion and improve power quality.

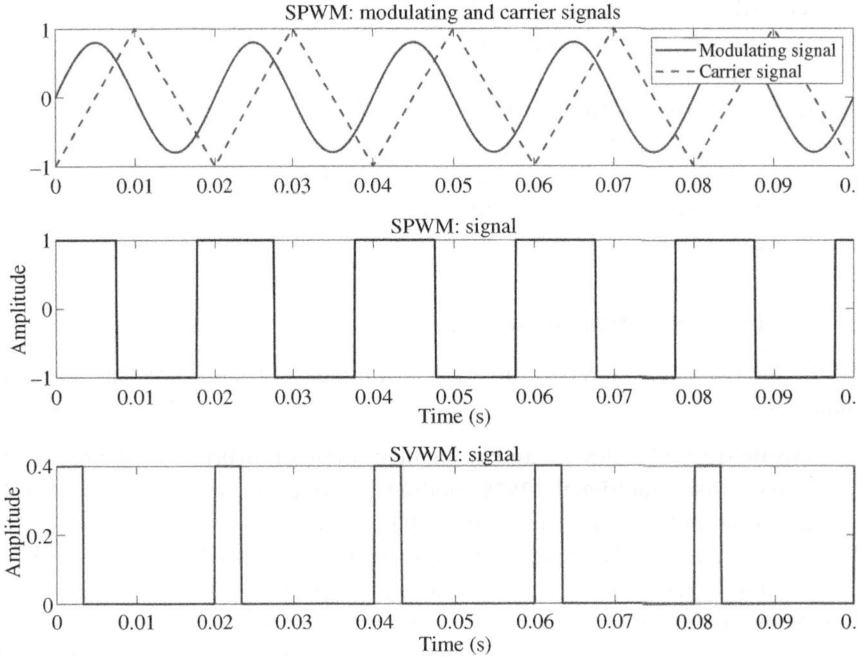

Figure 2.23 SPWM modulating signals in hybrid PWM.

2.15.3 Applications of Hybrid PWM

1) **Electric vehicles (EVs):** EV powertrain inverters benefit from hybrid PWM by enhancing efficiency, extending battery life, and reducing EMI.
2) **Renewable energy systems:** Solar inverters and wind turbine converters use hybrid PWM to maximize power extraction and ensure stable grid integration with minimized harmonics.
3) **Industrial drives:** Motor drives in industrial applications leverage hybrid PWM to achieve precise control, reduce losses, and improve reliability.
4) **Uninterruptible power supplies (UPS):** Hybrid PWM in UPS systems enhances power quality and efficiency, ensuring reliable backup power with minimal harmonic distortion.

2.15.4 Advantages of Hybrid PWM

1) **Enhanced efficiency:** By combining different PWM techniques, hybrid PWM can adapt to varying load conditions, optimizing switching patterns to reduce losses.
2) **Improved power quality:** Hybrid PWM minimizes harmonic distortion and EMI, ensuring cleaner power delivery to the load.

3) **Dynamic performance:** The ability to switch between modulation techniques enables faster response to load changes and transient conditions, improving overall system stability and performance.
4) **Flexibility and adaptability:** Hybrid PWM offers the flexibility to adapt modulation strategies based on real-time operating conditions, enhancing the robustness and versatility of the power conversion system.
5) **Reduced EMI:** By managing the switching frequency and pattern, hybrid PWM can mitigate EMI, which is crucial for sensitive applications.

2.15.5 Implementation Challenges

While hybrid PWM offers numerous advantages, it also presents certain challenges:

1) **Complexity:** The design and implementation of hybrid PWM are more complex than traditional PWM methods. It requires sophisticated control algorithms and real-time processing capabilities.
2) **Cost:** The additional complexity and processing power needed for hybrid PWM can increase the cost of the power electronics system.
3) **Optimization:** Determining the optimal combination and switching criteria for different PWM techniques requires careful analysis and optimization, which can be time-consuming.

Hybrid PWM represents a significant advancement in power electronics, offering a versatile solution to enhance the performance, efficiency, and reliability of various power conversion systems. By integrating the strengths of multiple PWM techniques, hybrid PWM can meet the demanding requirements of modern applications such as EVs, renewable energy systems, and industrial drives. Despite its complexity, the benefits of hybrid PWM make it a valuable tool in the quest for more efficient and effective power management.

2.16 MATLAB Exercises

1) Implement a triangle comparison-based PWM for a single-phase inverter. Compare a sinusoidal reference signal with a triangular carrier wave to generate the PWM signal. Analyze the output voltage and harmonic content.
2) Simulate bus clamping PWM for a three-phase inverter. Implement the technique to clamp one of the phases to the DC bus during certain periods of the switching cycle. Evaluate the inverter's efficiency and harmonic performance.

3) Develop a unipolar PWM scheme for an H-bridge inverter. Compare two sinusoidal reference signals (one positive and one negative) with a triangular carrier to generate the PWM signals. Analyze the output voltage waveform and its harmonics.
4) Simulate a bipolar PWM for a single-phase full-bridge inverter. Use a single sinusoidal reference signal and compare it with a triangular carrier wave to switch the inverter's transistors. Evaluate the output voltage waveform and harmonic distortion.
5) Design a voltage-controlled PWM for a buck converter. Use a PI controller to regulate the output voltage by adjusting the duty cycle of the PWM signal. Simulate the system under varying load conditions and observe the voltage regulation performance.
6) Implement a current-controlled PWM for a DC motor drive. Use hysteresis control to maintain the motor current within a specified band around the reference current. Analyze the motor's speed response and current waveform.
7) Simulate a compensated PWM for a boost converter with load variations. Implement a feedforward compensation to improve the transient response. Compare the performance with and without compensation under step changes in load.
8) Implement SVPWM for a three-phase inverter. Use space vector theory to generate the PWM signals and drive an induction motor. Analyze the motor's speed, torque, and harmonic content of the output voltage.
9) Design a hybrid PWM scheme combining SPWM and hysteresis PWM for an inverter. Use SPWM at low frequencies and switch to hysteresis PWM at higher frequencies. Evaluate the system's efficiency and dynamic response under varying load conditions.

2.17 Detailed Step-by-Step Exercise for Unipolar PWM

Objective: To design and simulate a unipolar PWM scheme for a single-phase H-bridge inverter and analyze the output voltage waveform and harmonic distortion.

Steps

1) **Model setup:**
   - Open MATLAB/Simulink and create a new model.
   - Add an H-bridge inverter circuit using power electronic switches (e.g., IGBTs or MOSFETs).

2) **Reference signals:**
   - Create two sinusoidal reference signals with the same frequency but 180 degrees out of phase.
   - Use sin function blocks or signal generators for the reference signals.
3) **Carrier signal:**
   - Generate a triangular carrier signal with a frequency much higher than the reference signals.
   - Use a Repeating Sequence block or similar to create the triangular waveform.
4) **PWM generation:**
   - Compare the reference signals with the carrier signal using Relational Operator blocks.
   - Generate the gating signals for the H-bridge switches based on the comparison.
5) **Control logic:**
   - Implement the control logic to ensure proper switching of the H-bridge inverter.
   - Ensure the switches operate in complementary pairs to avoid short-circuiting the DC supply.
6) **Simulation:**
   - Run the simulation for a few cycles of the reference signal.
   - Use Scope blocks to observe the output voltage waveform.

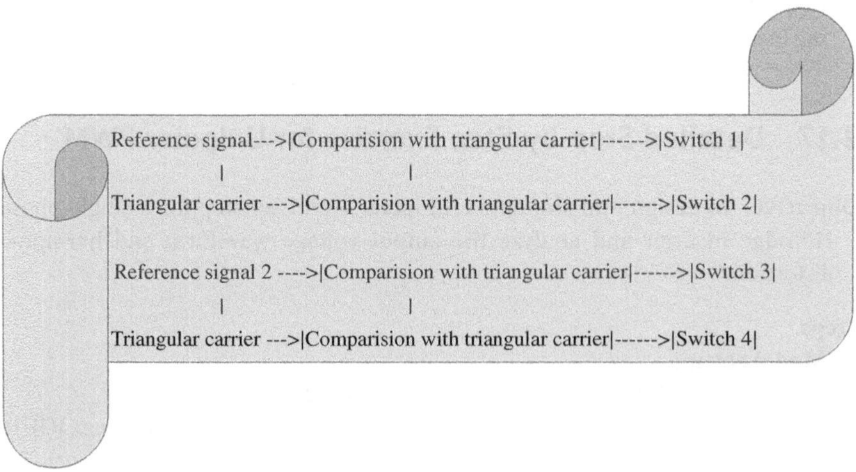

Figure 2.24 Conceptual diagram. ***[Hint: Single-phase H-bridge inverter simulation model is already shown in figure 7].***

7) **Analysis:**
   - Analyze the output voltage waveform using FFT analysis to determine the harmonic content.
   - Compare the harmonic distortion of the unipolar PWM with other PWM techniques (e.g., bipolar PWM).
8) **Parameter tuning:**
   - Experiment with different carrier frequencies and modulation indices.
   - Observe the impact on the output voltage waveform and harmonic distortion.

2.17.1 Simulation Diagram

Here is a conceptual diagram for generating a Simulink model of the unipolar PWM setup (Figure 2.24):

This detailed step-by-step approach can be similarly applied to other PWM techniques and exercises.

2.18 Experimental Setup and Analysis

1) Triangle Comparison-Based PWM

Experimental Setup

Components: Microcontroller (e.g., Arduino, STM32), DC power supply, oscilloscope, function generator, H-bridge inverter circuit, and load (e.g., resistor or motor).

Steps

1) Generate a triangular carrier signal using a function generator.
2) Generate a sinusoidal reference signal using the microcontroller.
3) Use the microcontroller to compare the reference and carrier signals to generate PWM signals.
4) Drive the H-bridge inverter with these PWM signals.
5) Connect the load to the inverter.

Analysis

1) Use the oscilloscope to measure the output voltage waveform.
2) Perform FFT analysis on the waveform to determine harmonic content.
3) Bus Clamping PWM.

Experimental Setup

Components: Similar to the above setup, with added logic to implement bus clamping.

Steps

1) Modify the microcontroller code to include bus clamping logic.
2) Generate PWM signals and drive the H-bridge inverter.
3) Connect the load.

Analysis

Measure output voltage waveform and harmonics.
Compare efficiency and harmonic performance with standard PWM.
3) Unipolar PWM

Experimental Setup

Components: Same as triangle comparison-based PWM.

Steps

1) Generate two sinusoidal reference signals 180 degrees out of phase.
2) Compare each reference with a triangular carrier using the microcontroller.
3) Generate PWM signals for the H-bridge inverter.
4) Connect the load.

Analysis

Measure and compare the output voltage waveform with bipolar PWM.
Analyze harmonic distortion.
4) Bipolar PWM

Experimental Setup

Components: Same as above.

Steps

1) Generate a single sinusoidal reference signal.
2) Compare it with a triangular carrier using the microcontroller.
3) Generate PWM signals for the H-bridge inverter.
4) Connect the load.

Analysis

Measure the output voltage waveform.
Perform harmonic analysis.
5) Voltage-controlled PWM

Experimental Setup

Components: Buck converter circuit, microcontroller, DC power supply, oscilloscope, and variable load.

Steps

1) Implement a PI controller in the microcontroller to adjust the duty cycle based on the output voltage.
2) Generate PWM signals to control the buck converter.
3) Connect the load.

Analysis

Measure output voltage stability under varying load conditions.

Analyze transient response.

6) Current-controlled PWM

Experimental Setup

Components: DC motor, H-bridge inverter, current sensor, microcontroller, DC power supply.

Steps

1) Use the current sensor to measure motor current.
2) Implement hysteresis control in the microcontroller to adjust PWM signals.
3) Drive the motor with the H-bridge inverter.

Analysis

Measure motor speed and current waveform.

Analyze control accuracy and response time.

7) Compensated PWM

Experimental Setup

Components: Boost converter circuit, microcontroller, voltage sensor, oscilloscope, and variable load.

Steps

1) Implement feedforward compensation in the microcontroller.
2) Generate PWM signals to control the boost converter.
3) Connect the load.

Analysis

Measure output voltage stability with and without compensation.

Analyze transient response and regulation performance.

8) Space vector PWM (SVPWM)

Experimental Setup

Components: Three-phase inverter, microcontroller, oscilloscope, and three-phase motor.

Steps

1) Implement the SVPWM algorithm in the microcontroller.
2) Generate PWM signals for the three-phase inverter.
3) Connect the three-phase motor.

Analysis

Measure motor speed, torque, and output voltage waveform.

Perform harmonic analysis on the output voltage.

9) Hybrid PWM

Experimental Setup

Components: Similar to unipolar/bipolar PWM setups, depending on the hybrid method used.

Steps

1) Implement hybrid PWM logic (e.g., combining SPWM and hysteresis) in the microcontroller.
2) Generate PWM signals to drive the inverter.
3) Connect the load.

Analysis

Measure and compare efficiency, harmonic distortion, and dynamic response.

General Analysis Techniques

For each PWM technique, the following analysis techniques can be applied:

1) **Waveform analysis:** Use an oscilloscope to capture and analyze the output voltage waveform.
2) **Harmonic analysis:** Perform FFT analysis to determine the harmonic content of the output waveform.
3) **Efficiency measurement:** Measure the input and output power to determine the efficiency of the inverter or converter.
4) **Transient response:** Analyze the system's response to step changes in load or input conditions.
5) **Stability analysis:** Examine the stability of voltage and current under different operating conditions.

By following these steps and using the specified components, you can experimentally implement and analyze the various PWM techniques.

2.19 Summary

PWM is a broadly used technique in the area of power electronics for limiting power delivery to various electrical devices. Each PWM method has exclusive characteristics and applications.

1) **Triangle comparison-based PWM:** This technique compares a triangular carrier waveform with a reference signal to produce the PWM signal. It is simple and widely used for its ease of execution.
2) **Bus clamping PWM:** This technique decreases switching losses by clamping the bus voltage to zero through certain intervals, thereby reducing the number of switching events.
3) **Unipolar PWM:** In Unipolar PWM, the output switches between zero and a positive or negative voltage level, which decreases switching losses and improves efficiency.
4) **Bipolar PWM:** Here, the output switches between positive and negative voltage levels. It's easier but results in greater switching losses compared to unipolar PWM.
5) **Voltage-controlled PWM:** This technique adjusts the duty cycle of the PWM signal based on a control voltage to regulate the output voltage.
6) **Current-controlled PWM:** Like voltage-controlled PWM, but it regulates the output current by adjusting the duty cycle based on a control current signal.
7) **Compensated PWM:** This method balances nonlinearities and other imperfections in the system to improve accuracy and performance.
8) **Space vector PWM:** An advanced technique that represents three-phase voltages as vectors in a two-dimensional plane, optimizing the inverter's output waveform and reducing harmonic distortion.
9) **Hybrid PWM:** Combines multiple PWM techniques to leverage their respective advantages, optimizing performance and efficiency for specific applications.

The choice of PWM technique depends on the specific requirements of the application, such as the desired control precision, efficiency, complexity, and cost. Each PWM method has unique characteristics that make it suitable for certain applications. Triangle comparison PWM and bipolar PWM offer simplicity and robustness, making them ideal for basic motor control and high-power applications. Unipolar PWM and SVPWM provide higher efficiency and lower harmonic distortion, suitable for motor drives and efficient inverters. Voltage-controlled PWM and current-controlled PWM offer precise control for regulated power supplies, while compensated PWM and hybrid PWM provide improved stability and tailored performance for advanced applications. Understanding these differences helps engineers select the most appropriate PWM technique for their needs.

Appendix: Comparative Analysis of Different PWM Techniques

A comparative analysis of triangle comparison PWM, bus clamping PWM, unipolar PWM, bipolar PWM, voltage-controlled PWM, current-controlled PWM, compensated PWM, SVPWM, and hybrid PWM based on their components, applications, advantages, disadvantages, switching frequency, control complexity, efficiency, voltage/current ratings, and cost is made.

Triangle Comparison PWM

- **Components needed:** Few (Comparator, Triangular Wave Generator, and Control Logic)
- **Applications:** Basic motor control, simple inverters
- **Advantages:** Simple to implement, well-suited for low-frequency applications
- **Disadvantages:** Less efficient at high frequencies, higher harmonic content
- **Switching frequency:** Low to moderate
- **Control complexity:** Low
- **Efficiency:** Moderate
- **Voltage/Current ratings:** Low to moderate
- **Cost:** Low

Triangle Comparison PWM is known for its simplicity. It involves comparing a reference signal with a triangular carrier wave to generate the PWM signal. This method is straightforward and suitable for low-frequency applications, making it ideal for basic motor control and simple inverters. However, its efficiency drops at higher frequencies due to increased switching losses and harmonic distortion.

Bus Clamping PWM

- **Components needed:** Moderate (Controller, Bus Clamping Circuit)
- **Applications:** Motor control, inverters with reduced switching losses
- **Advantages:** Reduced switching losses, lower harmonic distortion
- **Disadvantages:** Increased control complexity, may require specialized hardware
- **Switching frequency:** Moderate
- **Control complexity:** Moderate
- **Efficiency:** High
- **Voltage/Current ratings:** Moderate to high
- **Cost:** Moderate

  Bus Clamping PWM offers reduced switching losses by clamping the bus voltage to avoid unnecessary switching. This results in lower harmonic distortion and improved efficiency, making it suitable for motor control and inverter applications. However, the increased control complexity and potential need for specialized hardware can be a drawback.

Unipolar PWM

- **Components needed:** Moderate (H-Bridge Inverter, Control Logic)
- **Applications:** Motor drives, power inverters, where EMI is a concern
- **Advantages:** Lower switching losses, reduced EMI, higher effective switching frequency
- **Disadvantages:** More complex control logic compared to bipolar PWM
- **Switching frequency:** High
- **Control complexity:** Moderate
- **Efficiency:** High
- **Voltage/Current ratings:** Low to high
- **Cost:** Moderate

Unipolar PWM switches the voltage applied to the load between zero and a positive or negative value, leading to lower switching losses and reduced EMI. This technique is beneficial for motor drives and power inverters. The higher effective switching frequency improves performance, although it involves more complex control logic.

Bipolar PWM

- **Components needed:** Few (H-Bridge Inverter, Control Logic)
- **Applications:** Motor drives, power inverters for high-power applications
- **Advantages:** Simpler control logic, robust, and reliable
- **Disadvantages:** Higher switching losses, higher harmonic distortion
- **Switching frequency:** Low to moderate
- **Control complexity:** Low
- **Efficiency:** Moderate
- **Voltage/Current ratings:** Moderate to high
- **Cost:** Low

Bipolar PWM alternates the output voltage between positive and negative supply voltages, resulting in a robust and reliable system. Its simplicity in control logic makes it suitable for high-power applications like motor drives and power inverters. However, it suffers from higher switching losses and harmonic distortion.

Voltage-Controlled PWM

- **Components needed:** Moderate (Voltage Control Circuit, Comparator)
- **Applications:** Motor drives, power supplies needing a stable voltage
- **Advantages:** Precise voltage control, stable output voltage

- **Disadvantages:** Requires additional control circuitry, potential for voltage overshoot
- **Switching Frequency:** Moderate
- **Control Complexity:** Moderate
- **Efficiency:** High
- **Voltage/Current Ratings:** Low to moderate
- **Cost:** Moderate

Voltage-controlled PWM provides precise voltage regulation by modulating the duty cycle based on the desired voltage level. This technique is ideal for applications needing s voltage, such as motor drives and power supplies. The main disadvantages are the need for additional control circuitry and the potential for voltage overshoot.

Current-Controlled PWM

- **Components needed:** Moderate (Current Sensor, Control Logic)
- **Applications:** Motor drives, power supplies with current regulation
- **Advantages:** Precise current control, inherent overcurrent protection
- **Disadvantages:** Requires current-sensing components, more complex control logic
- **Switching frequency:** Moderate
- **Control complexity:** High
- **Efficiency:** High
- **Voltage/Current ratings:** Low to moderate
- **Cost:** Moderate

Current-controlled PWM adjusts the duty cycle to maintain a desired current level. This technique offers precise current control and inherent overcurrent protection, making it suitable for motor drives and regulated power supplies. The need for current-sensing components and complex control logic can be challenging.

Compensated PWM

- **Components needed:** Moderate to high (Feedback Loop, Compensation Network)
- **Applications:** Precision motor drives, regulated power supplies
- **Advantages:** Improved system stability, better transient response
- **Disadvantages:** Requires additional feedback and compensation components, complex design
- **Switching frequency:** Moderate
- **Control complexity:** High
- **Efficiency:** High
- **Voltage/Current ratings:** Low to moderate
- **Cost:** High

Compensated PWM uses feedback and compensation networks to improve system stability and transient response. This technique is ideal for precision motor drives and regulated power supplies. However, the additional components and complex design increase the overall cost and complexity.

Space Vector PWM (SVPWM)

- **Components needed:** High (DSP or Microcontroller, Complex Control Logic)
- **Applications:** High-performance motor drives, efficient inverters
- **Advantages:** Optimizes switching patterns, reduces harmonic distortion, improves efficiency
- **Disadvantages:** Complex control algorithm, requires powerful processing capabilities
- **Switching frequency:** High
- **Control complexity:** High
- **Efficiency:** Very High
- **Voltage/Current ratings:** Moderate to high
- **Cost:** High

SVPWM optimizes the switching patterns of the inverter to minimize harmonic distortion and improve efficiency. This technique is widely used in high-performance motor drives and efficient inverters. The main drawback is the complexity of the control algorithm, which requires powerful processing capabilities such as DSPs or microcontrollers.

Hybrid PWM

- **Components needed:** High (Combination of Multiple PWM Techniques, Control Logic)
- **Applications:** Advanced motor control, power conversion applications
- **Advantages:** Combines the advantages of multiple techniques, tailored performance for specific applications
- **Disadvantages:** Increased system complexity, potential for higher cost
- **Switching frequency:** Variable
- **Control complexity:** Very high
- **Efficiency:** Very high
- **Voltage/Current ratings:** Moderate to high
- **Cost:** Very high

Hybrid PWM combines the strengths of various PWM techniques to achieve tailored performance for specific applications. It is used in advanced motor control and power conversion systems, offering high efficiency and optimized performance. However, the increased system complexity and higher cost can be significant drawbacks.

3

PWM for Power Converters

3.1 Introduction

Pulse width modulation (PWM) is an essential control methodology employed in power electronic converters to properly regulate voltage, current, and power flow. By altering the duty cycle of switching devices while maintaining a fixed switching frequency, PWM facilitates accurate regulation of average output values with minimal power loss. PWM is exceptionally appropriate for contemporary applications, including motor drives, renewable energy systems, and power supplies.

In PWM-based converters, semiconductor devices like MOSFETs and IGBTs predominantly function in switching mode instead of linear mode. This markedly decreases conduction and switching losses, enhancing overall system efficiency. Furthermore, PWM facilitates compact converter designs as elevated switching frequencies diminish the dimensions of passive components such as inductors and capacitors.

A significant benefit of PWM is its adaptability in contouring output waveforms. Effective modulation schemes will reduce harmonic distortion and enhance power quality. This is especially important in grid-connected systems where stringent harmonic requirements must be adhered to.

Overall, PWM works as the backbone of modern power electronics, providing intelligent, efficient, and adjustable energy conversion across a wide variety of voltage and power levels.

3.2 Controlled Rectifiers

A controlled rectifier converts AC power into controllable DC power by substituting diodes with controllable semiconductor devices such as thyristors. Unlike uncontrolled rectifiers, the output voltage of controlled rectifiers will be modified by altering the firing angle of the thyristors. This controllability makes them appropriate for applications like DC motor drives and battery charging systems.

The mechanism of operation relies on delaying the turn-on of transistors with respect to the AC supply voltage. By increasing the firing angle, the average DC output voltage drops. This offers smooth and continuous control of DC output without changing the supply or load conditions.

Controlled rectifiers will function in several modes, such as rectification mode and inversion mode, depending on the firing angle. This directional power flow capacity is useful in regenerative braking applications where power is supplied back to the source.

Despite their advantages, controlled rectifiers introduce substantial harmonics and have poorer power factors compared to diode rectifiers. Hence, PWM-based rectification approaches are increasingly used in new high-performance systems.

3.2.1 Continuous Firing Angle Regulation

Pulse output for controlled rectifiers focuses on creating accurate and synchronized firing pulses for thyristors. These pulses must be properly timed with regard to the AC supply to achieve correct conduction intervals. Early pulse generation methods used analog circuits such as RC phase shift networks.

With developments in electronics, digital pulse generation techniques using microcontrollers, DSPs, and phase-locked loops have become increasingly more common. These approaches offer improved accuracy, versatility, and noise immunity. Synchronization with the supply voltage is critical to avoid misfiring and commutation failure.

Different topologies, such as single-pulse control, multiple-pulse control, and PWM-based control, are applied depending on application needs. A multiple-pulse system will minimize harmonics and increase output waveform quality compared to single-phase solutions.

Modern pulse generation topologies also integrate production and fault detection features. These features guarantee reliable operation under abnormal conditions such as overcurrent, voltage dips, and device failures.

3.2.2 Integral Cycle Control

Integral cycle control, also known as burst firing or on–off control, is a power control system where the supply to the load is switched ON and OFF for entire cycles of the input AC waveform. Instead of modulating the duty cycle inside a single cycle as in typical PWM, ICC manages power by altering the number of complete cycles applied to the load over a set control interval. This approach is extensively utilized in AC voltage controllers and resistive heating applications.

In integral cycle control, the switching device conducts an integral number of complete cycles and remains OFF for another set of full cycles. The ratio of ON cycles to the total number of cycles determines the average power delivered to the load. Since switching occurs at zero crossings of the AC waveform, switching losses and electromagnetic interference (EMI) are greatly minimized compared to phase angle control.

One of the key advantages of ICC is its simplicity of deployment. The control circuitry is less advanced, and synchronization with the supply frequency is straightforward. Additionally, because the waveform inside each cycle stays undistorted, harmonic generation is lower than that of phase-controlled rectifiers, making ICC suited for applications with relaxed dynamic response requirements.

However, integral cycle control suffers from poor output power resolution at the low frequencies and introduces low-frequency power pulsations. These pulsations will generate heat cycling and mechanical stress in certain loads. As a result, ICC is best suited for systems with considerable thermal inertia such as industrial heaters and furnaces, where fast dynamic response is not necessary (Table 3.1).

Table 3.1 Comparison of power electronic control strategies.

Parameter	Phase angle control	PWM control	Integral cycle control
Control principle	Delays the firing angle within each AC cycle	Varies pulse width at high switching frequency	Switches complete AC cycles ON and OFF
Switching frequency	Line frequency	High frequency (kHz range)	Line frequency
Output waveform	Highly distorted	Near sinusoidal (with filtering)	Undistorted within ON cycles
Harmonic content	High	Low to moderate	Low (but low-frequency pulsations)
Switching losses	Moderate	Higher due to high frequency	Very low
EMI	High	Moderate to high	Very low
Control resolution	Continuous within a cycle	Very fine and precise	Coarse at low power levels
Dynamic response	Moderate	Fast	Slow
Circuit complexity	Simple	Complex	Very simple
Typical applications	Light dimmers, AC regulators	Motor drives, SMPS, inverters	Industrial heating, furnaces

3.3 Pulse Generation Topologies for Controlled Rectifiers

To ensure effective performance of switching devices such as thyristors and power transistors, controlled rectifiers widely employed in power electronic systems for AC to DC conversion with regulated output voltage require accurately timed gate pulses. The performance, reliability, and controllability of these rectifiers greatly depend on the pulse generation technology chosen. Several pulse generation topologies have been devised and are routinely employed in control rectifier systems, depending on the needed precision, complexity, and application domain.

3.3.1 Phase-Locked Loop-Based Pulse Generation

In phase-locked loop (PLL)-based topologies, the generated firing pulses are synchronized with the input AC supply using phase-locked loop circuits. PLL continually tracks the phase of the AC waveform, ensuring perfect alignment of firing pulses with zero crossing points of the supply voltage. This fine synchronization is important for maintaining a uniform firing angle and stable rectification during supply frequency changes. Such topologies are particularly suited for high-precision control systems where reliable synchronization with the AC mains is required.

3.3.2 Microcontroller and Digital Signal Processor-Based Topologies

Microcontrollers and digital signal processors (DSPs) are extensively employed for pulse generation in modern controlled rectifiers due to their flexibility and computational capability. These digital platforms create gate pulses depending on measured system variables and predefined control settings. Advanced control techniques, including PID control and adaptive control schemes, will be employed to ensure accurate voltage and current regulation. These techniques are currently employed in advanced power electronics systems such as motor drives, renewable energy interfaces, and smart power converters.

3.3.3 Zero-Crossing Detector-Based Logic Circuits

Zero-crossing detectors are employed to identify the instants at which the AC input voltage crosses zero. Using this information, logic circuits such as counters, flip-flops, and delay generators produce approximately timed firing pulses to achieve phase angle control. This technique offers a simple and cost-effective solution for pulse production, making it suited for low-cost controlled rectifiers where ease of implementation and reliability are major design factors.

3.3.4 Analog Timing Circuit-Based Pulse Generation

Analog pulse generation approaches involve timing circuits created from operational amplifiers, RC networks, or integrated timing devices such as the 555 timer IC. By inserting a regulated time delay with respect to the AC reference waveform, firing pulses of the design with timing will be created, although limited in flexibility and precision compared to digital alternatives. Analog timing circuits will be applied in legacy systems and in no applications where a digital application is impractical.

3.3.5 FBGA-Based Topologies

FPGA-based pulse generation techniques provide a very flexible and high-performance option for controlled rectifiers. FBGA offers accurate timing control, parallel processing, and the construction of many control loops simultaneously. Their ability to handle high-frequency switching and real-time control makes them particularly ideal for industrial-grade converters, high-power drives, and advanced power electronic systems demanding deterministic performance.

3.3.6 Direct Digital Synthesis-Based Circuits

Direct digital synthesis (DDS) techniques leverage digital methods to generate waveforms with carefully regulated frequency and phase characteristics. In controlled rectifier applications, a DDS-based system provides precise control of firing angle and switching sequences. These topologies are suitable for applications demanding high-frequency operation and fine phase precision such as high-speed motor drives and specialized power conversion systems.

3.4 Single-Phase Inverter

The primary objective of a single-phase inverter is to generate an AC output waveform that under optimal conditions closely approaches the standard sinusoidal waveform of AC power supplied by the utility grid, exhibiting

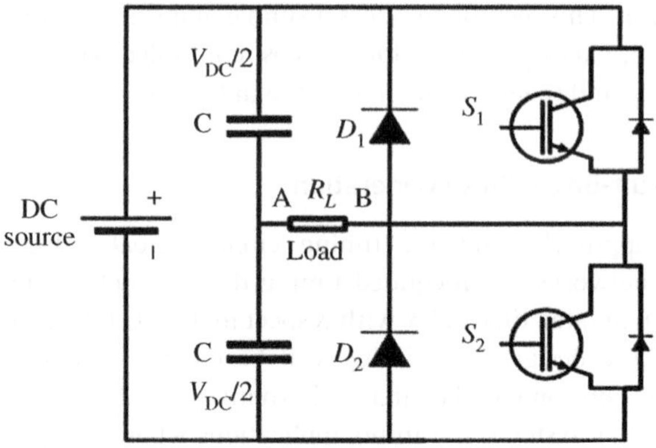

Figure 3.1 Half H-bridge inverter. *Source:* Monolithic Power Systems (2025).

minimal harmonic distortion. A high-quality sinusoidal waveform is essential to prevent harmonic distortion and enable the proper functioning of various loads, including electric motors and sensitive electrical devices.

3.4.1 Circuit Topologies

The circuit configurations for single-phase inverters are as follows:

Inverter Half-Bridge (Figure 3.1):

The half-bridge inverter represents the basic design of a single-phase inverter. It comprises two controlled switching devices, typically MOSFETs, IGBTs, or bipolar transistors, arranged in series over a DC voltage source. Each switch is equipped with an antiparallel feedback diode to enable current flow during freewheeling periods. Two capacitors are arranged in series across a DC source to create a split DC bus with the load placed between the midpoint of the switching devices and the midpoint of the capacitors.

This architecture generates an alternating output voltage across the load through the complementary switching actions of the two power devices. When one switch is activated, the load is linked to the positive half of the DC supply. Conversely, when the other switch is activated, the load is connected to the negative half. For inductive or partially inductive loads, feedback diodes offer a channel for the load current during the nonconductive intervals of the switches, providing continuous current flow and reducing undue voltage stress on the devices. Although the half-bridge converter is physically simple and cost-effective, it requires either a split-capacitor arrangement or a center-tapped DC source to obtain the requisite output voltage levels. This requirement increases the size and voltage-balancing complexity of the DC-link capacitors, particularly in higher-power applications.

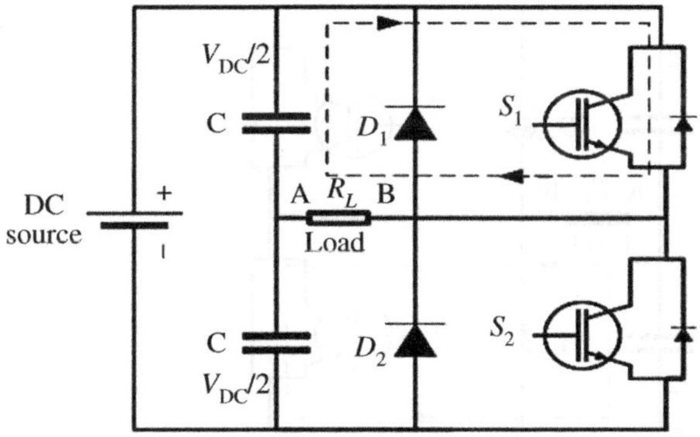

Figure 3.2 Mode 1 for R load in a half H-bridge inverter. *Source:* Adapted from Monolithic Power Systems (2025).

3.4.1.1 Working with R Load

For a half-bridge inverter with a resistive load, the two modes of operation are (Figure 3.2).

Mode 1 of the half H-bridge inverter is applicable for resistive loads. In this mode, the output voltage is half of the DC voltage source, current flows to the load, and the upper switch (S_1) and lower switch (S_2) are turned off. The output voltage across the load will be calculated as follows:

$$V_O = \frac{V_{dc}}{2}$$

The output current is determined by (Figure 3.3)

$$V_O/R_L$$

The current output of the half H-bridge inverter operating in mode 2 is similar for resistive loads. In this mode, the output voltage is equal to the negative half of the DC source voltage, allowing current to flow through the load and the bottom switch (S_2), while the top switch (S_1) remains off. At the load, the output voltage is

$$V_O = V_{dc}/2$$

Similarly, the current output is

$$V_O/R_L$$

The half-bridge inverters' output voltage, the current that flows through the switches, and the waveforms of the switching signals are considered when connected to a resistive load. It is straightforward to demonstrate that the RMS value of the output's fundamental component is 0.45 V_{DC}.

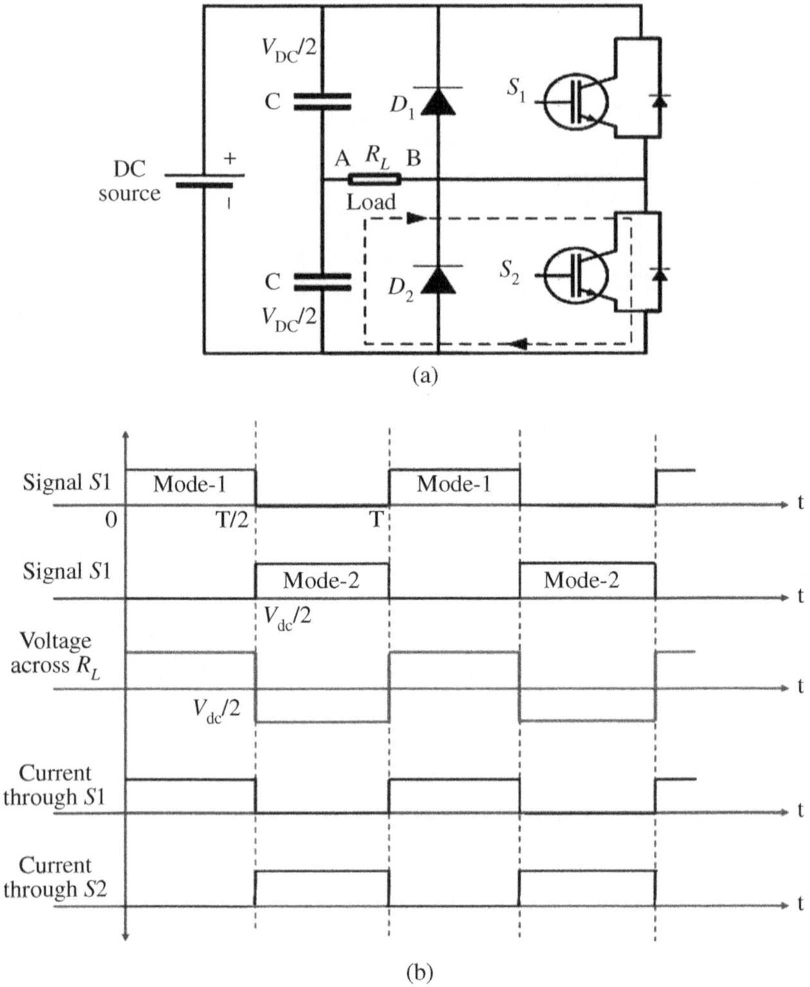

Figure 3.3 (a). Mode 2 for R load in a half H-bridge inverter. (b). Output waveform of a half H-bridge inverter. *Source:* Adapted from Monolithic Power Systems (2025).

3.4.1.2 R_L Load Operation

There are four active half-bridge inverters (Figure 3.4).

When operating in mode 2, an R_L load in a half H-Bridge inverter causes D2 to turn on, resulting in the output voltage being the negative half of the DC source voltage. Consequently, current flows through the load in the same direction as in mode 1, due to the inductor's resistance, sudden current fluctuations, and the current progressively declines as the inductor discharges until the load current reaches zero (Figure 3.5).

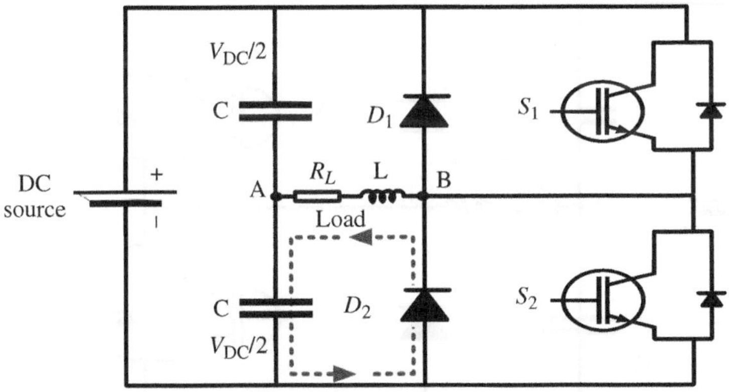

Figure 3.4 Mode 2 in a half H-bridge inverter with R_L load. *Source:* Adapted from Monolithic Power Systems (2025).

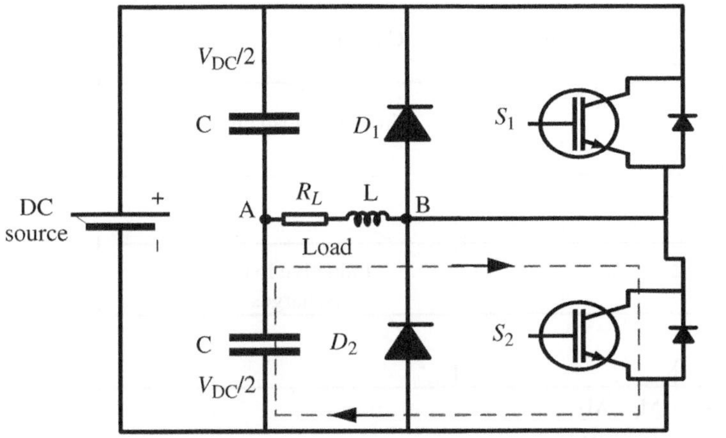

Figure 3.5 Mode 3 of the half H-bridge inverter for R_L load. *Source:* Adapted from Monolithic Power Systems (2025).

The load R_L operates in half-bridge inverter mode 3. When the output voltage is equal to the negative half of the DC source voltage, current passes through the load and the lower switch (S_2). The lower switch (S_2) activates once the inductor has fully discharged through the load. The current flowing through the load reverses direction compared to modes 1 and 2. After S_2 discharges, the current starts to increase and ultimately hits a negative peak. The inductor stores the charge once again (Figure 3.6).

For the R_L load of the half H-bridge inverter, select mode 4. Due to the inductor's resistance to sudden changes in current, when D_1 is activated, the output voltage in

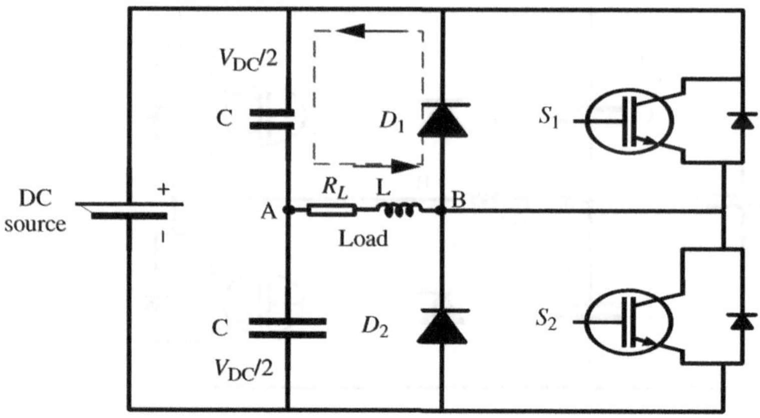

Figure 3.6 Mode 4 for half H-bridge inverter with R_L load. *Source:* Adapted from Monolithic Power Systems (2025).

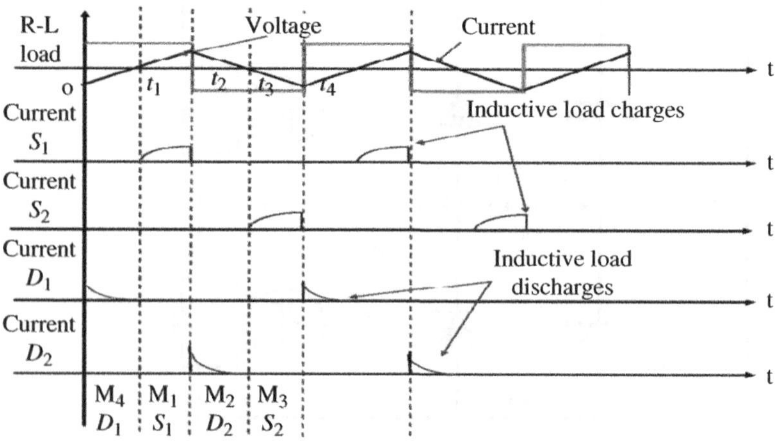

Figure 3.7 Waveforms of a half H-bridge inverter with R_L load.

this mode is half of the DC source voltage, and the current flows through the load in the same direction as in mode 3. The current gradually decreases as the inductor discharges, until the load current I_L drops to zero (Figures 3.7 and 3.8).

The H-bridge configuration in a full-bridge inverter features four switching devices and four feedback diodes. This topology provides a higher output voltage capability when compared to the half-bridge architecture. Despite having more components and increased complexity, full-bridge inverters are known for their superior performance and are frequently employed in various single-phase inverter applications such as motor drives, solar inverters, and UPS systems.

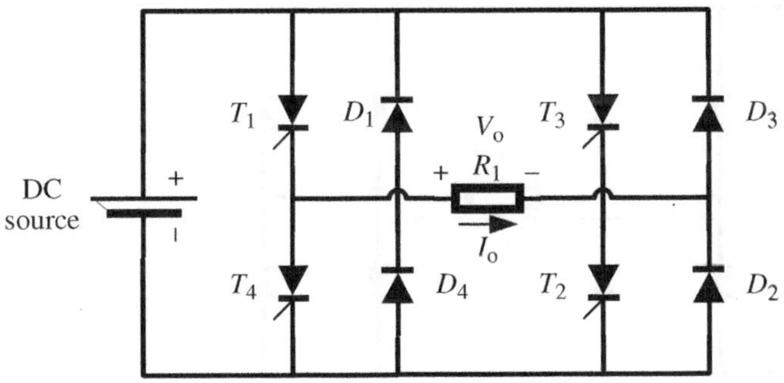

Figure 3.8 Full-bridge inverter. *Source:* Adapted from Monolithic Power Systems (2025).

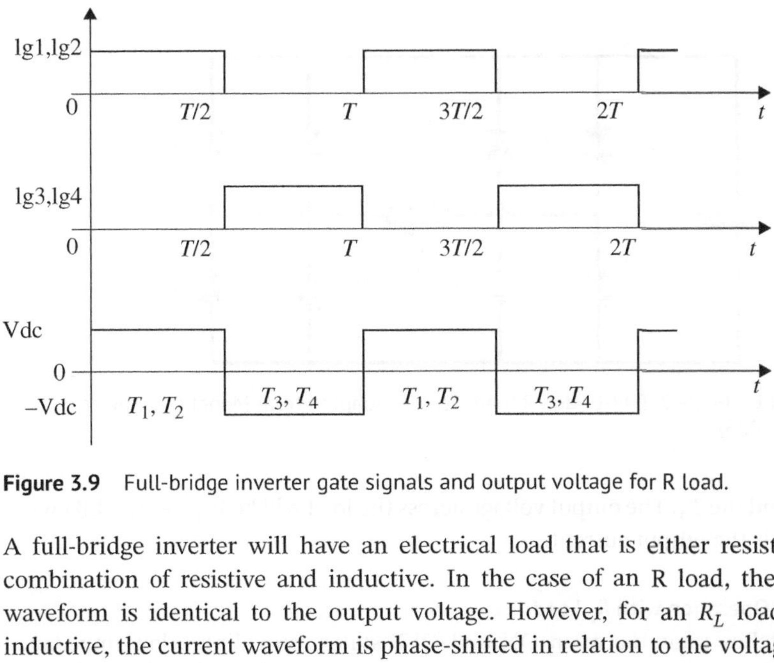

Figure 3.9 Full-bridge inverter gate signals and output voltage for R load.

A full-bridge inverter will have an electrical load that is either resistive or a combination of resistive and inductive. In the case of an R load, the current waveform is identical to the output voltage. However, for an R_L load that is inductive, the current waveform is phase-shifted in relation to the voltage waveform. The degree of this phase shift is influenced by the power factor of the load (Figures 3.9 and 3.10).

R load in a full-bridge inverter operating in mode 1. When T_1 and T_2 are turned on, T_3 and T_4 are turned off, and the output voltage equals the DC source voltage. The current passes by load, T_1, and T_2. The output voltage across the load is $V_O = V_{dc}$. Likewise, the output current is V_O/R_L (Figure 3.11).

During the second mode, for a resistive load in a full-bridge inverter, T_3 and T_4 are turned ON and T_1 and T_2 are turned OFF; the output voltage equals the negative DC source voltage. In this scenario, the current travels through the load,

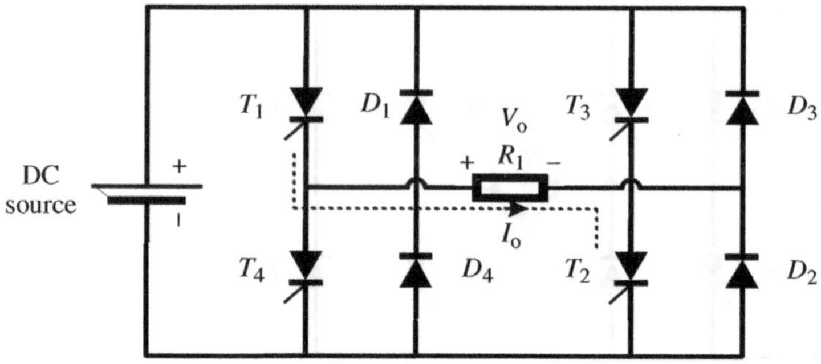

Figure 3.10 Mode 1: Full-bridge inverter R load. *Source:* Adapted from Monolithic Power Systems, 2025).

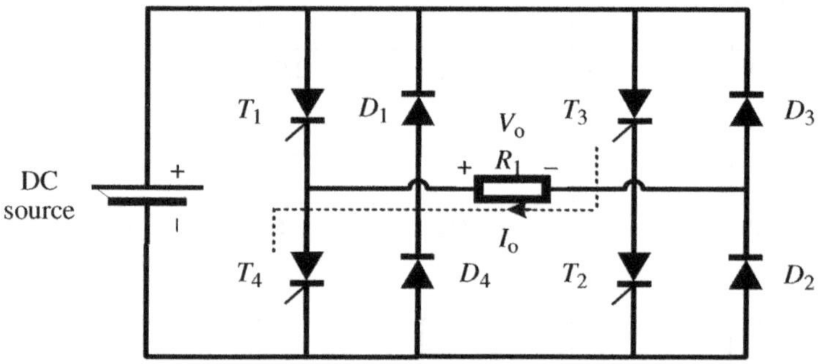

Figure 3.11 Mode 2: Full-bridge R load. *Source:* Adapted from Monolithic Power Systems (2025).

the T_3, and the T_4. The output voltage across the load will be $V_O = -V_{dc}$. Likewise, $V_O = R_L$ is the output current.

3.4.1.3 Operation with R_L Load

The switches rotate between ON and OFF states, providing a low-impedance channel for the inductive load current. The freewheeling diodes increase the inverter's overall performance and reliability by protecting the switches from voltage spikes caused by the inductive load while also allowing current to flow continuously through the load during switching transitions. This will operate in four different modes (Figure 3.12).

The R_L load in the full-bridge inverter operates in mode 1. In this situation, the switches T_1 and T_2 deliver current from the source to the load. The load maintains a positive polarity throughout the entire input voltage V_{DC}, and the current steadily increases until it hits the peak (Figure 3.13).

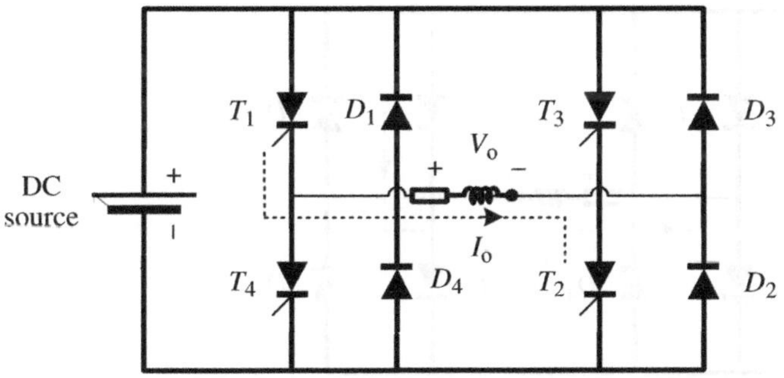

Figure 3.12 Mode 1: Full-bridge with R_L load. *Source:* Adapted from Monolithic Power Systems (2025).

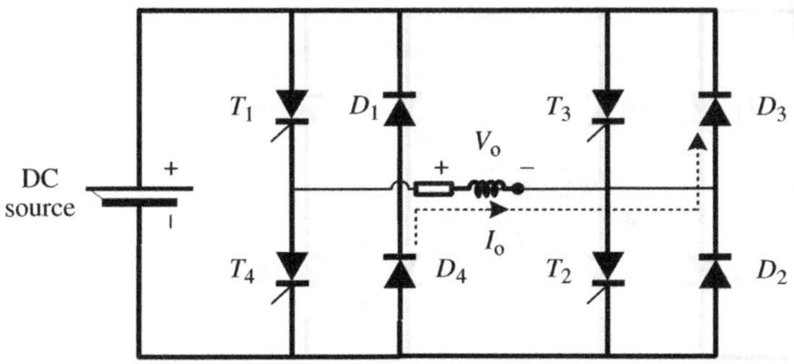

Figure 3.13 Mode 2: Full-bridge with R_L load. *Source:* Adapted from Monolithic Power Systems (2025).

A full-bridge inverter utilizes mode 2 for R_L loads. In mode 2, the R_L load operates within a full-bridge inverter in mode 1. Here, the switches T_1 and T_2 transmit transport current from the source to the load. The load maintains a positive polarity across the complete input voltage V_{dc}, and the current steadily increases until it hits its peak. The inductor accumulates energy as a voltage and current in this mode and shares the same polarity.

The feedback diodes will cause the thyristors T_1 and T_2 to immediately begin conduction. Furthermore, negative voltage is generated across the load, and the di/dt spike is reduced by these diodes. However, the polarity of the current remains constant as it gradually decreases until it reaches zero (Figure 3.14).

In mode 3, the T_3 and T_4 will be conducted immediately after the inductive load has been completely dissipated. As the current flows against it, it accelerates and eventually reaches a negative peak (Figure 3.15).

For an R_L load in a full-bridge inverter, utilize mode 4. Mode 4 uses the feedback diodes D_1 and D_2, which start to conduct, and T_3 and T_4 will get commutated.

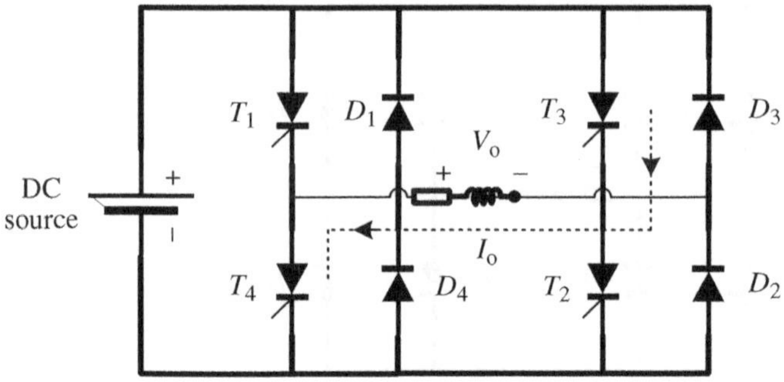

Figure 3.14 Mode 3: Full-bridge with R_L load. *Source:* Adapted from Monolithic Power Systems (2025).

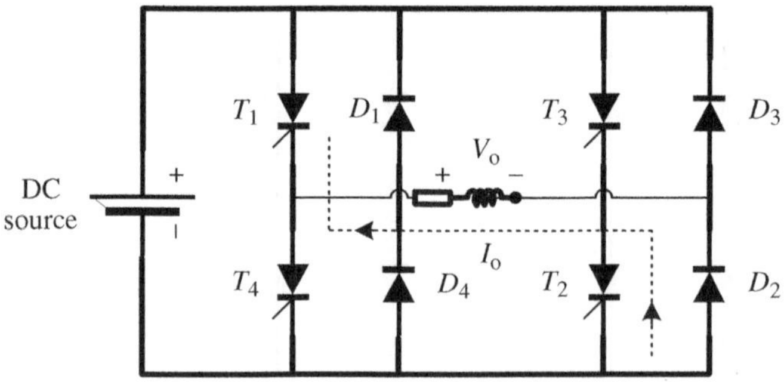

Figure 3.15 Mode 4: Full-bridge with R_L load. *Source:* Adapted from Monolithic Power Systems (2025).

Figure 3.16 shows the voltage and current waveforms for inductive and R_L loads. Current lags after the voltage's basic component. Each mode has active diodes, and the image depicts thyristors.

3.5 Three-Phase Inverter

Three-phase inverters are responsible for converting DC electricity into the three-phase AC power, which will be configured for phase difference, amplitude, and frequency. They are essential in a variety of applications, such as power distribution networks, renewable energy systems, and industrial motor drives (Figure 3.17).

3.5.1 Circuit Topology

Three-phase full-bridge inverters have two fundamental operating modes: the 120° and the 180° conduction modes (Figure 3.18).

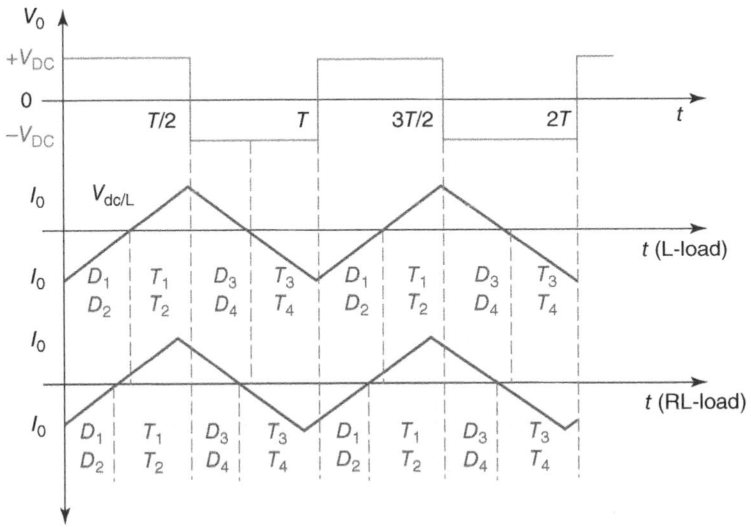

Figure 3.16 Full-bridge inverter's output voltage and current with L and R_L loads.

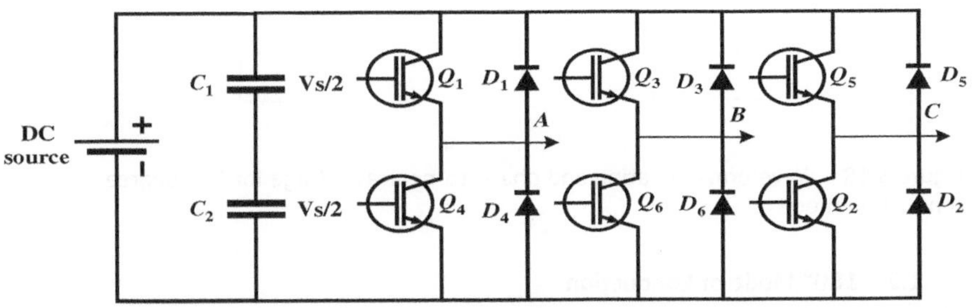

Figure 3.17 Topology of three-phase full-bridge inverter. *Source:* Adapted from Monolithic Power Systems (2025).

3.5.1.1 120° Conduction Mode

Whenever a Y-connected resistive load is employed, the gating signals and phase-to-neutral voltages are generated in the 120-degree conduction mode. The sequence in which transistors conduct is as follows: 61, 12, 23, 34, 45, 56, 61. Each switch operates at a 120-degree angle. Only two switches are active at any given time in each phase leg, and these switches will conduct for ($1/3^{rd}$) of the fundamental period. V_s emerges between nodes a and b when g1 and g6 are active during the first time interval (0 to $\pi/3$), making $V_{an} = V_s/2$ and $V_{bn} = -V_s/2$. Similarly, it is simple to extrapolate and understand various voltage patterns (Figure 3.19).

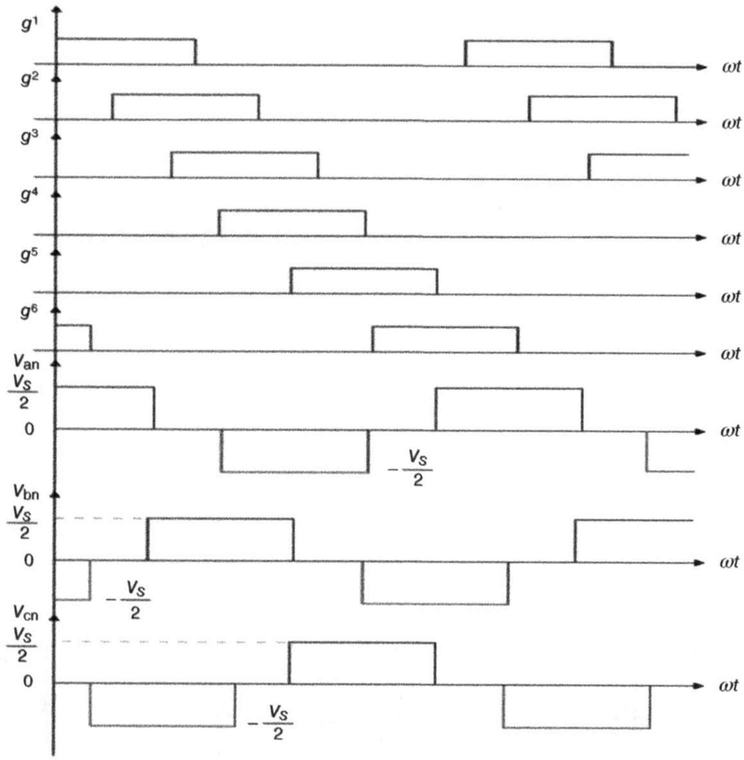

Figure 3.18 Waveforms of gating and phase-to-neutral voltage for 120-degree conduction mode.

3.5.1.2 180° Mode of Conduction

The gating signals and phase-to-phase voltages of a balanced system load in the 180° conduction mode. Three transistors are perpetually active, with each switch conducting for 180° of the primary output voltage waveforms. The output phase-to-phase voltage pattern in the 180° connection mode is easy to comprehend. Consider the first time interval (0 to π/3) when G1, G5, and G6 are active, which results in the establishment of V_S between notes A and B as well as between nodes C and B; thus, $V_{AB} = V_S$ and $V_{BC} = -V_S$. Similarly, various voltage patterns will be readily extrapolated and interpreted. The phase-to-neutral voltages for balanced star connector loads will be computed using the gating signal patterns, which eventually transform into step square waveforms because three transistors are simultaneously active. The step sizes are VS by three when G1, G5, and G6 are active. VS appears between nodes A/C and B, and during the initial time interval of 0 to π/3, $V_{an} = V_s/3$ (Figures 3.20 and 3.21).

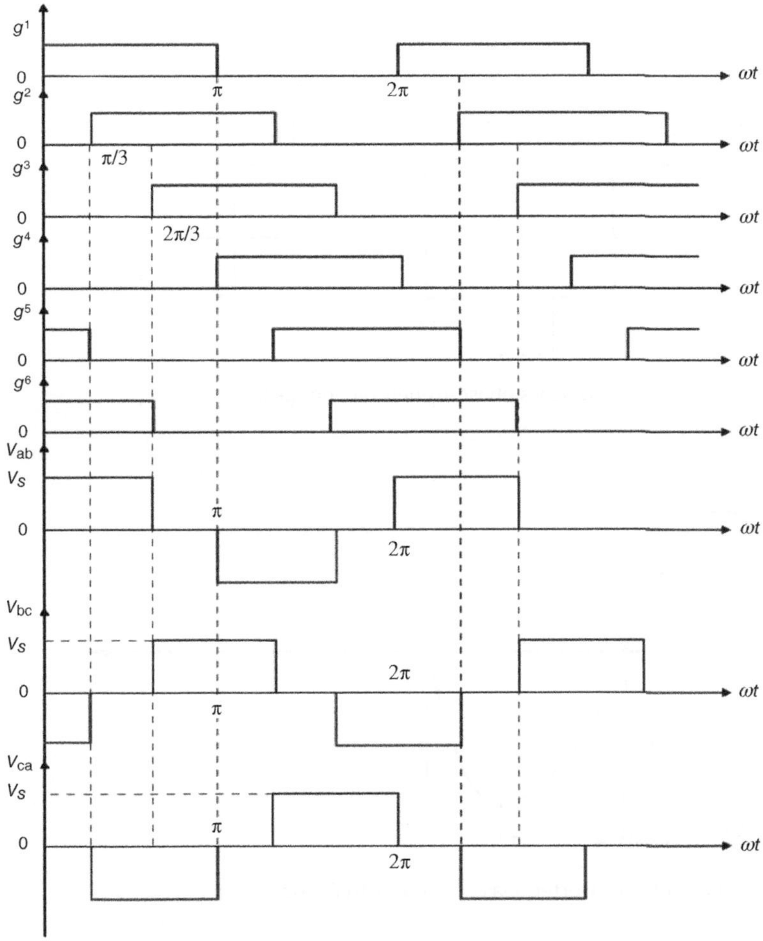

Figure 3.19 Voltage and gate signals for 180-degree conduction mode.

In three-phase inverters with R_L loads, freewheeling diodes are required. Each inverter arm's current takes longer to reach its fundamental voltage, as shown in the waveform. Even though the polarity across the voltage has reversed, the current will not vary immediately because of its inductive nature. For example, if Q1 is turned on, a positive load current will occur. Because D4 will act as a conduit for the load current's flow and is inductive, it does not collapse immediately. It should be emphasized that the transistors must be constantly gated since the load power factor influences the conduction time of both diodes and transistors.

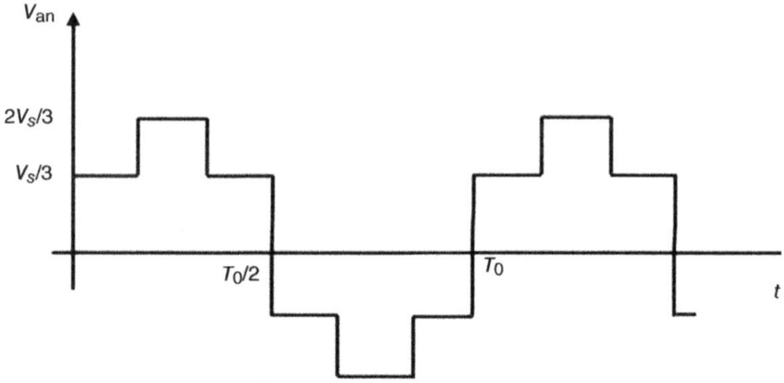

Figure 3.20 Three-phase inverter's phase-to-neutral voltages.

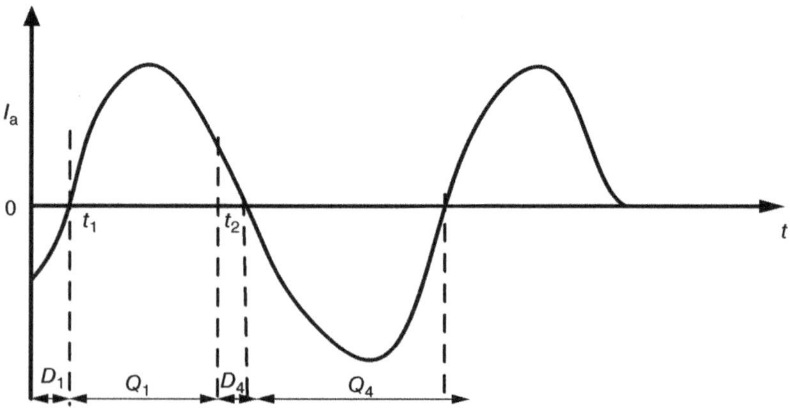

Figure 3.21 Three-phase inverter phase current with R_L load.

3.5.2 Conduction Modes: 120-Degree Versus 180-Degree

In the 120° conduction mode, the third load terminal is kept open while the other two are connected to the input supply. The characteristics of the load reveal the surprising capabilities of the open terminal. When assessed against the 180° conduction mode for the same load conditions, the 120° conduction mode for each transistor shows underutilization. For three-phase inverters, the 180° conduction mode is recommended for various reasons.

3.6 Pulse Generation Topologies for Inverter

There are many methodologies and strategies used to generate switching signals that regulate the operation of power electronic switches (such as MOSFETs or

IGBTs) in inverter circuits, which are known as pulse generation topologies for inverters. These signals determine the inverter's output voltage and frequency. Here are a few common pulse generation topologies seen in inverters.

3.6.1 Pulse Width Modulation (PWM)

- **Sinusoidal PWM (SPWM):** This method produces a sinusoidal output voltage by comparing a triangle carrier signal to a sine waveform, which will be used as a reference signal. Due to this, the circuit produces a series of pulses of varied widths that correspond to the sine wave's amplitude.
  - **Space vector PWM (SVPWM):** Improves performance by optimizing the DC bus voltage. Three-phase voltage vectors are represented in two dimensions.
  - **Hysteresis band PWM:** Adjusts output current within a hysteresis band. When the current exceeds the top or lower limits, the inverter switches on or off to bring it back inside the band.
  - **Selective Harmonic Elimination (SHE):** Pre-calculate switching angles to remove specified harmonics from the output waveform. It is especially beneficial for applications that require extremely low harmonic distortion.
  - **Carrier-based PWM comes in two types:** Bipolar and unipolar. Bipolar PWM alternates between positive and negative DC bus voltage, whereas unipolar PWM alternates between zero and either positive or negative voltage.
  - Multilayer PWM technique reduces harmonic content while improving output waveform quality by producing a large number of voltage levels.
  - Modern inverters utilize DSPs or microcontrollers to generate PWM signals. These digital controllers may perform complicated algorithms, allowing for precise control over inverter functioning.
  - Random PWM spreads harmonics across a larger frequency spectrum, decreasing noise and interference. Each of these pulse-generating topologies has advantages and is selected based on the application's specific requirements, such as effectiveness, harmonic distortion, control complexity, and cost.

3.7 Duty Cycle Generation for Choppers

3.7.1 Pulse Width Modulation

This technique modifies the output waveform's pulse width (t_{ON}) while maintaining a constant chopping frequency (f) and, as a result, a constant chopping period (T). As a result, adjusting the ON time (t_{ON}) will produce the desired output voltage. Figure 3.22 shows the voltage output for varied ON times.

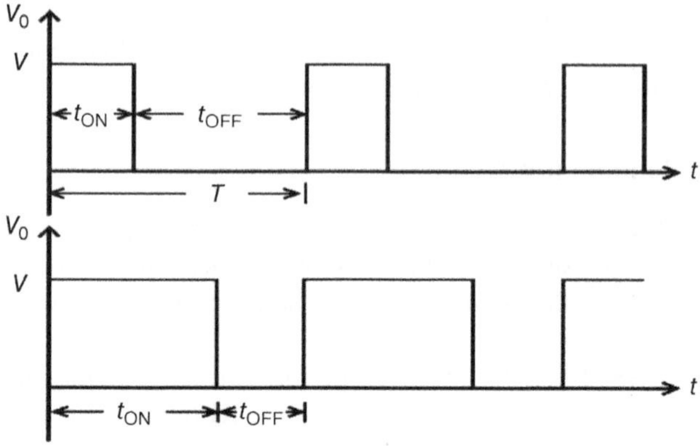

Figure 3.22 PWM control.

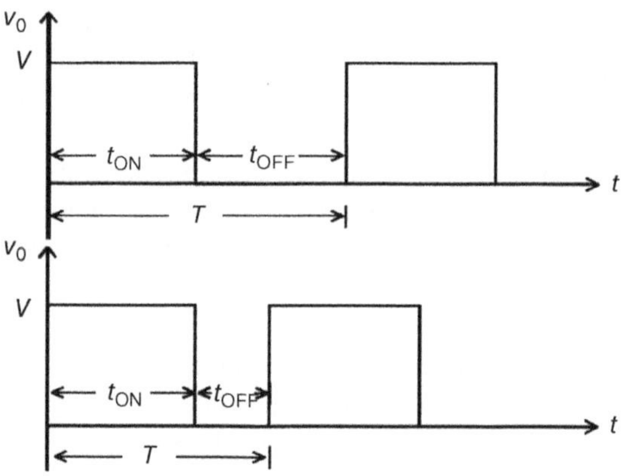

Figure 3.23 Output voltage waveforms for time ratio control.

3.7.2 Variable Frequency Control

In this approach, chopping frequency alone gets varied while maintaining the ON/OFF time as constant. This is also known as FM (frequency modulation). Displays output voltage waveforms with a variable chopping period T and a constant t_{ON}. This method creates harmonics, and the current will become discontinuous for substantial t_{OFF} loads (Figure 3.23).

3.8 Soft-Switching PWM

A variety of soft-switching pulse width modulated converter topologies have been developed to combine the advantages of conventional PWM and resonant converters while minimizing their respective drawbacks. Among these, the zero-voltage switched quasi-square wave converter represents a key category of the zero-voltage soft-switching PWM converters alongside zero-voltage switching (ZVS)-PWM and zero-voltage transition (ZVT-PWM) converters. In parallel, zero current soft-switching approaches, including zero current switched PWM and zero current transition converters, have been investigated to evaluate their operational benefits and limitations. Experimental results obtained from several converter prototypes validate the performance characteristics and the practical applicability of each soft-switching PWM topology.

3.8.1 Zero-Voltage Switching (ZVS)

When the output voltage reaches zero, the power to the load (heater, cooler, or other device) will turn on or off. This is known as ZVS. The zero-voltage switch is designed to switch the thyristors (TRIACs or SCRs) as close to the moment where the supply voltage waveform crosses zero or the zero line as possible. This disadvantage of phase control is eliminated if a thyristor is engaged at the zero crossing, as the RFI created is negligible. By turning on the load in small bursts, the average load power can be managed. For example, turning the thyristor on for five cycles and then off for twenty cycles can result in low load power.

A zero-voltage switch consists of an input, also known as the control circuit. This is the circuit that controls how the thyristors switch. It could be a simple on/off switch, such as a resistance bridge or thermostat, which employs thermistors to provide “proportional” temperature control.

A detector for zero crossing. This circuit determines the precise timing of the trigger pulses transmitted to the thyristors.

- A pulse generator. This circuit generates the trigger pulse for the thyristor.
- Zero-voltage switches can be created using discrete components. The technique is no longer used and is out of date.

Integrated circuits (ICs) are designed specifically for this function.

A ZVS is mostly used to control heating loads. Because most heating elements have a high thermal inertia, the burst firing of thyristors will not significantly affect them. A ZVS is not suitable for controlling motor speed or illumination due to pulsing motor torque and lamp output at low levels. This would cause an unpleasant flicker in incandescent lights. When basic on-off control is required, a ZVS can be used. To accomplish this, a solid-state relay (SSR) was developed. An SSR consists of the zero-crossing controller and the primary thyristor.

3.8.2 Zero Current Switching (ZCS)

Modern power electronics significantly rely on ZCS converters, which allow for efficient conversion of electrical energy while minimizing switching losses.

The term "ZCS" refers to the approach of switching the power semiconductor device at zero current, which reduces switching losses and increases system efficiency.

3.8.2.1 Principle of ZCS Converters

Because of the nature of inductive loads, ZCS converters function fundamentally. When the current flows through the switches of a ZCS converter reaches zero, it shuts down. This function increases the system's overall efficiency by reducing power dissipation across switching devices.

3.8.2.2 Types of ZCS Converters

- **Resonant ZCS converters:** These converters use resonant circuits, which provide a natural zero-current situation at switching points. The resonant circuit causes the current to flow sinusoidally. As a result, the current naturally becomes zero at certain periods, allowing the switch to turn off with minimal losses.
- **Quasi-resonant ZCS converters:** Unlike resonant ZCS converters, quasi-resonant converters do not maintain a resonant condition during operation. The switching losses are reduced by turning off the switch at zero current.

3.8.3 Advantages and Applications of ZCS Converters

ZCS converters have significant advantages, which make them widely used in a variety of applications. They considerably reduce switching losses, increase power conversion system efficiency, and extend the life of power semiconductor components. Furthermore, ZCS converters produce less EMI, hence increasing the reliability of electronic systems.

Due to their high efficiency and low EMI, ZCS converters are widely utilized in a range of applications, including computer power supplies, renewable energy systems, telecommunications equipment, and electric car chargers.

3.8.4 Limitations of ZCS Converters

One of the key issues is that the ZCS state can only be attained at specified operational moments. This indicates that the ZCS state may not be maintained under all load levels, resulting in increased switching losses.

Furthermore, resonant circuits in ZCS converters add complexity to the design, making them more difficult to design and build compared to typical converters.

Also, these circuits will create large peak currents, which will lead to greater stress on the components and potentially limit their lifespan.

Reference

Monolithic Power Systems (2025). Single-phase inverters. https://www.monolithicpower.com/en/learning/mpscholar/power-electronics/dc-ac-converters/single-phase-inverters (accessed on December 15, 2025).

4

PWM Applications for System Control

4.1 Introduction

PWM plays a major role in system control. Especially in power engineering, it finds its applications in voltage regulation, grid synchronization, harmonic analysis, and drive control.

4.2 Voltage Regulation Using PWM

The capacity of a system to provide nearly constant voltage under a variety of load conditions is known as voltage regulation, or VR. It is a measurement of how much the voltage changes between a component's transmitting and receiving ends, like a transmission or distribution line. Many VR issues are caused by the power system's excessive impedance, which prevents the load from being adequately utilized. Another way to put it is that the power system isn't strong enough to support the load. As a result, when the load is high, the voltage decreases too low. On the other hand, an overvoltage condition may arise if the load falls too low after the source voltage is increased to overcome the resistance.

The input terminals of AC loads may need a steady or variable voltage. When inverters are used to feed such loads, it is crucial that the inverters' output voltage be regulated to meet the needs of AC loads. To keep the voltage at the AC load terminals constant at the specified level, any fluctuations in the DC input voltage must be appropriately adjusted. Inverter output voltage can be controlled in a number of ways, but there are some restrictions on how this can be done.

4.2.1 AC Voltage Control

It requires an AC voltage controller between the inverter and load terminals, which gives rise to the higher harmonic content in the output voltage, and it is much suitable for low-power applications only.

4.2.2 Series Inverter Control

This method involves the series connection of two or more inverters, which can be summed up with the help of transformers to obtain an adjustable AC voltage. The drawback of this method is that it does not enhance the harmonic content even at low output voltage levels.

4.2.3 External Control of DC Voltage

In this method, the DC input voltage to the inverter is controlled by means of a chopper, which requires a filter circuit for reducing the ripple content. It results in more losses and reduced efficiency due to the increased number of power-handling stages.

4.2.4 Internal Control

Additionally, PWM control can be used to regulate the output voltage without the need for external peripherals. Modified PWM with an advanced control architecture is necessary to lower the overall THD in order to improve the power quality, in contrast to ordinary PWM schemes.

Equations (4.1) and (4.2) provide the equations for line and load regulation systems that can be used to maintain a constant voltage. However, low-power DC circuits can benefit greatly from these approaches.

$$\text{Line regulation} = \frac{(\Delta V_{\text{out}}/V_{\text{out}})}{\Delta V_{\text{in}}} \times 100\% \tag{4.1}$$

$$\text{Load regulation} = \left(\frac{V_{\text{NOLOAD}} - V_{\text{FULLLOAD}}}{V_{\text{FULLLOAD}}}\right) \times 100\% \tag{4.2}$$

Maintaining the balanced DC input voltage for multilevel inverters requires careful consideration of capacitor voltage balancing. However, these techniques work well for DCMLI and FCMLI topologies, which keep these inverters from being widely used in real-world applications. Since each DC link is fed by a separate DC source (a solar PV array), the main benefit of CMLI is that it has no imbalance issues. Considering that advantage, it presents a significant challenge to both comply with electrical grid rules and preserve a steady output voltage sent to the customers. The DC voltage magnitudes at the maximum power point are near to one another for PV modules with varying temperatures or levels of radiation. Typically, the variance is under 15%. Therefore, an appropriate controller is needed to regulate the output voltage at the inverter end before it is supplied to the customers. The primary structural drawback of the voltage control topology utilizing vector transformation is its bidirectional power switches, which raise the inverter's overall cost and switch count.

4.3 Problem Statement

Subsequently, it is vital to make sure that the output voltage of the required magnitude and frequency is attained when AC loads are fed through inverters. By changing the input DC voltage while keeping the inverter's gain constant, a variable output voltage can be produced. However, if the DC input voltage is set and uncontrollable, a variable output voltage can be achieved by adjusting the inverter's gain, which is often achieved using PWM control inside the inverter.

The amount of energy absorbed from solar radiation varies periodically. A solar panel's rated voltage is 48 V with a current of roughly 7 A; during the day, it fluctuates between 48 and 20% and 48 + 20%. The design of suitable power electronic circuits is required to transform this energy into the grid's requirements. The following are the goals:

- To change the DC voltage from unstable to stable.
- To transform a steady DC voltage into an AC voltage.
- To connect the main grid to the solar panel system.

According to the objectives, the technical features are set as follows:

- Provide an inverter with a 1% variance to match the grid data.
- Provide a DC link voltage with a 1% fluctuation to supply the inverter with output.
- A DC-DC converter with a large voltage transfer gain is necessary since the input voltage is 48 V with a ±20% fluctuation.

In conventional methods, the VR is accomplished using tap-changing under-load transformers, set voltage regulators, and fixed or switched capacitors. The operation of these devices is mechanical in nature and effective in managing slow variations in voltage. The lifetime is dramatically reduced by the increased number of operations needed to handle fast voltage variations due to the sudden changes in the supply. Also, the setting of conventional devices can be optimized every hour in anticipation of the overall change in load. For example, air conditioners are turned on during a hot afternoon. A few examples of isolation devices are motor-generator sets, ferro-resonant transformers, and UPS systems. These are devices that, by some kind of energy conversion, effectively separate the load from the power supply. As a result, the device's load side can be independently controlled and kept at a steady voltage regardless of what happens at the power source. These devices have the drawbacks of being expensive, adding to losses, and potentially causing harmonic distortion issues for the power supply system. Also, the filter, transformer, and booster circuit lead to complexity of the circuit and increase the cost.

Hence, a control scheme to maintain constant voltage magnitude at the point where a sensitive load under system disturbances is connected. The controller input is an error signal obtained from the reference voltage and the terminal voltage measured. Such an error signal is processed by a controller to generate a reference signal, which is provided to the PWM signal generator.

4.4 Fuzzy Logic-Based Controller

Basic set-theoretic operations provide the foundation of a fuzzy control system. In contrast to classical or digital logic, which operates on discrete values of either 1 or 0, it can also be described as a mathematical system that analyzes analog input values in terms of logical variables that take on continuous values between 0 and 1. More design choices than usual are needed when designing a fuzzy controller, such as those pertaining to the rule base, inference engine, defuzzification, and pre- and post-processing of data. Fuzzy logic is an effective approach to solve problems. Fuzzy offers a straightforward method of deriving conclusions from ambiguous, imprecise, or obscure data.

Fuzzy logic's capacity to derive accurate solutions from approximations of data is somewhat like human decision-making. In contrast to classical logic, it necessitates accurate numerical values, correct equations, and a thorough comprehension of a system. A different way of thinking is incorporated into fuzzy logic, which makes it possible to represent complicated systems using a higher level of abstraction derived from human knowledge and experience. This knowledge can be expressed using fuzzy logic, which maps subjective concepts like "very hot," "bright red," and "long time" into precise numerical ranges.

4.4.1 Problem Formulation and Parameterization

Figure 4.1 demonstrates the structure of the Fuzzy Logic Controller (FLC) utilized to carry out VR in a 15-level solar-fed inverter. This figure compares the output voltage V_o, which is obtained from a 15-level inverter, to the reference voltage V_{ref}, which is the voltage that the inverter is intended to attain. The FLC uses the following error, $e = V_{ref} \div V_o$, and the error's rate of change, de/dt, as input attributes. There are five main block sets in the FLC. These include rule bases, databases, prediction systems, fuzzifiers, and defuzzifiers. Fuzzification uses a lookup in any number of membership functions to translate every component of input data to degrees of membership. This block determines how well each rule's condition matches that specific input instance by comparing the input data with the rules'

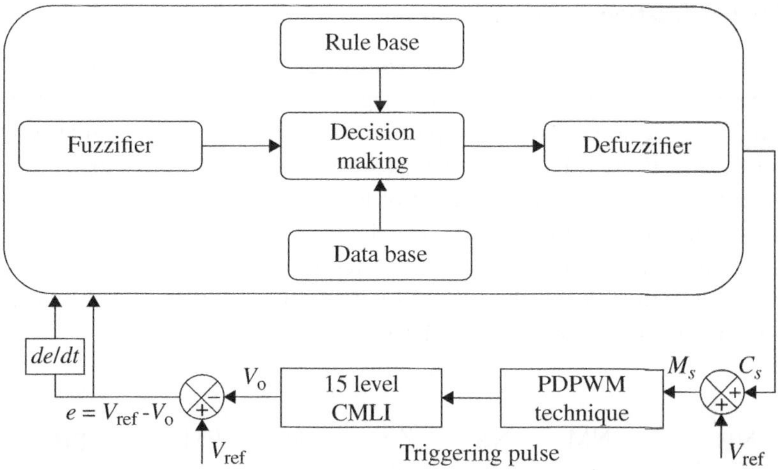

Figure 4.1 Structure of fuzzy logic controller.

conditions. Each linguistic phrase that pertains to that input has a degree of membership. The modulating signal M_s needed for PWM production is then produced by comparing the commanding signal C_s produced by the FLC with V_{ref}. This process yields the appropriate gating signals for the inverter switches.

The input error, change in error signal, and output modulating signal are the parameters considered when creating rules. Both in the condition and the conclusion, the rules may make use of multiple variables. Thus, the controllers can be used for both Single-Input Single-Output (SISO) and Multi-Input Multi-Output (MIMO) problems. The goal is to use an error signal to regulate a control signal. Although the controller may really require the error, change in error, and accumulated error as inputs, it might be referred to as single-loop control since, in theory, all three are derived from the error measurement. In order to keep things simple, this section assumes that the control goal is to regulate some process output around a specified set point or reference. As a result, the display is restricted to single-loop control. To provide the fuzzy set to the process as a control signal, it needs to be transformed into a number. Defuzzification is the term for this process. It fuzzifies the resultant fuzzy set into a clear control signal by extracting the control signal from the rule set controller. The rules represent the fuzzy proportional-derivative controller technique, which calls for the control signal to be a combination of the reference error and the change in error. The vertical lines on the first and second columns of the rules chart show the instances of the error and the change in error. The inference engine determines the membership values in each rule's condition, and 49 rules are created to control the voltage for the decision-making process.

4.4.2 Implementation Procedure

The problem is formulated with the error and its derivative membership function. The membership function for the error signal is shown in Figure 4.2. In this figure, *N* indicates negative, *P* indicates positive, and *Z* indicates zero. Similarly, *B* indicates Big, *M* indicates Medium, *S* indicates Small, and *E* indicates Error. The derivative of the error signal for the input of FLC and its membership function is shown in Figure 4.3. Figure 4.4 illustrates the fuzzy logic output, which is the reference signal created from a membership function.

Table 4.1 exhibits the rules for the rule table with two inputs (error and its derivative signal) and one output (reference signal).

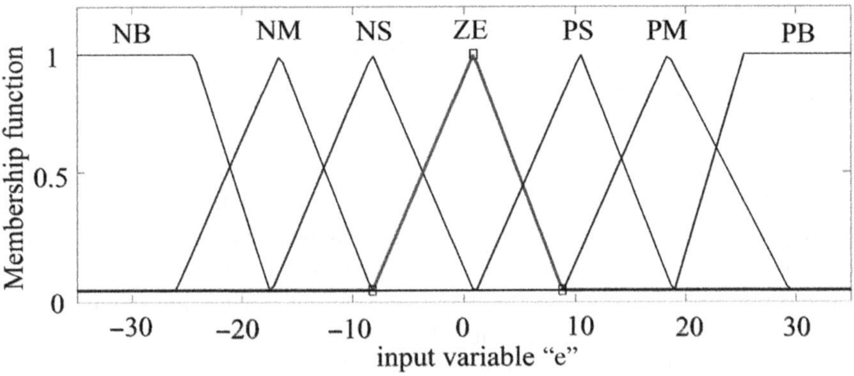

Figure 4.2 Membership function for the error signal.

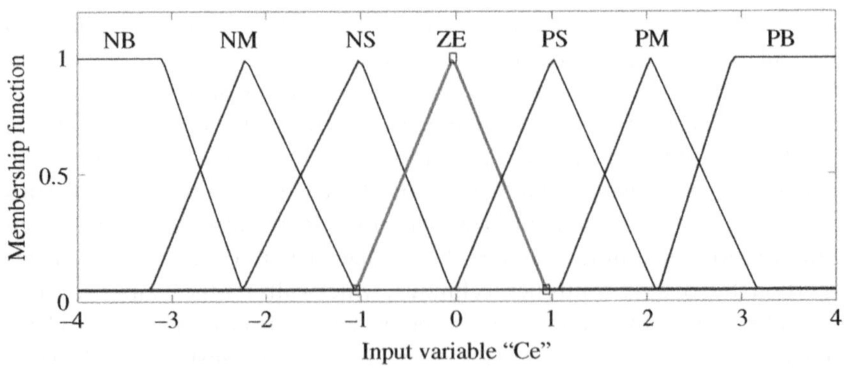

Figure 4.3 The error signal's membership attribute.

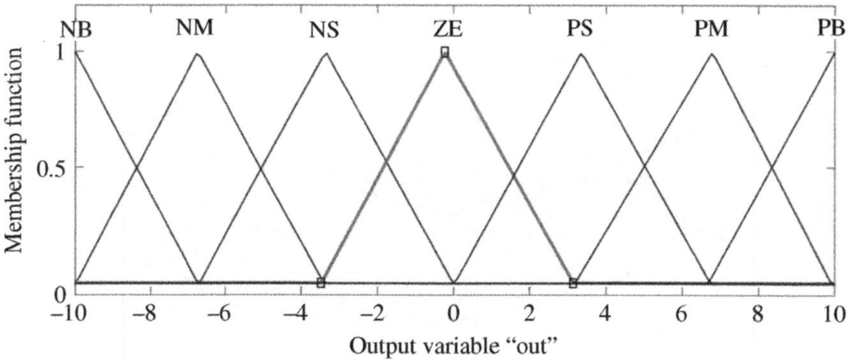

Figure 4.4 The intent of membership for the reference indication.

Table 4.1 The rule for fuzzy logic controllers.

e Ce	NB	NM	NS	ZE	PS	PM	PB
NB	PB	PB	PB	PB	PM	PS	ZE
NM	PB	PB	PB	PM	PS	ZE	NS
NS	PB	PB	PM	PS	ZE	NS	NM
ZE	PB	PM	PS	ZE	NS	NM	NB
PS	PM	PS	ZE	NS	NM	NB	NB
PM	PS	ZE	NS	NM	NB	NB	NB
PB	ZE	NS	NM	NB	NB	NB	NB

The defuzzification to process transforms the fuzzy value into a crisp value. Center of Gravity (COG) conversion is the technique used. Figure 4.5 displays the input error and its derivative in three dimensions along with the output reference signal.

4.5 PI-Based Controller

A PI controller can control a weighted sum of the error (difference between the output and desired set point) and the integral of that value. The gain of the PI controller is chosen by trial and error. These PI increases don't change. However, when the PI controller gains are selected by trial and error, the overshoot for the reference step change and the steady-state error issues occur. A triangle pulse width

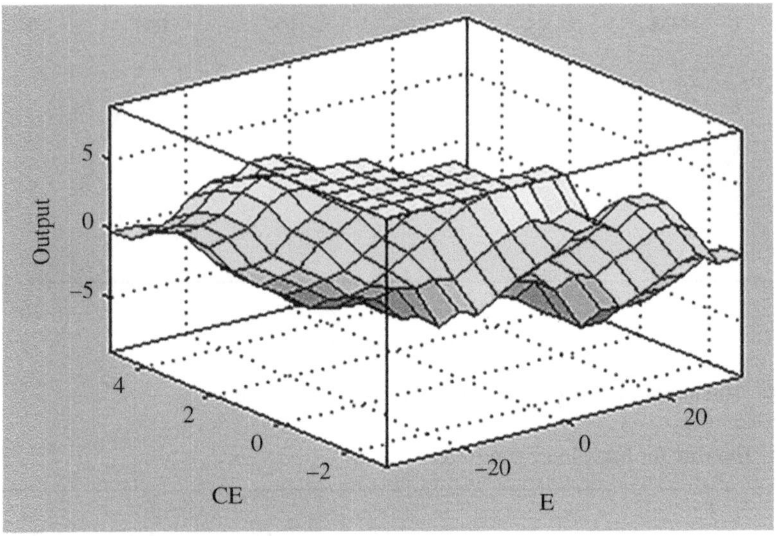

Figure 4.5 3D fuzzy rule visualization.

modulator receives the variable voltage value produced by the PI controller's integration of the error between the feedback and reference to produce the gate signal for the H-bridge converter. The PWM PI controller has shown several benefits over the conventional hysteresis controller. (i) a steady frequency of switching; (ii) closed-loop control; (iii) minimal current ripple; and (iv) minimal acoustic noise.

Because the measured value of the controlled variable is "fed back" to the comparator device, a PI controller is a closed-loop or feedback system. The controlled variable and the intended value or set point are compared in the comparator. An error is created if the measured variable and the set point vary in any way. The controller receives this error and modifies the final control element to bring the controlled variable back to the set point.

Negative feedback makes sure that the control element is adjusted based on the difference between the measured variable and the set point, which tends to lessen the error. The comparator would have an unstable positive feedback system if the two signals were combined to create the signal. Figure 4.6 illustrates how the PID controller is frequently utilized in many kinds of feedback systems.

4.5.1 Tuning of PI Controller

The issue at hand presents several difficulties that must be resolved, some of which are:

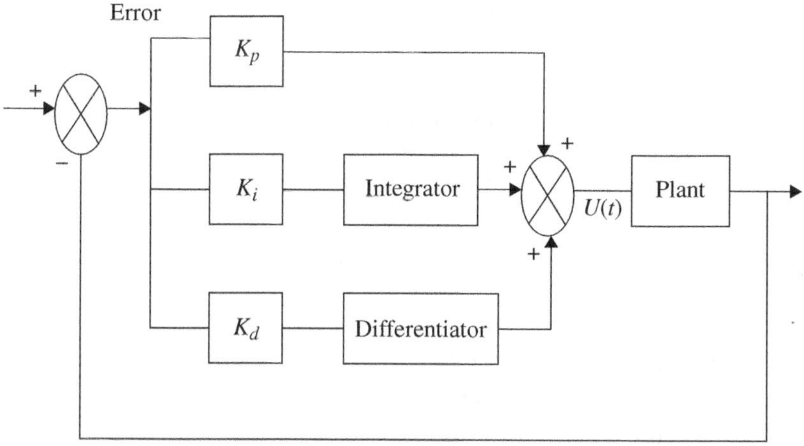

Figure 4.6 Structure of PID controller.

- Variations in the transport coefficients' orders of magnitude based on the radial positions that change with time.
- The actuators are subject to strict nonlinear form limitations, and the accessible parameters are saturated.
- Any suggested control scheme's resilience to disturbances and numerical issues, especially considering the disparity in the size of the transport coefficients.
- Linking the boundary condition and the control applied to the infinite-dimensional system.
- Real-time implementation of the control algorithms is required (with limitations on the computing cost).

The PI controller gain is adjusted in relation to the varying irradiance for different error signals. A reference signal for the several error signals produced by the fluctuations in the solar PV is produced by the PI controller. In order to control the output voltage, the PI controller is manually adjusted for various error values by considering the real-time solar data for the geographic area under consideration, which is indicated in the table (Table 4.2).

4.5.2 Implementation Procedure

Based on the tuned values for the controller, the error signal, which was generated based on the comparison between the actual output voltage obtained and the reference voltage, deteriorates. The major difference between the FLC and PI-based controller lies in the way of regulation, and the drawback resists in the lengthy process for finding the controller gains. The mathematical modeling for the given

Table 4.2 PI controller with tuned specifications for voltage control.

K_p	K_i	Output peak voltage (V_p)
0.355	0.5	315.8
0.1	0.5	322.4
.	.	.
.	.	.
.	.	.
0.05	0.2	323.7
0.45	0.2	313.35
0.6	0.3	309.5
0.2	0.3	320
0.03	0.3	324
0.01	0.5	325
.	.	.
.	.	.
.	.	.
0.03	0.1	324.2

system cannot be made, as the model needs to incorporate the changes in all the stages. These stages lie in the input solar, changes in the battery charging and discharging, and the unique property of the switching elements. Hence, the controller gains obtained from the method cited above using real-time data save more time and are more reliable. The controller provides the reference signal, which is then compared with the carrier to generate the complementary signal for the PWM signal suitable for the inverter switches.

4.6 ANN-Based Controller

Simple synthetic nodes known as "neurons," "neuroses," "processing elements," or "units" are connected in order to generate a synthetic neural network that resembles a biological neural network. Brain networks are comparable to biological brain networks in that their units carry out tasks concurrently and collectively rather than clearly defining the subtasks to which different units are assigned. Models used in artificial intelligence, psychological research, and statistics are typically referred to as "neural networks." Theoretical and computational neuroscience includes neural network models that mimic the central nervous system.

The control scheme's goal is to keep the voltage magnitude at the point where a sensitive load is connected during system disruptions constant. The error signal is processed by the neural controller, which then creates the necessary angle to drive the error to zero – that is, return the load RMS voltage to the reference voltage. The formula $V_{\text{error}} = V_{\text{ref}} - V_{\text{actual}}$ is used to determine the voltage error values. The ANN is trained using these error values. To train a neural network, perform these steps: (i) Set each neuron's input weight to its initial value; (ii) Give the network a training dataset. To process the error signal, the neural network is trained on a variety of samples at varying intervals.

4.6.1 Formulation of Problems Along with Parameterization

Applying any "learning method" for ANNs, expert learning, pattern recognition, or classical modeling is the first step. The optimal approach for the learning process is to split the available data into three data sets rather than two. The training set comes first, followed by the control (or fine-tuning) set and the final test set. With a propensity for the training set to be the smallest and the test set to be the largest, each of these three data sets should comprise roughly one-third of the total data.

In situations where the training is going well, the control set – which is typically disregarded – serves as a source of feedback information for adjusting the initially chosen ANN architecture. However, the outcomes on the control set – which was obviously not utilized throughout the training – are poor. Regretfully, this occurs frequently during any learning process. The learning and testing principle states that the test objects should not be involved in any way in acquiring the model, yet users typically leave this duty to the test set. Therefore, a fully "noncommittal" or impartial collection of facts should be used for the genuine test. Table 4.3 lists the various irradiance levels for which the dataset was gathered, ranging from $300\,\text{W/m}^2$ to $1000\,\text{W/m}^2$.

4.6.2 Algorithm

Algorithms are trained using a three-layer structure after gathering the necessary dataset. Error and change in error are the inputs, while the necessary modulating signal is the output. The algorithm utilized is called gradient descent backpropagation (traingd). The general characteristic is the generation of the deltas (error signals) for all output and hidden neurons by the backpropagation of output activations using the training pattern target. When compared to alternative techniques like Newton-Raphson, etc., gradient descent minimizes error and is simple to compute.

Table 4.3 Sample dataset for neural network training.

Irradiance (W/m^2)	Panel voltage (V)	Error value	Reference voltage (V)
900	47.6102	5.422	325
901	47.6143	5.442	325
906	47.6348	5.543	325
907	47.6389	5.564	325
.	.	.	.
.	.	.	.
.	.	.	.
908	47.6429	5.584	325
909	47.647	5.604	325
910	47.6511	5.624	325
1000	48	7.346	325

4.7 Simulation Results

Because the multilayer converters have many DC linkages, each string's MPP may be tracked and independently controlled. This feature can boost the PV system's efficiency in situations where the strings are mismatched because of uneven solar radiation, aging PV panels, differing cell types, or dust buildup on the panels' surfaces. Without VR, a 15-level inverter is an open-loop system. Every distinct stage of the CMLI is connected to a panel with a varied irradiance level. A series connection of seven cascaded H-bridges is made for the 15 levels. A comparison is made between a reference and carrier wave for the pulse signal. For the pulse signal, the carrier of a triangle wave and the reference signal of a sinusoidal wave are compared. For pulse generation, the bipolar PDPWM technique is used. A triangle wave is compared to a positive sinusoidal signal for one leg and a negative sinusoidal signal for the other leg for the pulse sequence. The modeling of a PV panel with changes in irradiance levels illustrating the shift in an inverter's output voltage is shown in Figure 4.7. The output voltage waveform produced by fluctuations in irradiance and partially shaded solar PV modules is shown in Figure 4.8. Voltage imbalance is the outcome of this uneven distribution of output voltage. This unequal change can be made up for by using VR techniques.

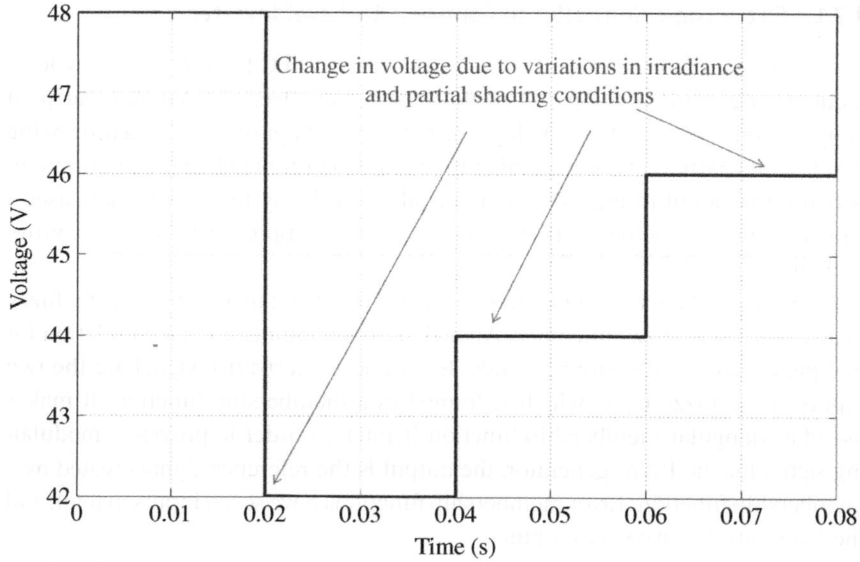

Figure 4.7 Output voltage variation in relation to irradiance.

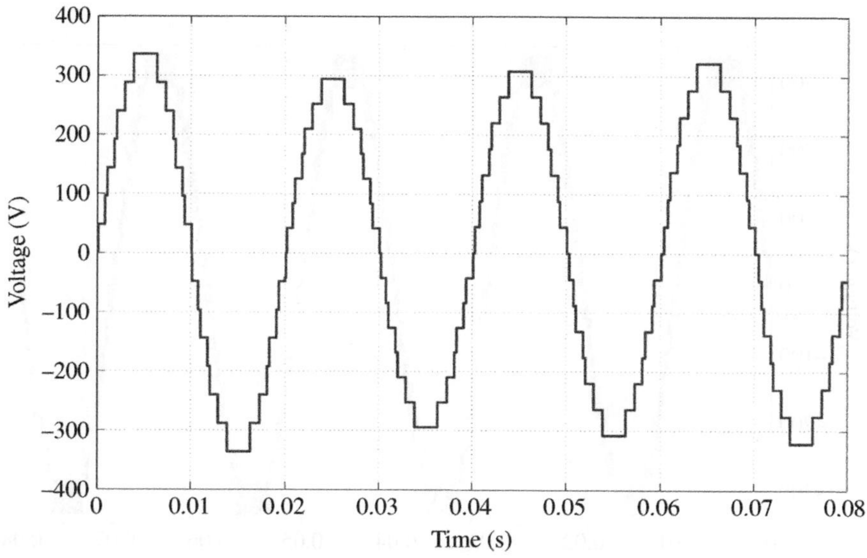

Figure 4.8 Output voltage with 15 levels and fluctuating irradiance.

4.7.1 Fuzzy Logic Controller to Control a 15-Level Inverter

The suggested method uses a 15-level inverter with FLC for VR. The fuzzy logic method to control problems is similar to human decision-making, except it happens considerably more quickly. Fuzzy logic is an approach to data processing that permits partial set membership instead of strict set membership or nonmembership. Instead of trying to mathematically model a system, fuzzy logic uses a straightforward, rule-based IF X AND Y THEN Z approach to solve a control problem.

The process of creating a mapping from a given input to an output using fuzzy logic is known as fuzzy inference. After that, the mapping offers a foundation for making choices or identifying trends. Error and its derivative signal are the two inputs of the fuzzy logic, which is framed as a membership function. It makes use of a triangular membership function (trimf). In order to provide a modulating signal for the PWM generator, the output is the reference signal created by a membership function. Error membership functions have seven inputs in total, and their derivatives have seven inputs.

In a similar manner, the output reference signal's seven membership functions are created. For improved VR and controller performance, a total of $7 \times 7 = 49$ rules are set. The regulated output voltage is depicted in Figures 4.9 and 4.10 displays the matching FFT analysis.

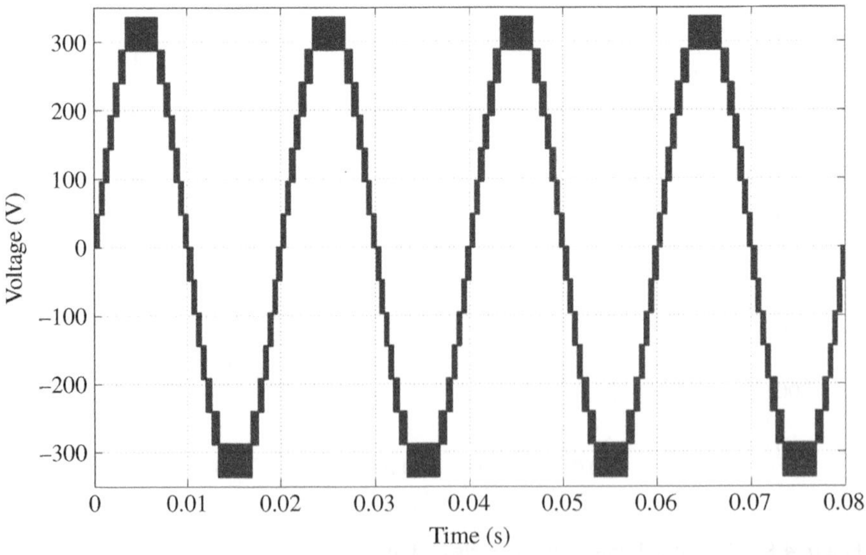

Figure 4.9 FLC-controlled output voltage with 15 levels.

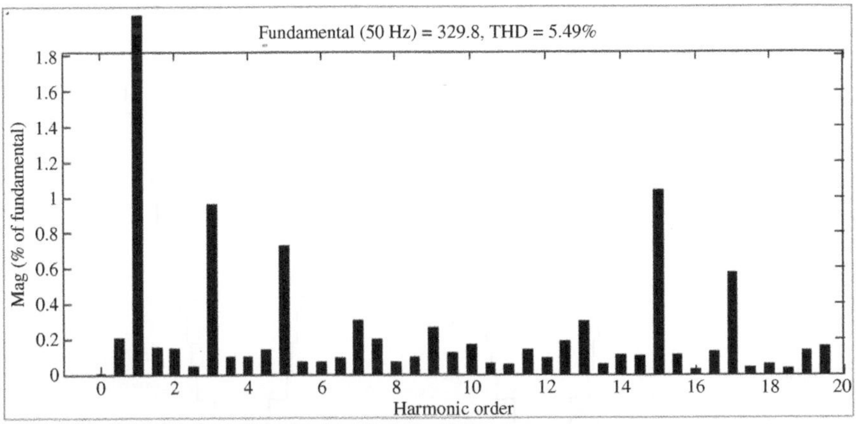

Figure 4.10 FFT analysis for voltage regulation based on FLC.

4.7.2 15-Level Inverter with Controller Based on PI

The DC input of CMLI changes according to the irradiance. The output obtained across the load of a solar-fed CMLI is compared with the reference signal, which produces an error voltage. The error signal is applied to the PI controller with gains $K_p = 0.04$ and $K_i = 0.5$. The output of the PI controller is the compensating signal, which is then added to the reference signal to yield the required modulating signal. Figures 4.11 and 4.12 display the regulated output voltage together with the associated FFT analysis.

4.7.3 An ANN-Based Controller for a 15-Level Inverter

Table 4.4 shows the specifications of the ANN used for the VR in a solar-fed 15-level inverter.

Figures 4.13 and 4.14 display the regulated output voltage together with the associated FFT analysis.

4.8 Hardware Outcomes

For the 7-stage, 15-level solar-fed CMLI with VR strategy, a 3 kWp solar PV power supply unit is constructed and put into operation. Each individual inverter stage has a 48 V voltage rating and a 7A current rating. Each module has a 115 W rated power and a voltage range of 16–21 volts (nominal 12 V) based on various working conditions, including light intensity. A nominal DC bus of 48 volts is obtained

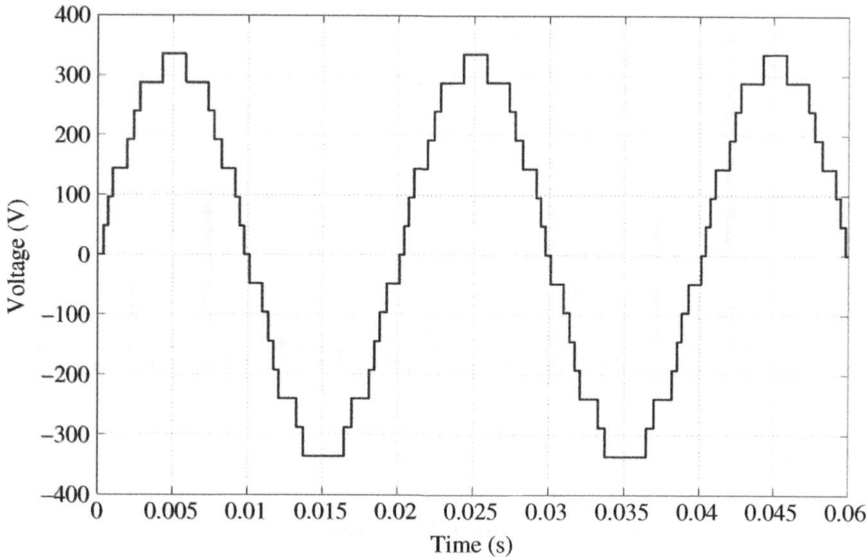

Figure 4.11 PI controller-controlled output voltage at 15 levels.

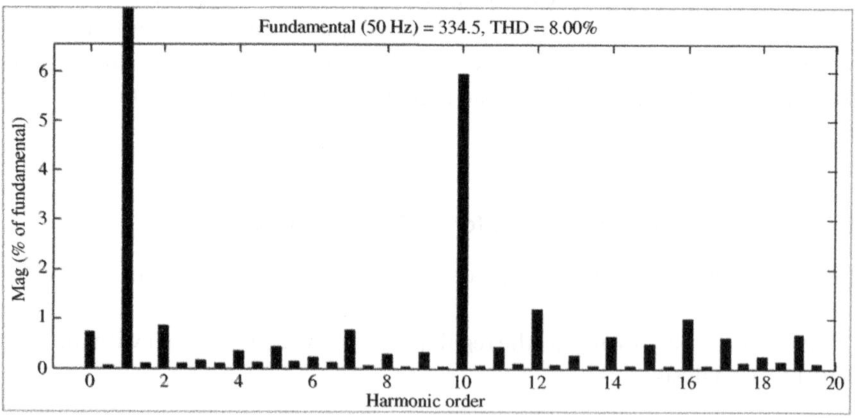

Figure 4.12 FFT analysis for PI-based voltage regulation.

by connecting four 12 V, 100 Ah battery packs in series. When the voltage across the battery setup is measured, variations between 37 and 58 volts are noted. At the inverter's output end, this variance results in an unequal voltage distribution. Therefore, the VR in the solar PV system is carried out by the microprocessor.

For comparison with the actual value produced, the necessary voltage is scaled down to the appropriate limits. The compensatory signal needed for pulse

Table 4.4 ANN specifications.

Parameters	Functions
Input layer	2
Hidden layer	10
Output layer	1
Network	Backpropagation
Algorithm	Gradient descent
Rate of learning	0.05
Objective: iteration count	$1e^{-5}$
Each iteration count	3000
Function of training	Purelin
Function of activation	Sigmoid and tangential

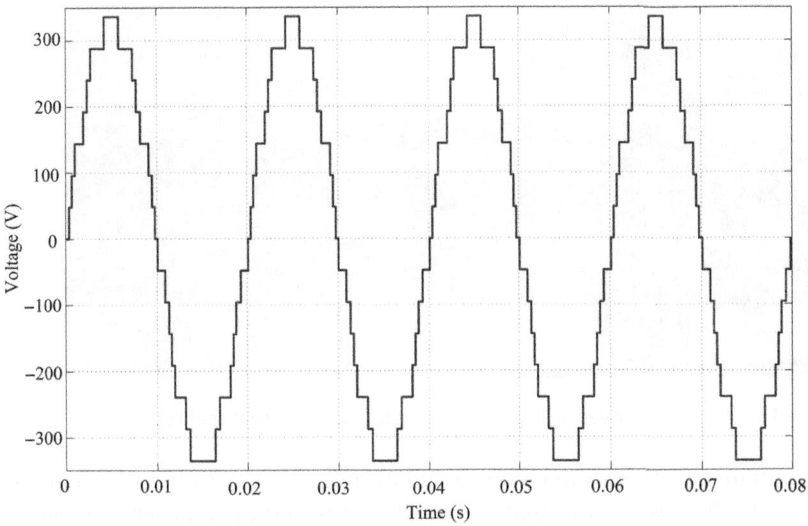

Figure 4.13 15-Level output voltage regulation using an ANN-based controller.

production is created based on this. Through an optocoupler, these pulses are sent to the MOSFET switches, resulting in the matching VR. The hardware configuration for a 15-level inverter is shown in Figure 4.15. The integration of the inverter with the solar PV system is shown in Figures 4.16–4.18 demonstrate the output voltage waveform in both open-loop and closed-loop operation.

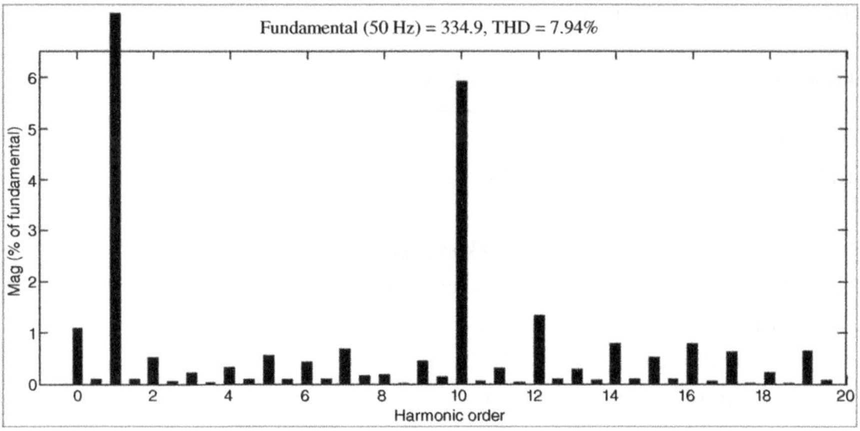

Figure 4.14 FFT analysis for voltage regulation using ANNs.

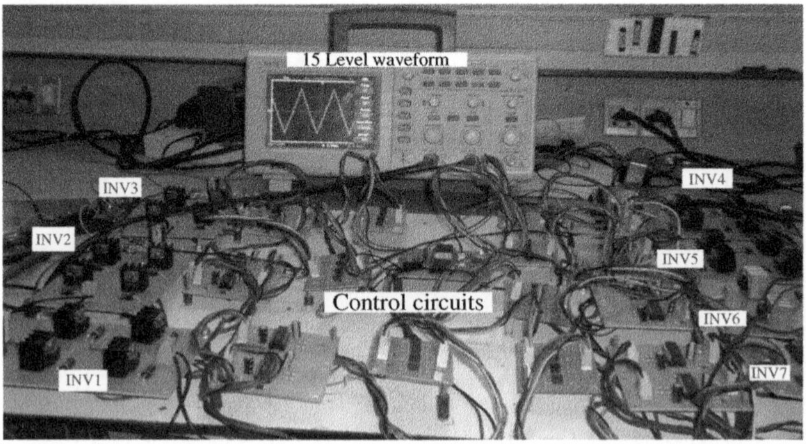

Figure 4.15 A 15-level inverter experimental setup using virtual reality.

The output waveforms are obtained and examined for the various situations. The output voltage waveform analysis, which shows all the parameters needed for regulation, is shown in Figures 4.19 and 4.20.

The THD obtained from power quality analysis using the PQA WT3000 is displayed in Figure 4.21.

Table 4.5 compares different controllers for output VR; FLC outperforms PI and ANN-based controllers in terms of voltage regulation.

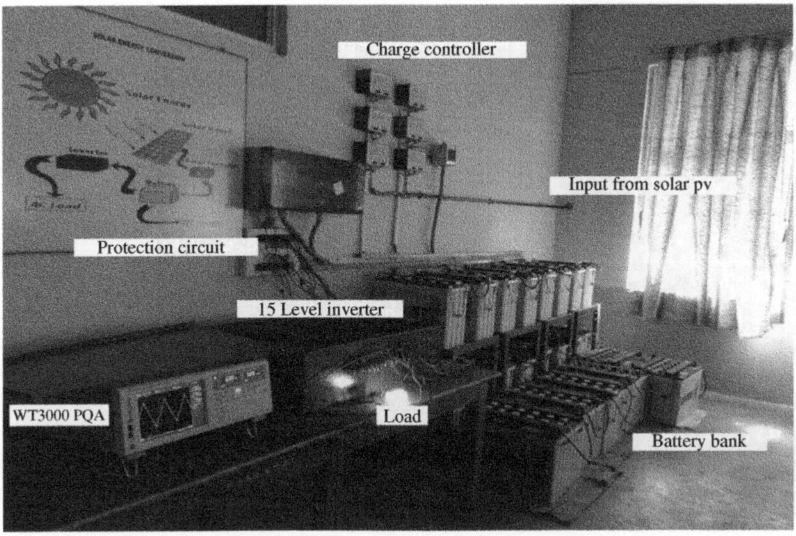

Figure 4.16 Combining solar PV with the suggested inverter.

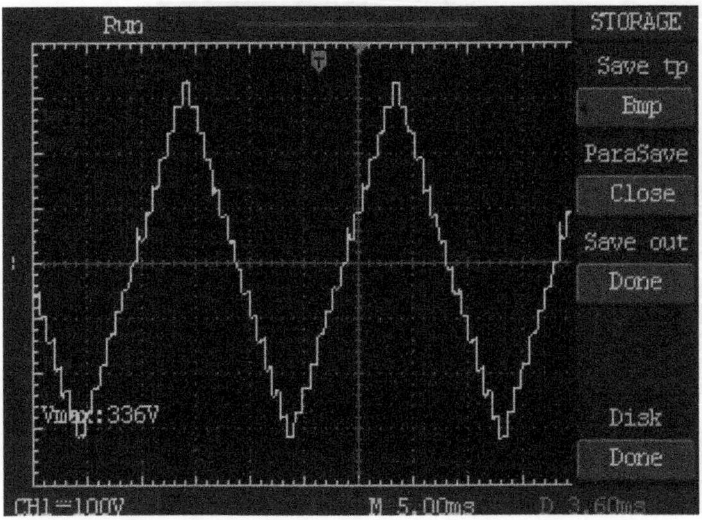

Figure 4.17 Waveform of the output voltage in an open loop.

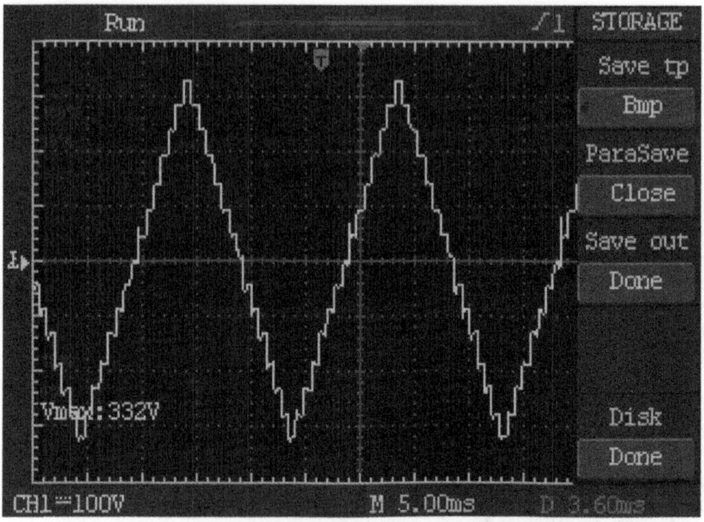

Figure 4.18 Output voltage waveform in closed-loop condition.

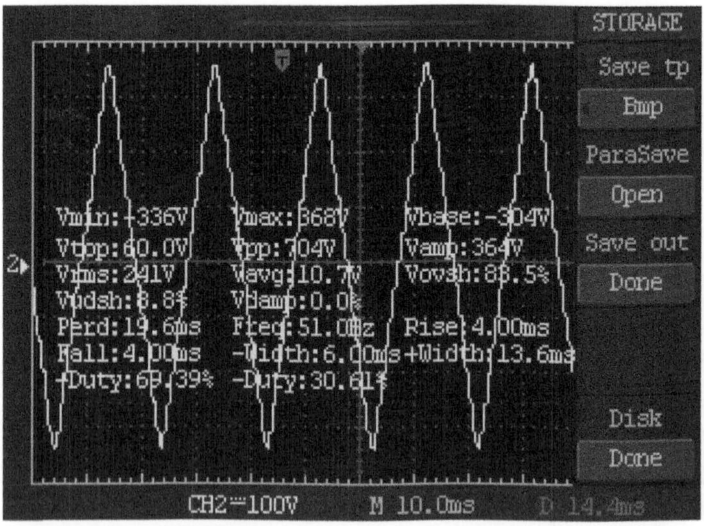

Figure 4.19 Analysis of the output voltage waveform ($V_{RMS} = 241$ V).

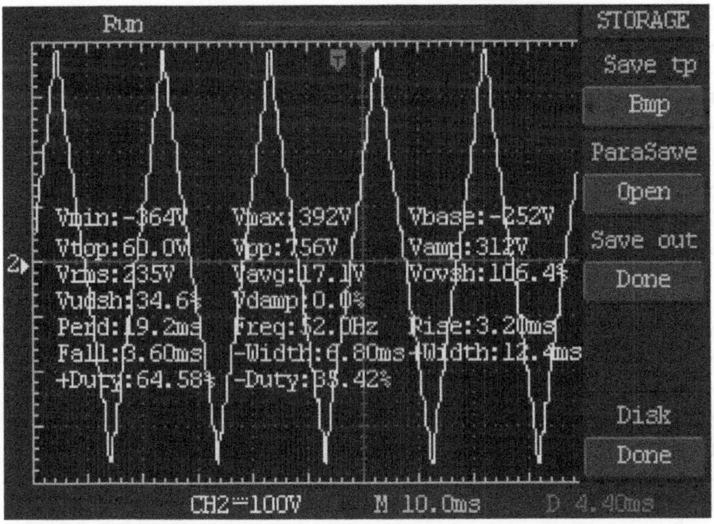

Figure 4.20 Output voltage waveform analysis (V_{RMS} = 235 V).

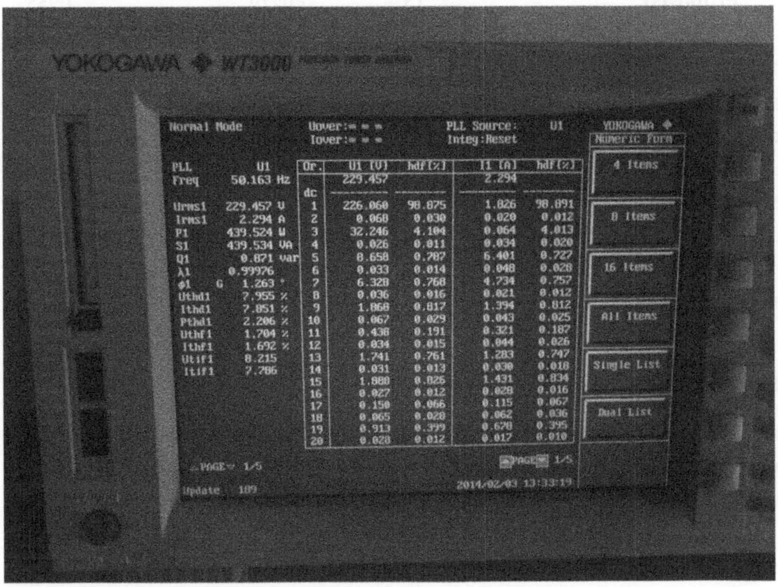

Figure 4.21 PQ analysis for the output voltage waveform.

Table 4.5 Results comparison.

S. no.	Method	Level	THD (%)	VR deviation
1	PI	15	8	9.3
2	ANN	15	7.94	9.7
4	FLC	15	**5.49**	**4.6**

Table 4.6 Compares the outcomes with other methods. Since there aren't many published research for the VR, the methodology and the levels taken into consideration were compared.

S. no.	Authors	Levels	Methodology	THD (%)
1.	Selvaraj and Rahim (2009)	5	PI	6.8
2	Ghazanfari et al. (2012)	5	Voltage balancing	6.13
3	Cecati et al. (2010)	9	Fuzzy	–
4.	Corzine et al. (2004)	7	PI	9
5.	**Proposed**	15	Fuzzy	7.955

The results are compared with alternative approaches in Table 4.6. The approach used and the levels taken into consideration were compared because there aren't many published studies for the VR.

The results show that, when considering the changes at the input solar PV, FLC offers superior outcomes for VR. Despite this, Cecati et al. (2010) contemplate FLC for the nine levels; nevertheless, the implementation is done using a DC power supply instead of solar panels.

4.8.1 Harmonic Analysis Using PWM

The equivalent carrier frequency is not raised by modulation techniques based on the vertical dispersion of carriers. Alternate Phase Opposition Disposition (APOD), Phase Opposition Disposition (POD), and Phase Disposition (PD) are further classifications for this method. Figures given below show the carrier arrangements for PD, POD, and APOD considering various inverter levels.

4.8.2 Phase Disposition

Three level

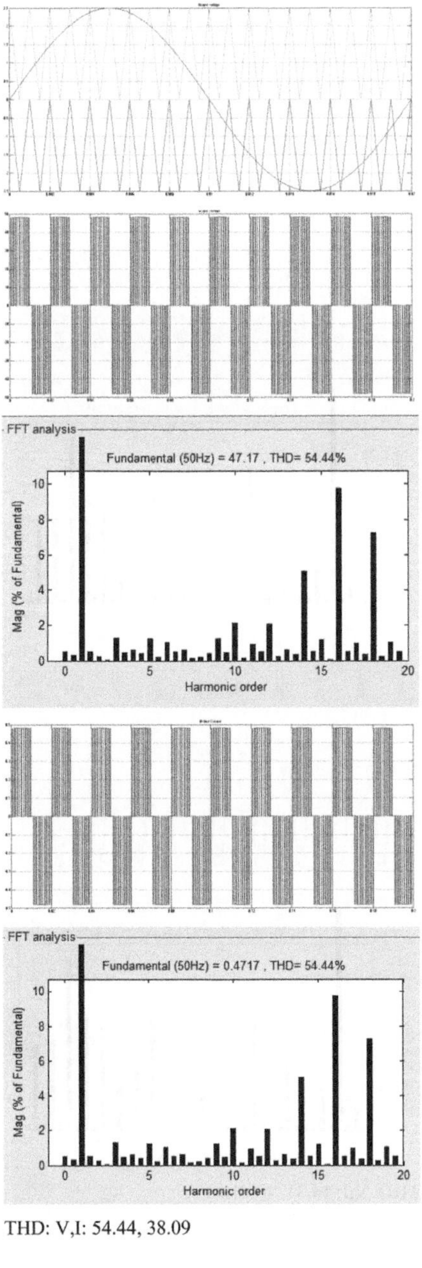

THD: V,I: 54.44, 38.09

Five level

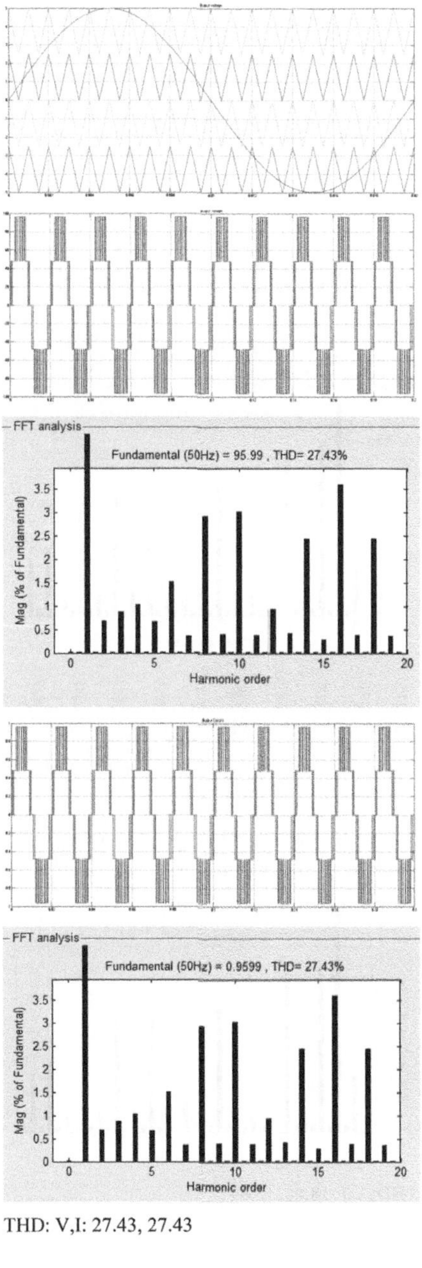

THD: V,I: 27.43, 27.43

Seven level

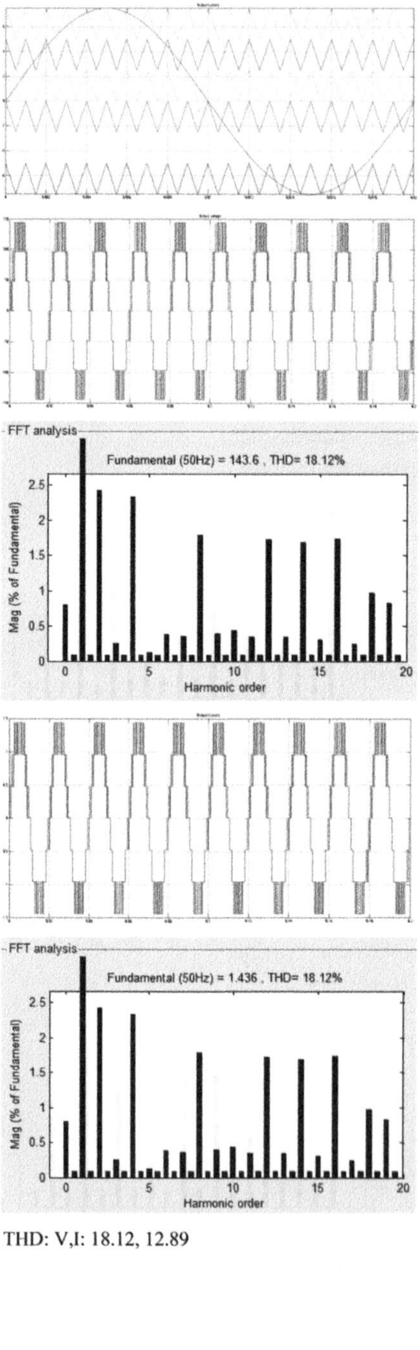

THD: V,I: 18.12, 12.89

Nine level

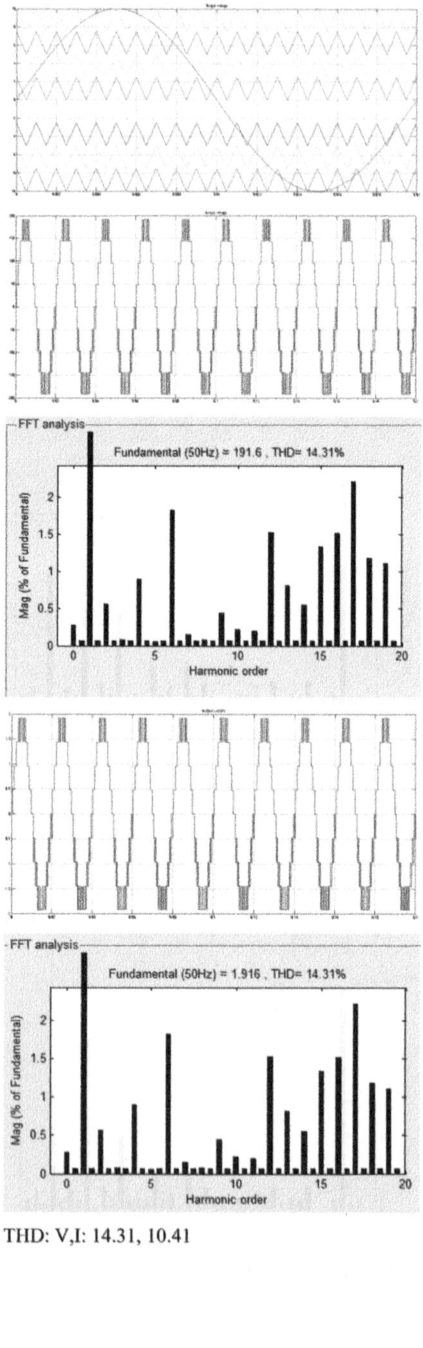

THD: V,I: 14.31, 10.41

Eleven level

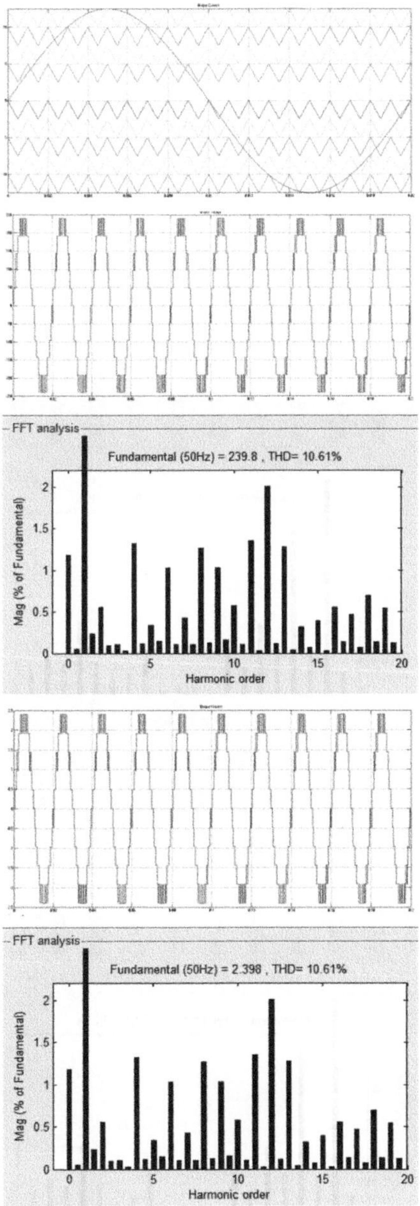

THD: V,I: 10.61, 7.76

Thirteen level

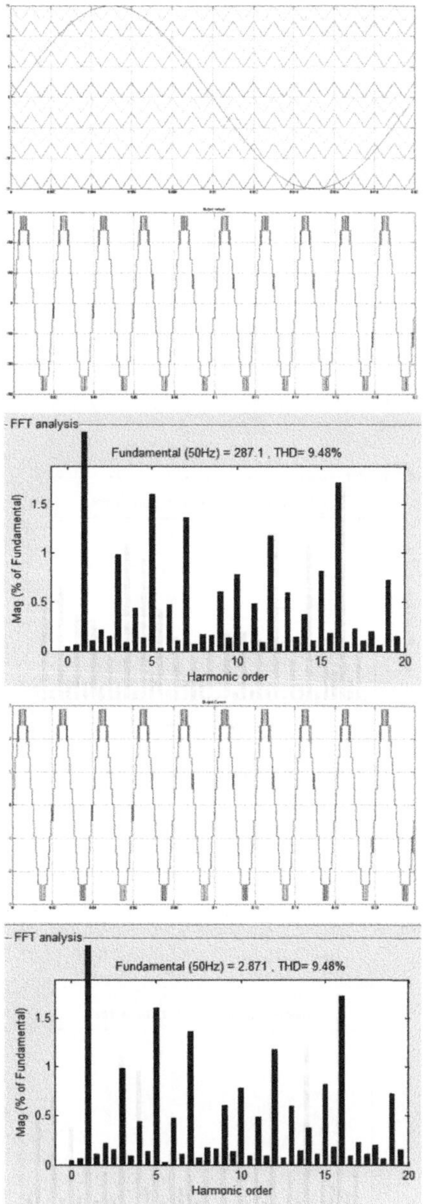

THD: V,I: 9.48, 7.01

Variable Amplitude

Fifteen level

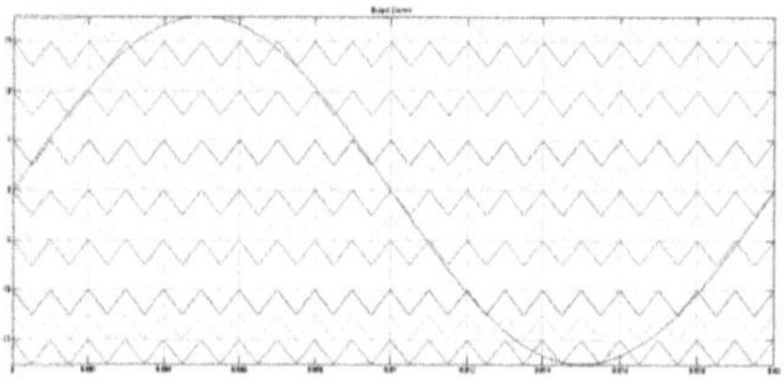

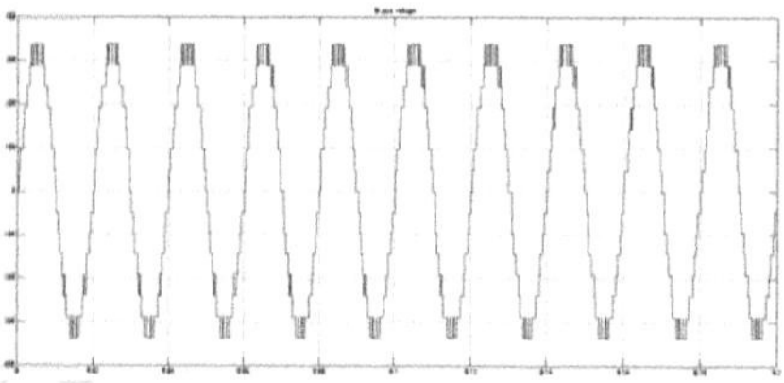

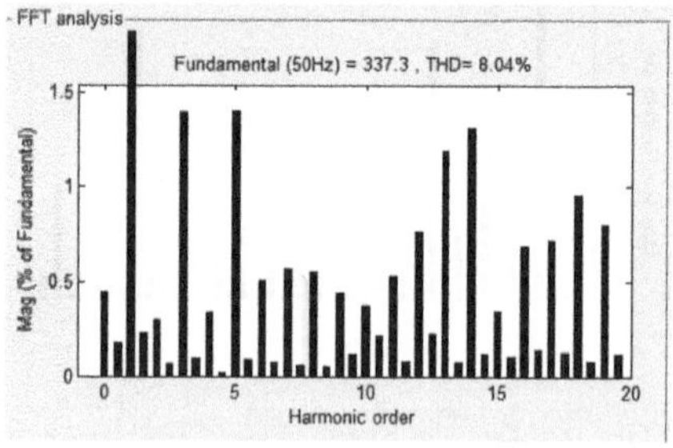

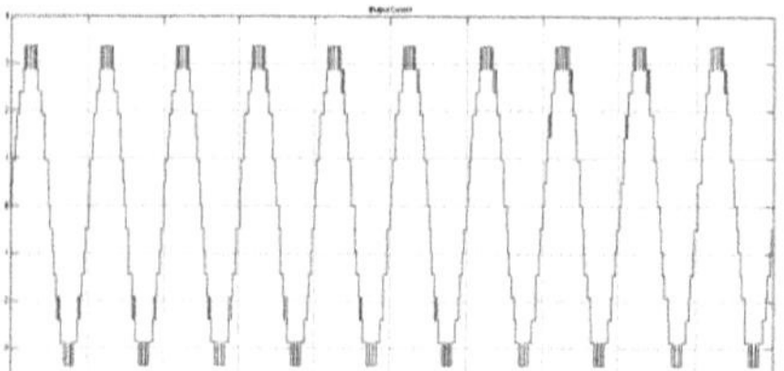

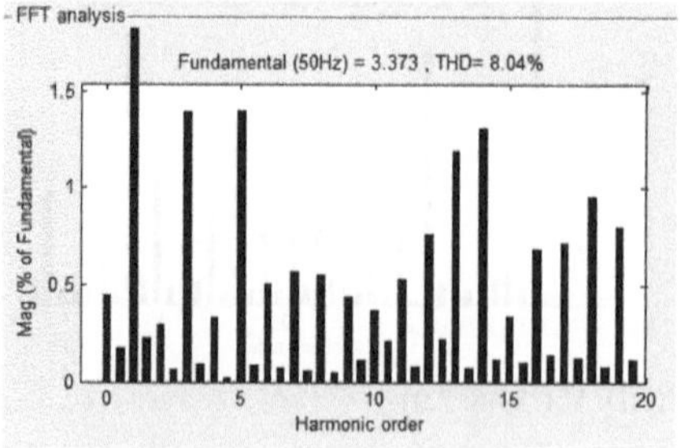

THD: V,I: 8.04, 5.96

Five level:

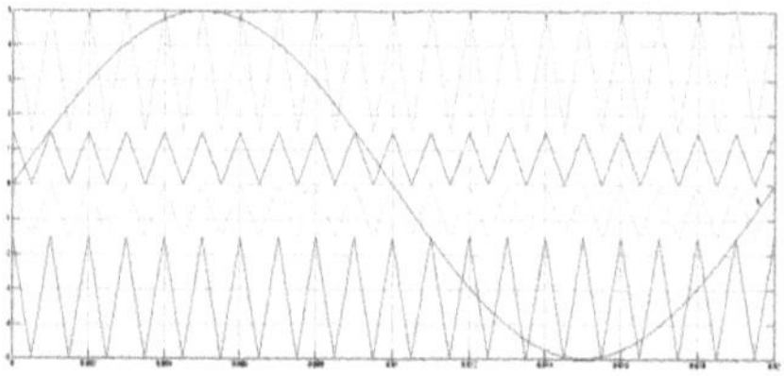

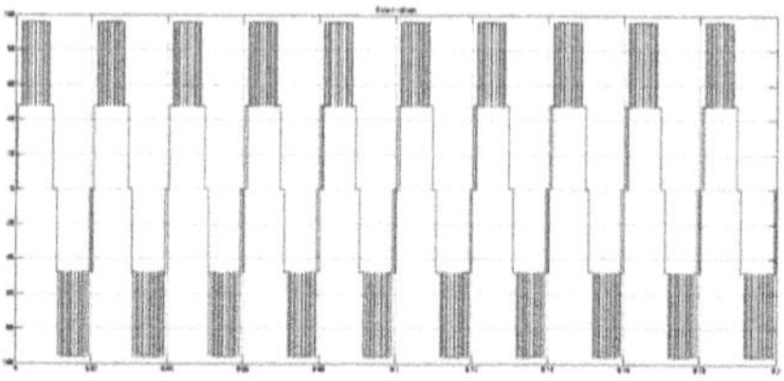

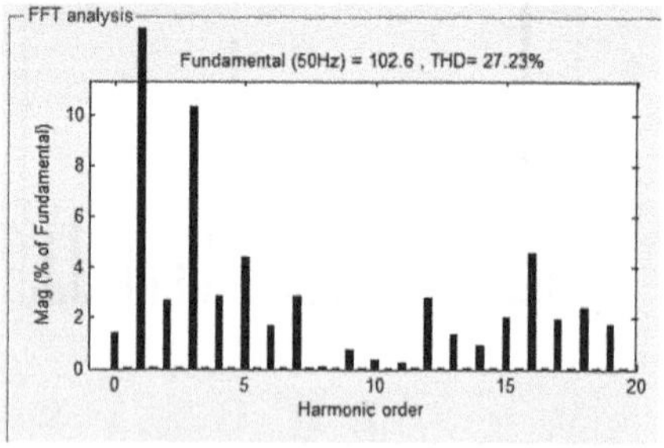

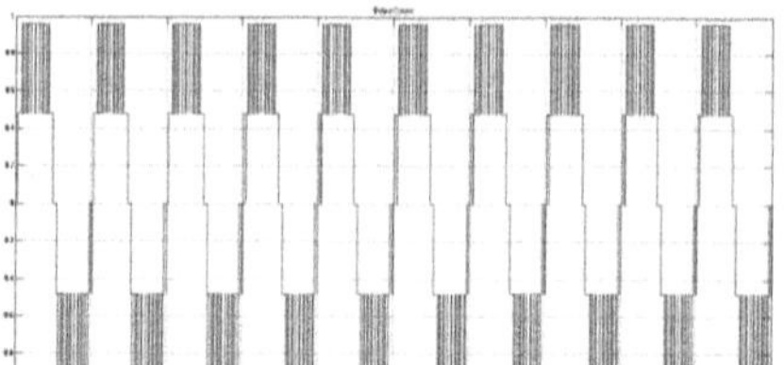

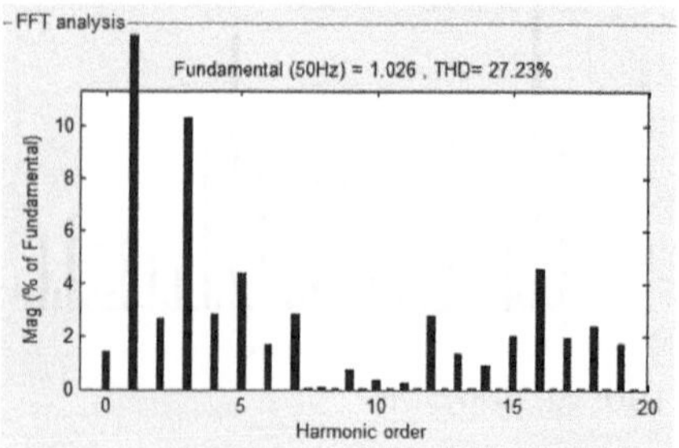

THD:V,I: 27.23, 20.93

Seven level:

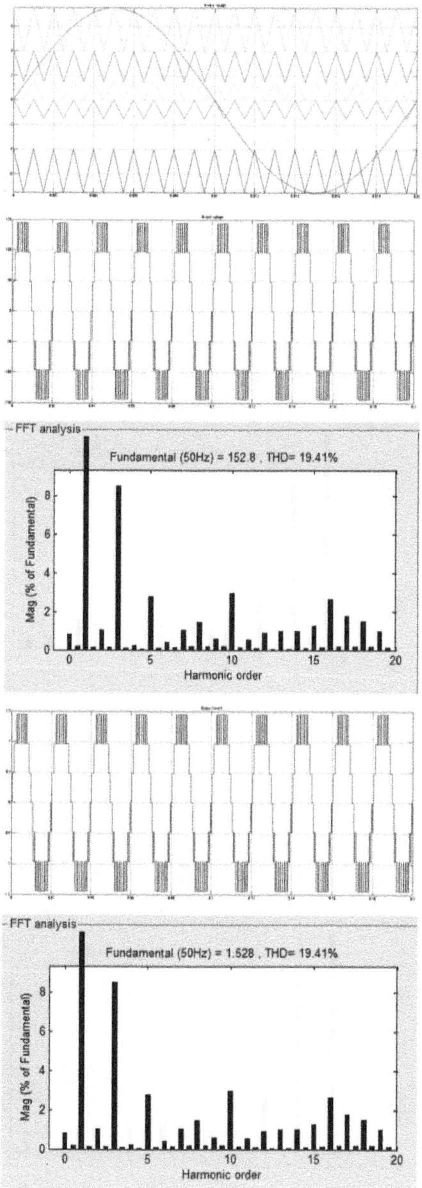

THD:V,I: 19.41, 15.22

Nine level

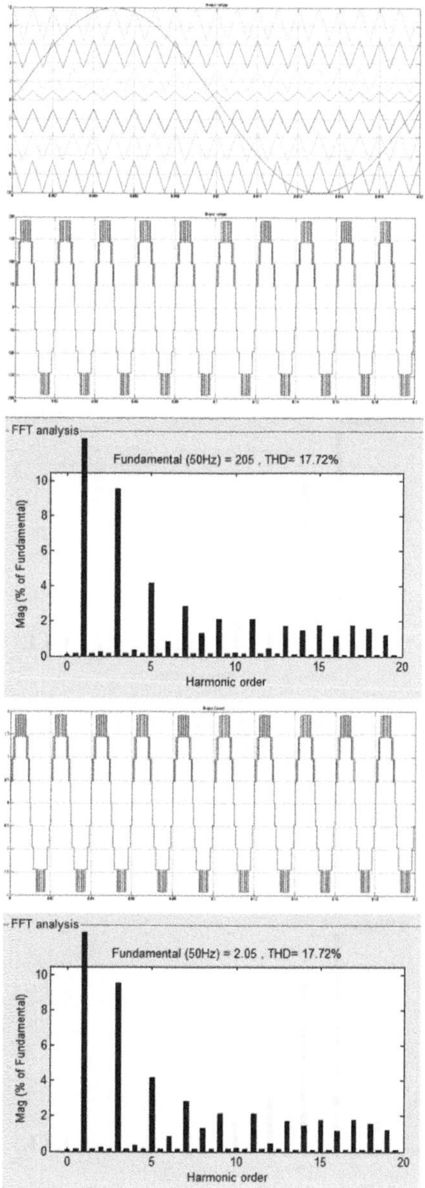

THD:V,I: 17.72, 14.76

Eleven level

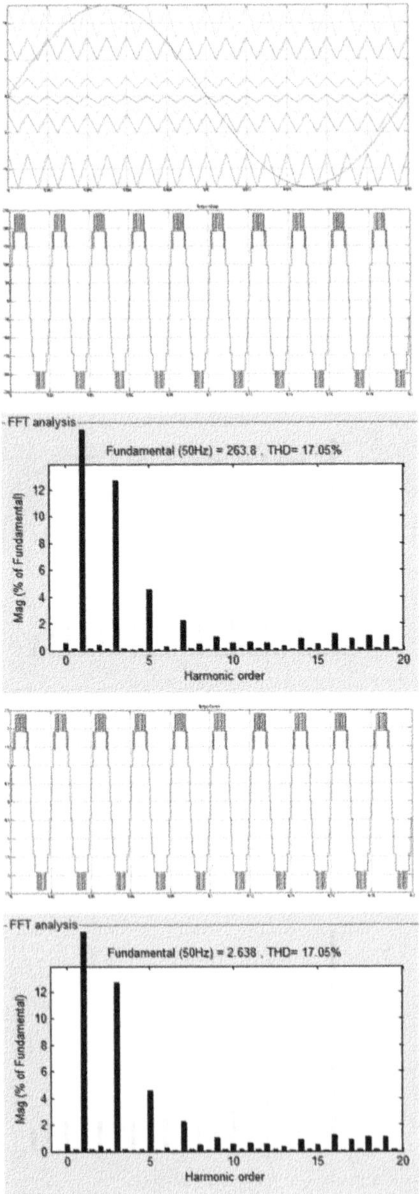

THD:V,I: 17.05,15.40

Thirteen level

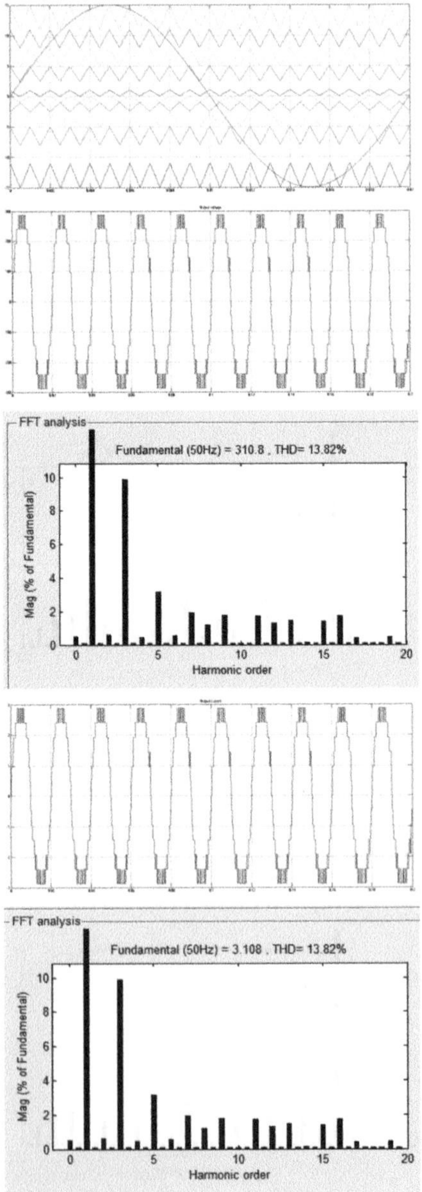

THD:V, I: 13.82, 12.38

Variable Time (inner f 500, 1000, 1500, 2000.....)

Fifteen level

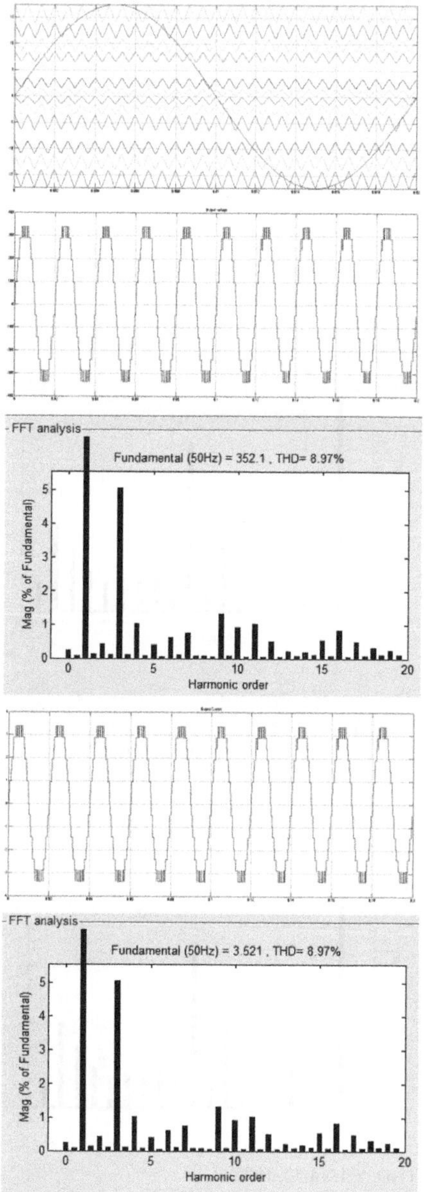

THD: V,I: 8.97, 7.39

Five level:

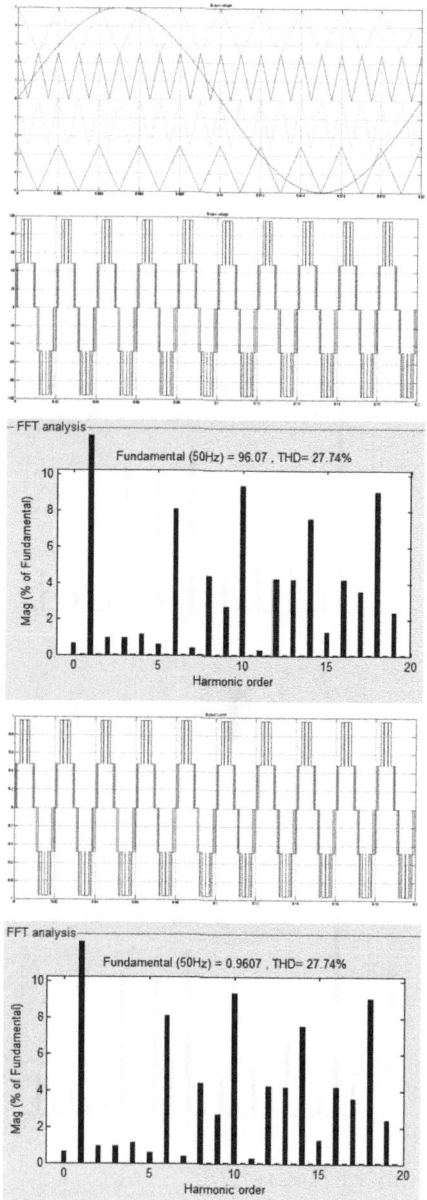

THD:V,I: 27.74, 22.06

Seven level

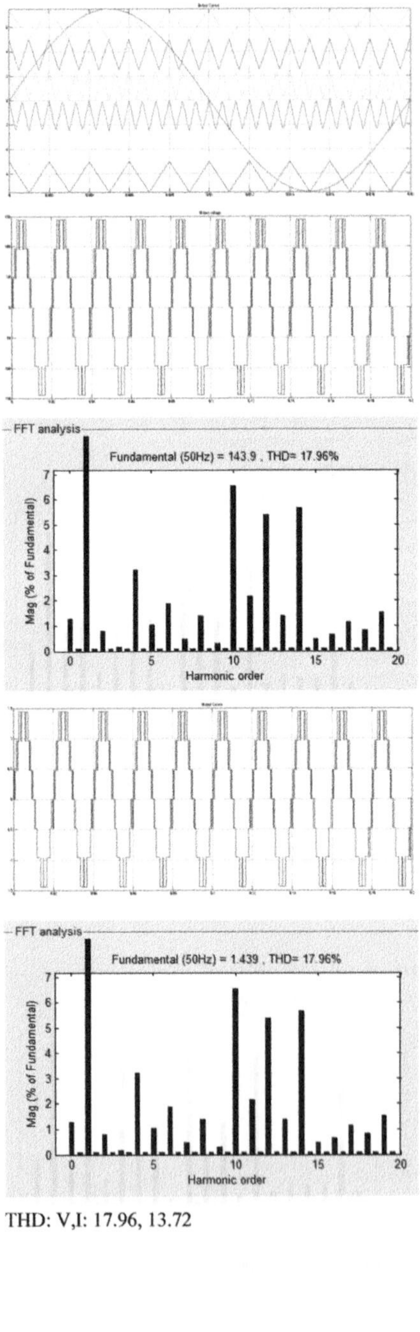

THD: V,I: 17.96, 13.72

Nine level

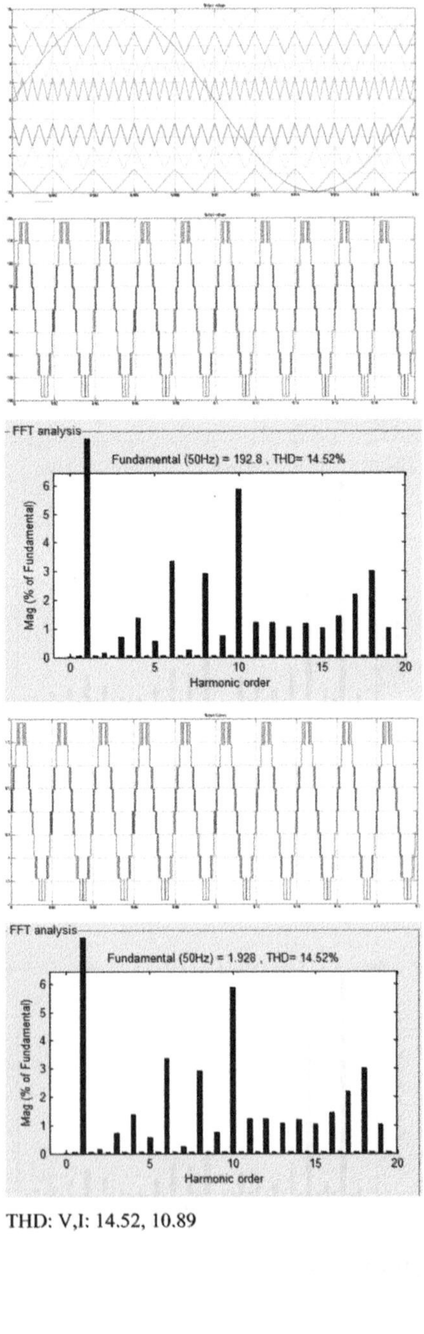

THD: V,I: 14.52, 10.89

Eleven level

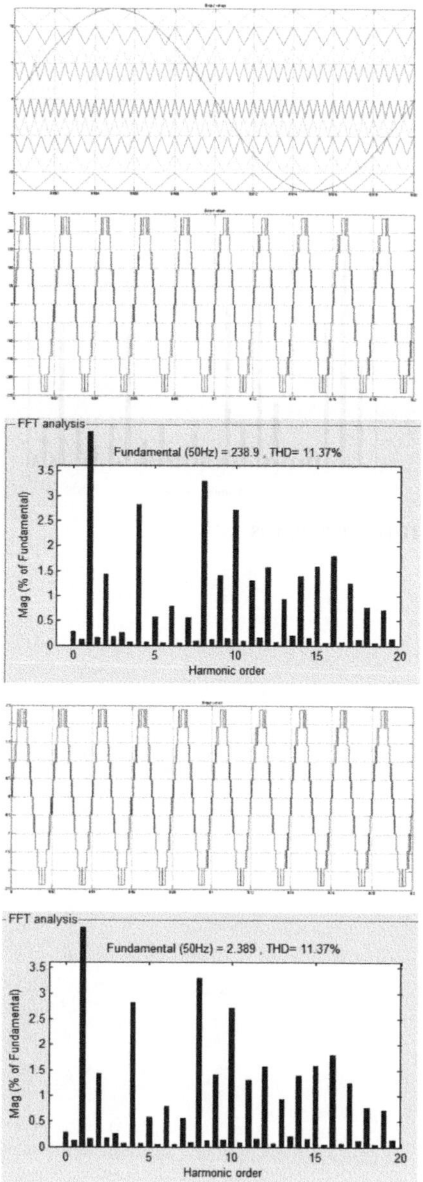

THD: V,I: 11.37, 8.25

Thirteen level:

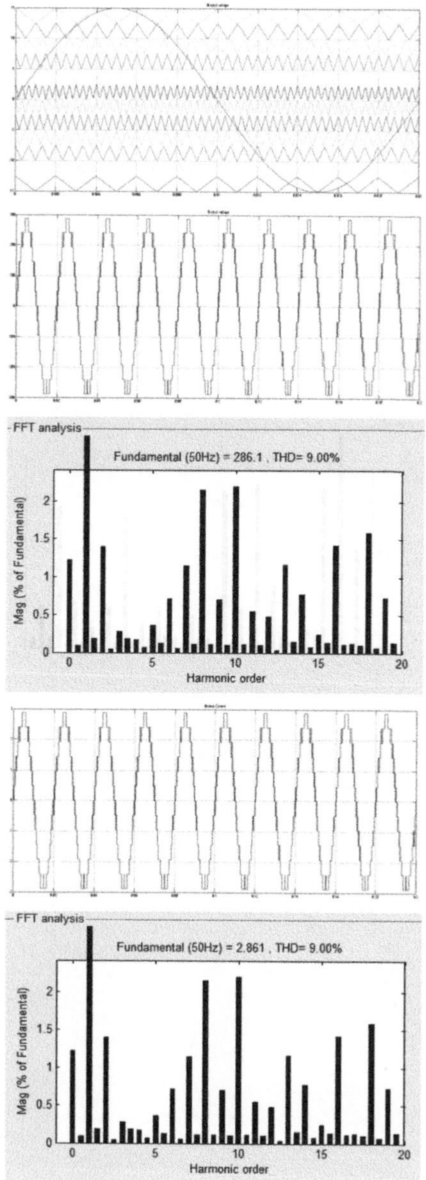

THD:V,I: 9.00, 6.10

Fifteen Level

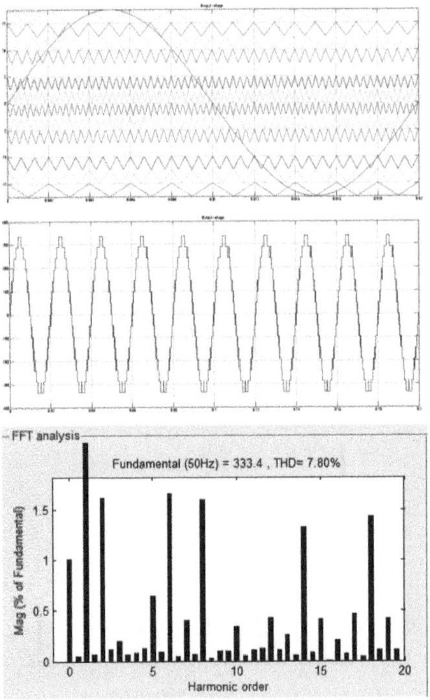

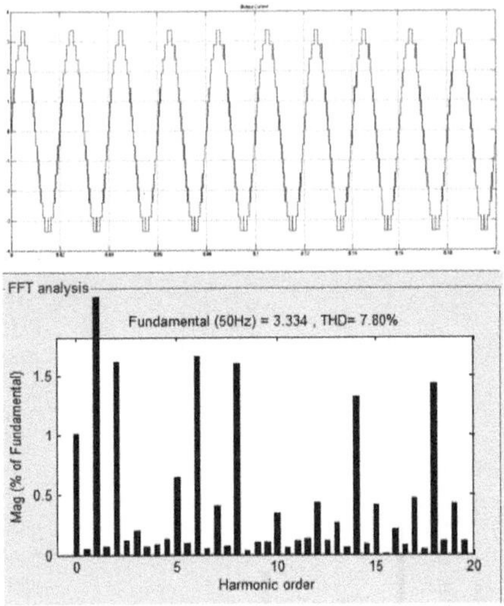

THD: V,I: 7.80, 4.98

Phase Opposition Disposition

Three level

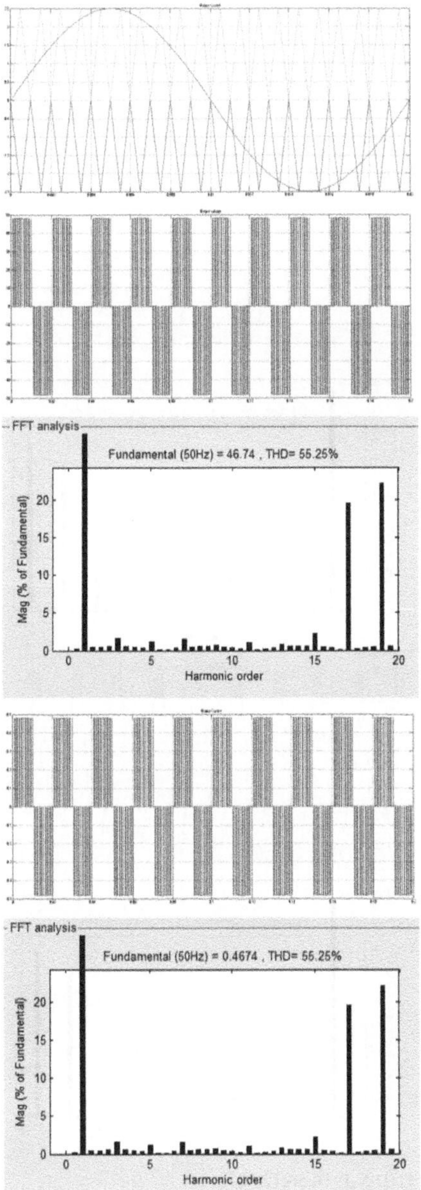

THD:V,I: 55.25, 38.95

Five level

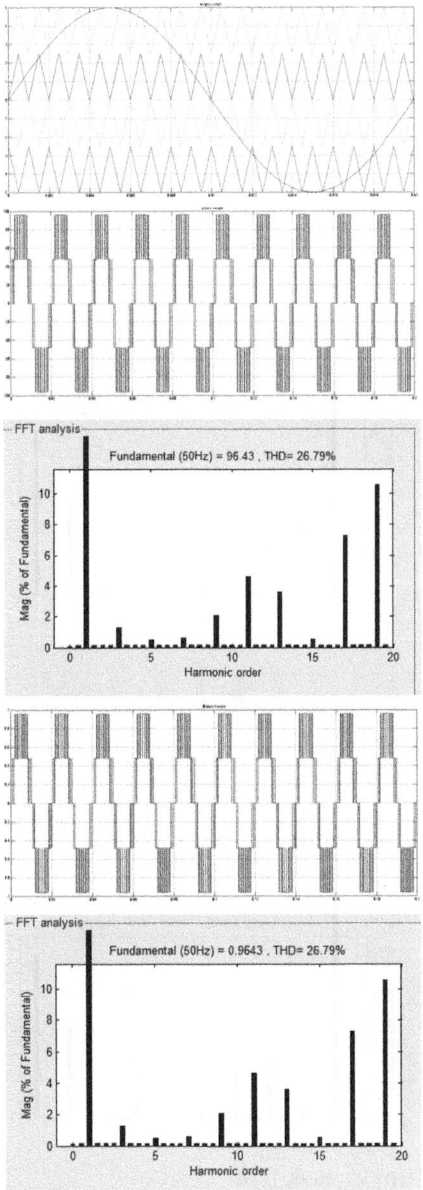

THD:V,I: 26.79, 18.45

Seven level | Nine level

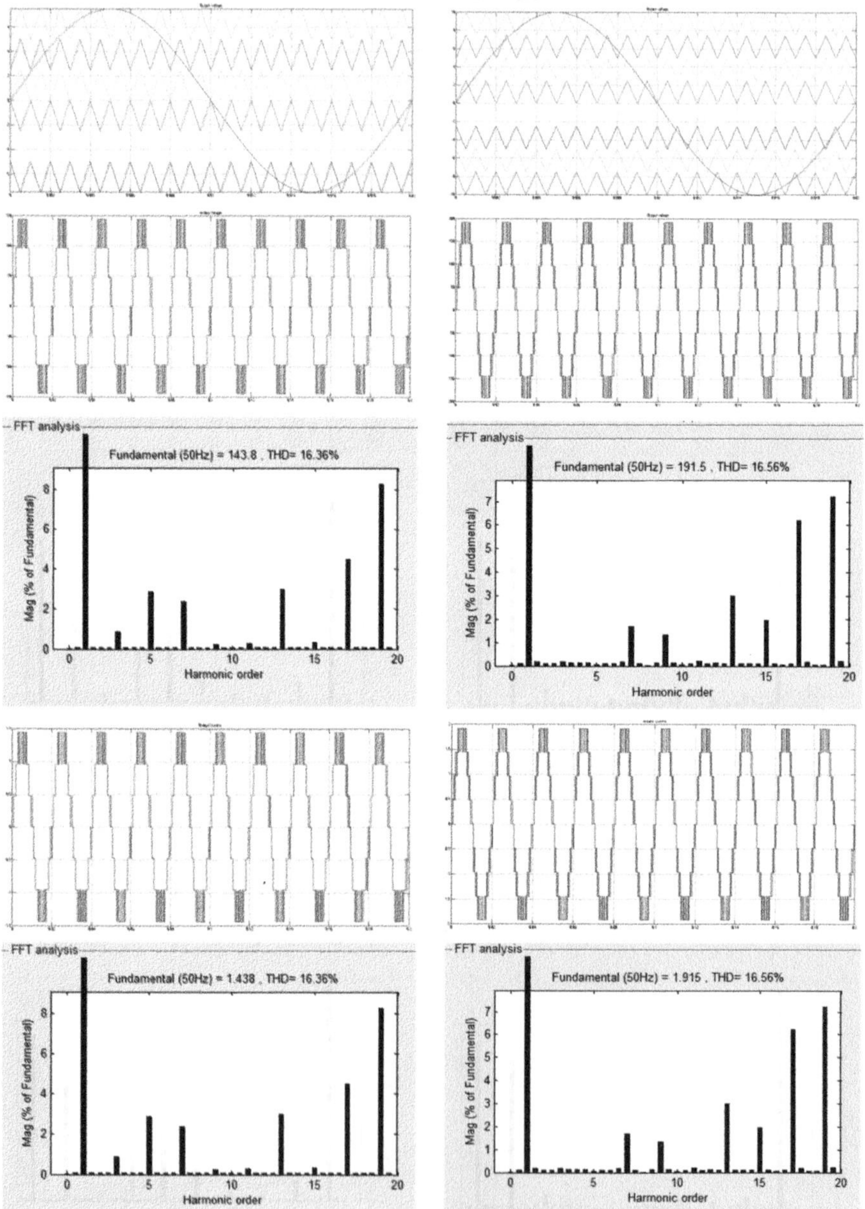

THD:V,I: 16.38, 12.02

THD:V,I: 16.56,12.30

Eleven level

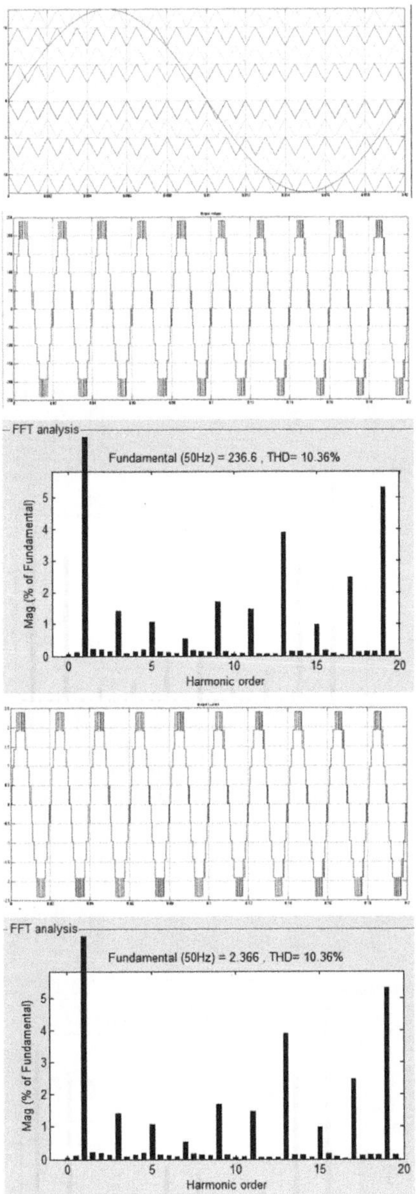

THD:V,I: 10.36, 7.74

Thirteen level

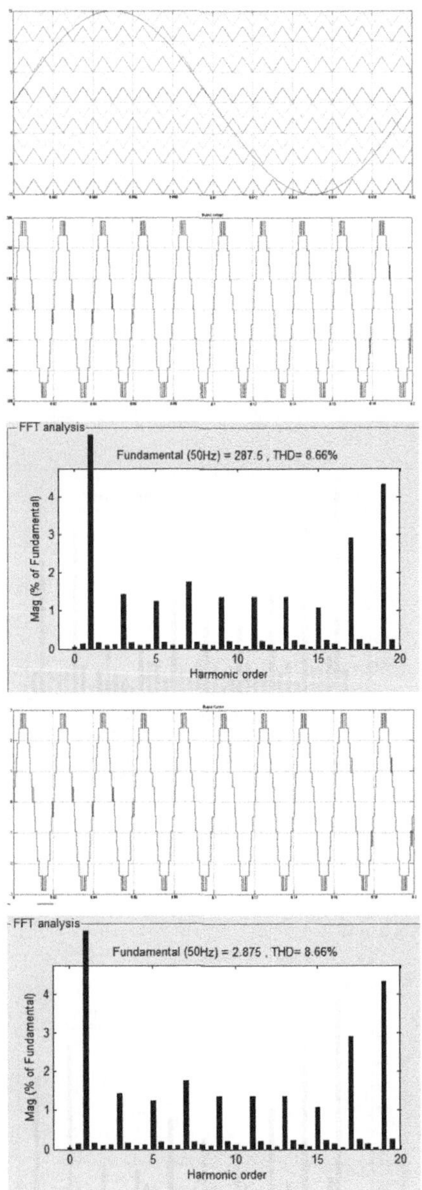

THD:V,I: 8.66, 6.48

Variable Amplitude

Fifteen level

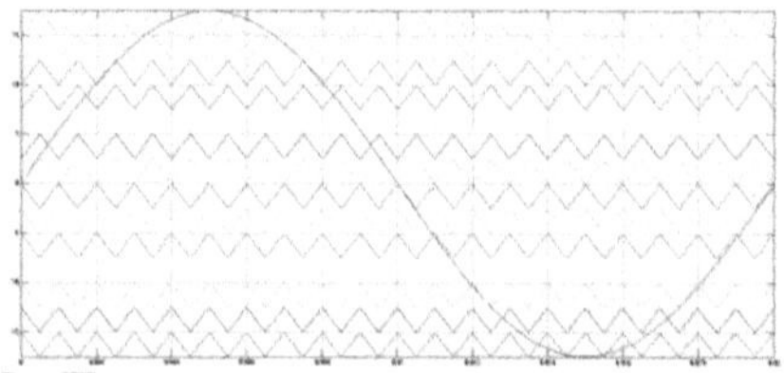

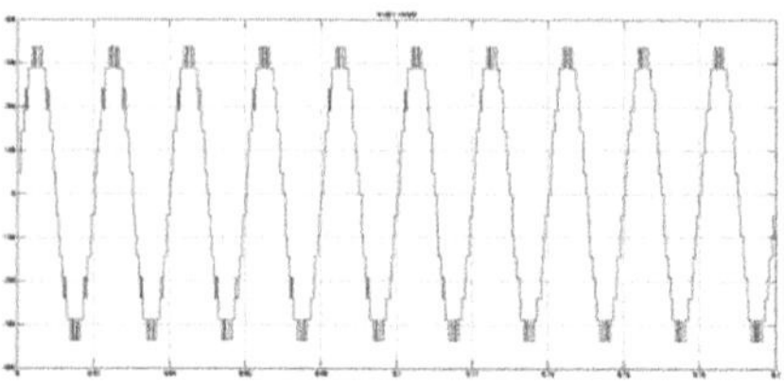

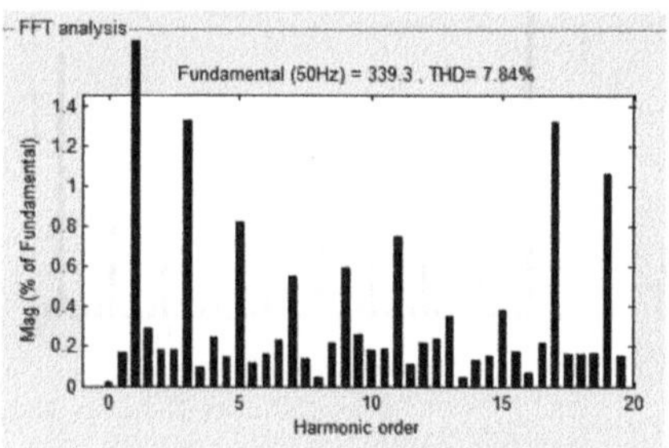

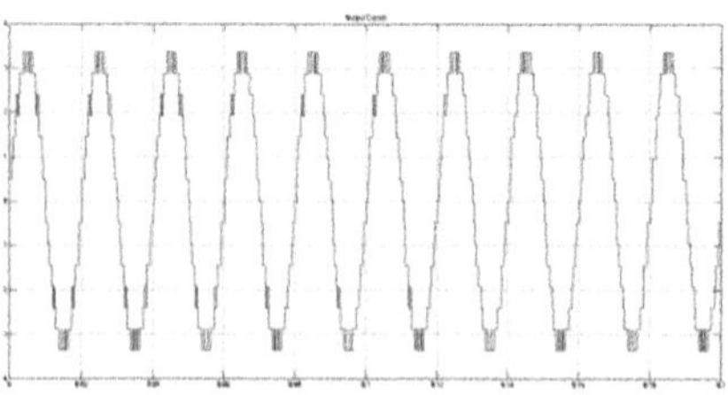

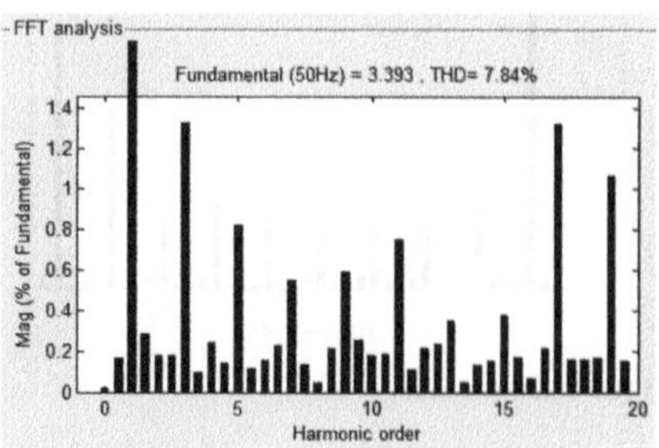

THD:V,I: 7.84, 5.63

Five level

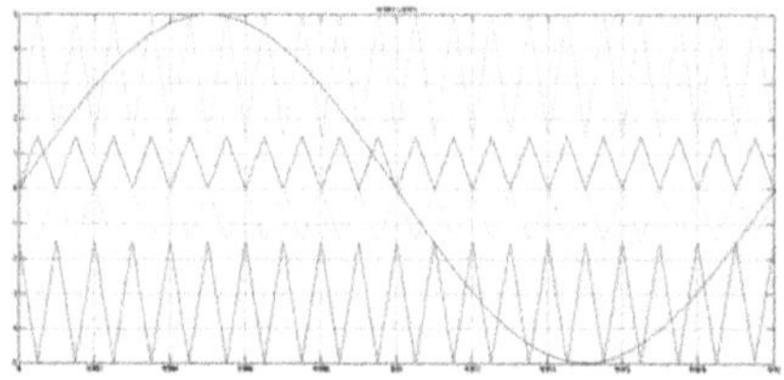

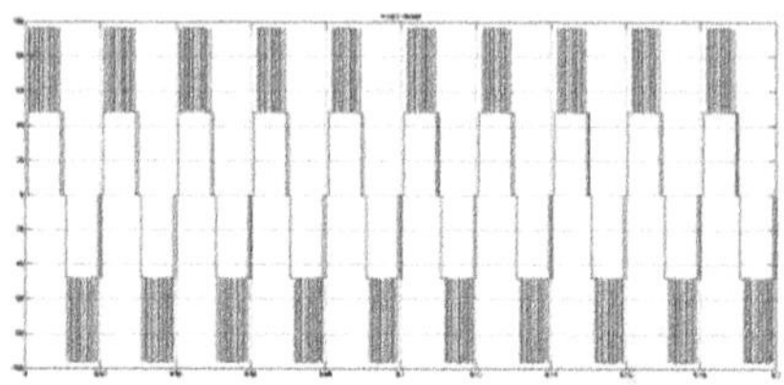

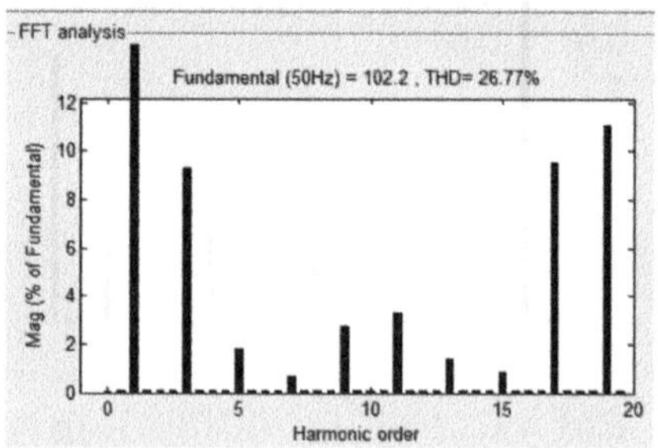

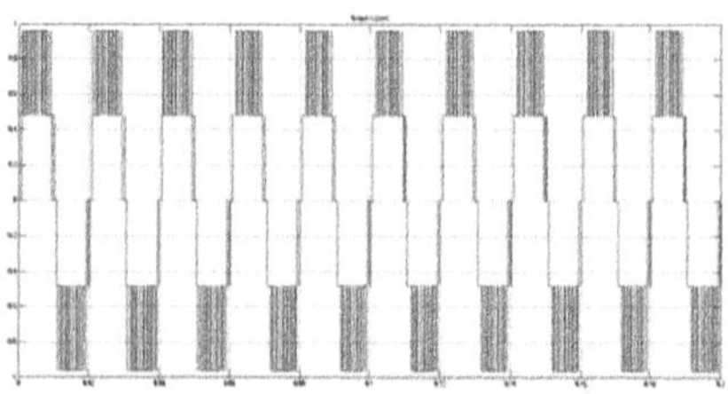

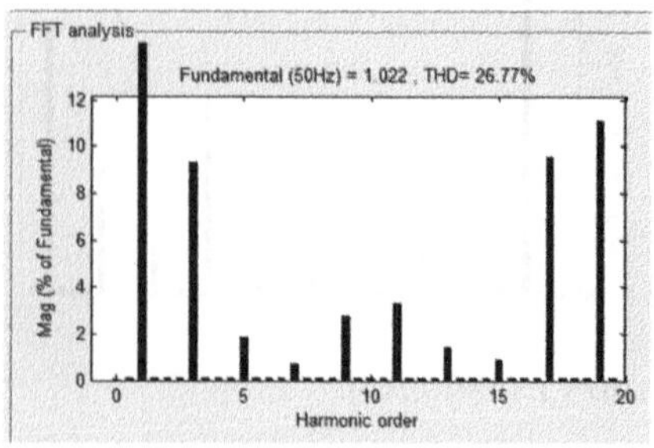

THD:V,I: 26.77, 20.00

Seven level

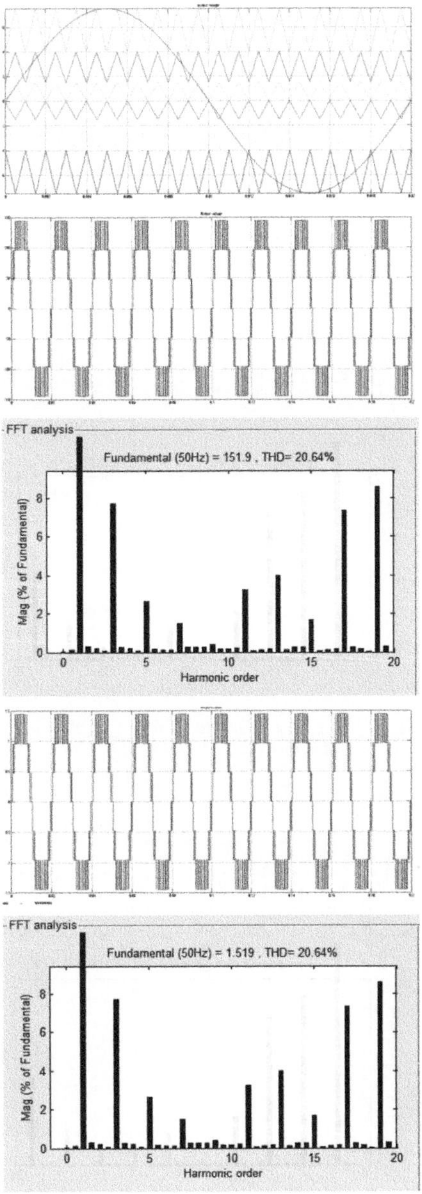

THD: V,I: 20.64, 16.02

Nine level

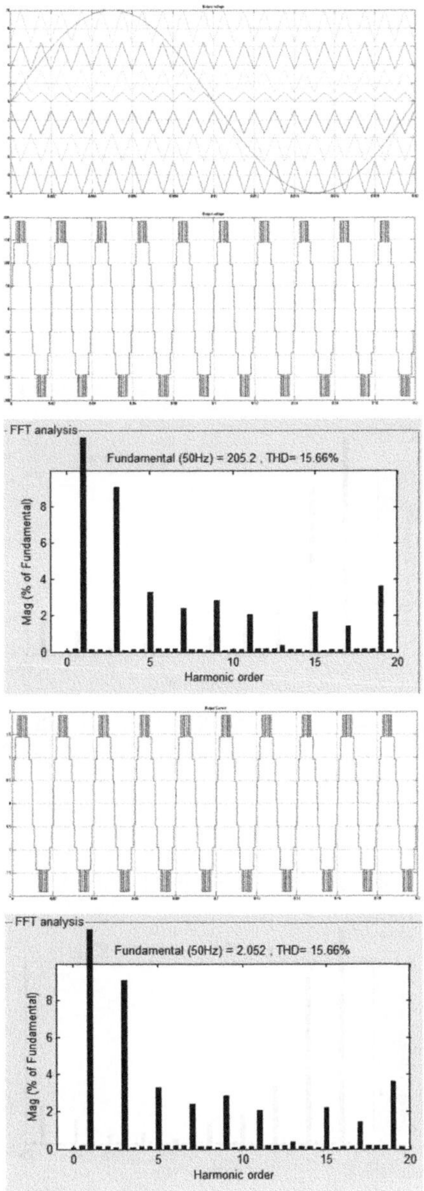

THD:V,I: 15.66, 13.14

Eleven level

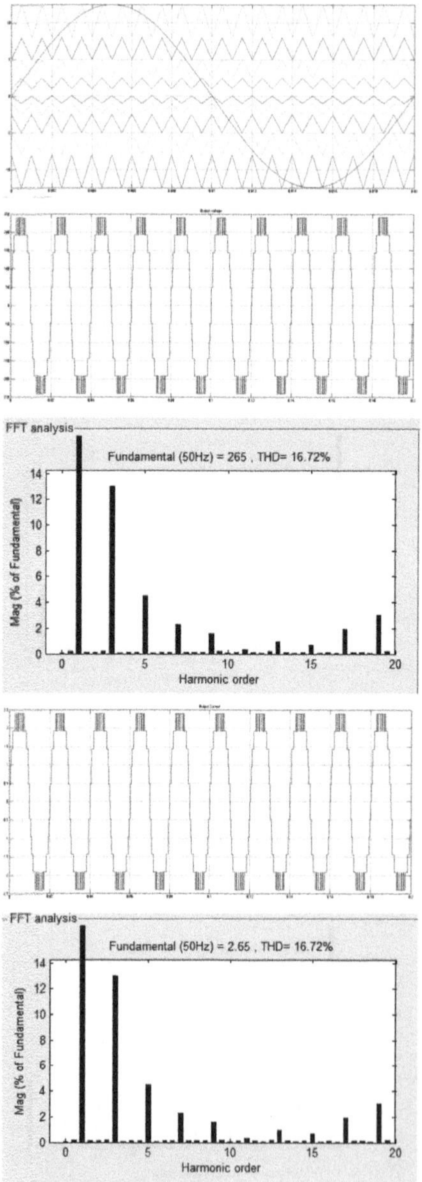

THD: V,I: 16.72, 15.28

Thirteen level

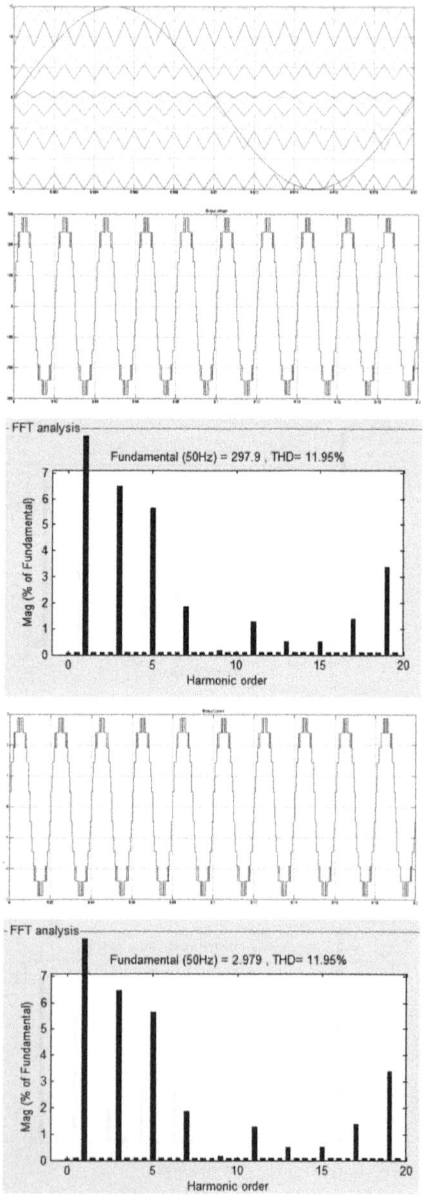

THD: V,I: 11.95, 10.26

Variable Time

Fifteen level

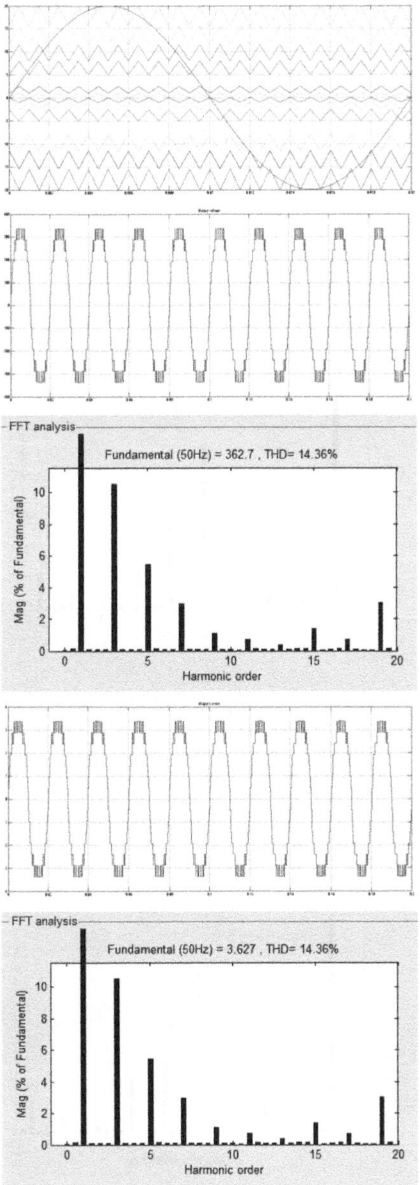

THD: V,I: 14.36, 13.18

Five level

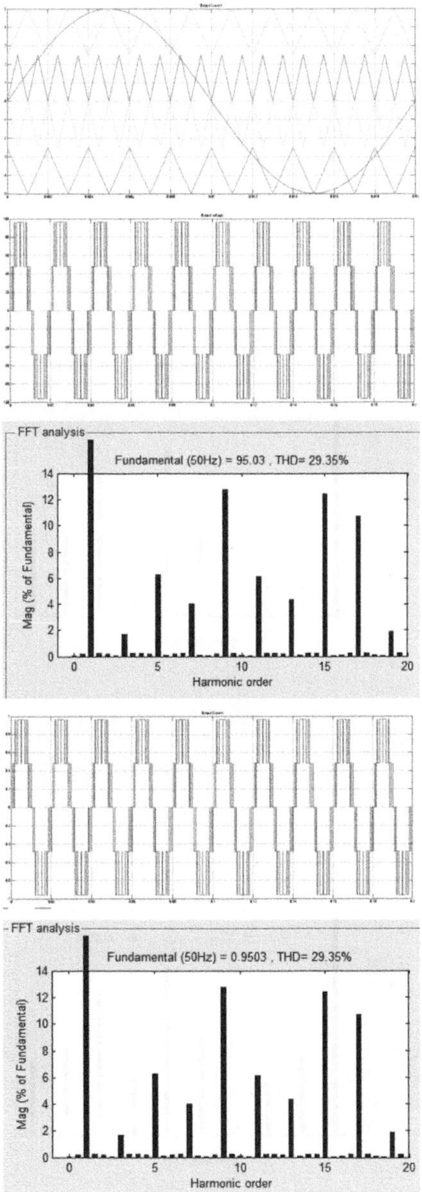

THD: V,I: 29.35, 23.89

Seven level

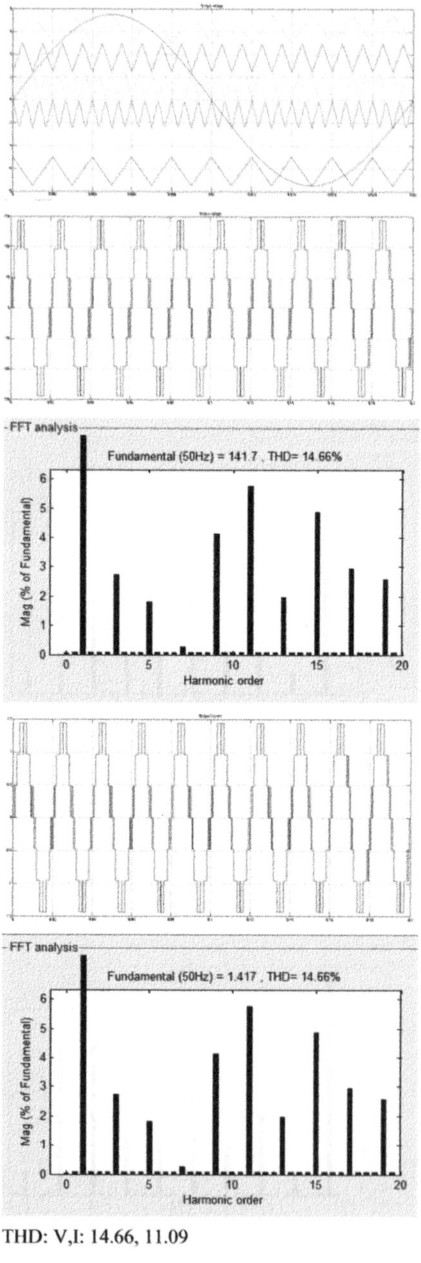

THD: V,I: 14.66, 11.09

Nine level

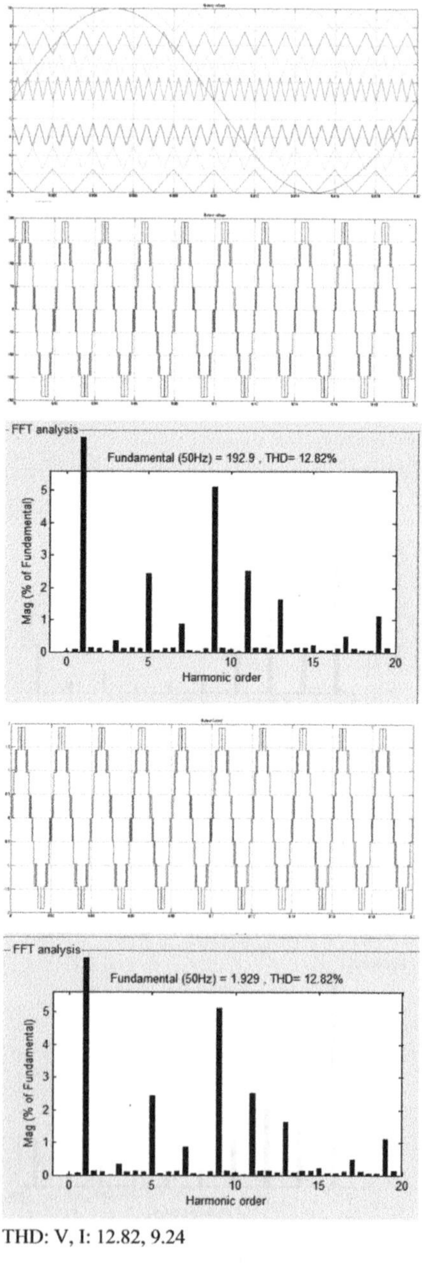

THD: V, I: 12.82, 9.24

Eleven level

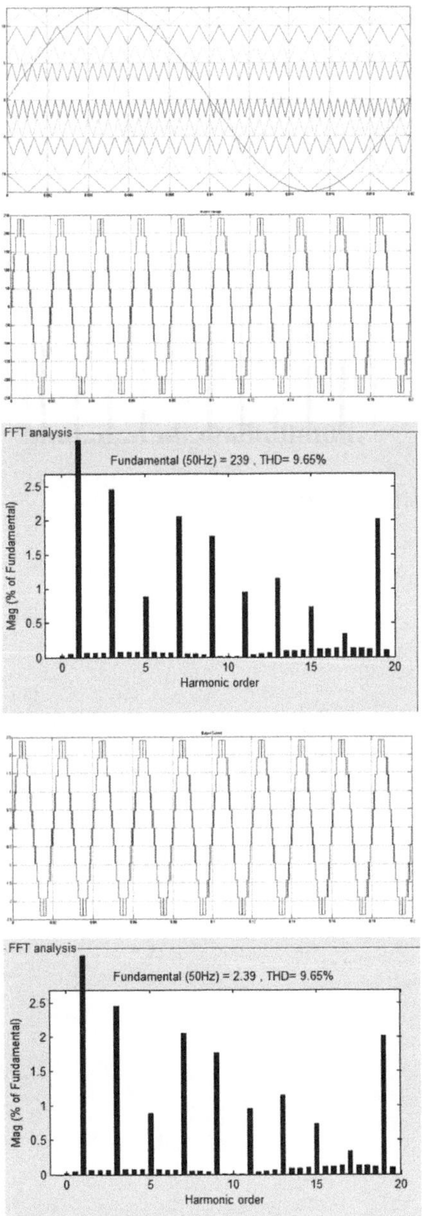

THD: V,I: 9.65, 6.37

Thirteen level

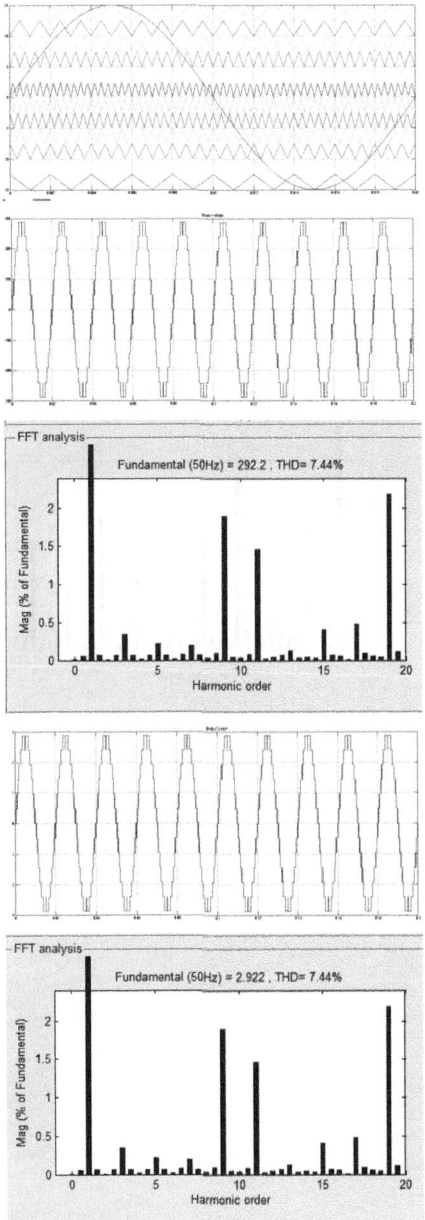

THD: V,I: 7.44, 4.58

Fifteen level

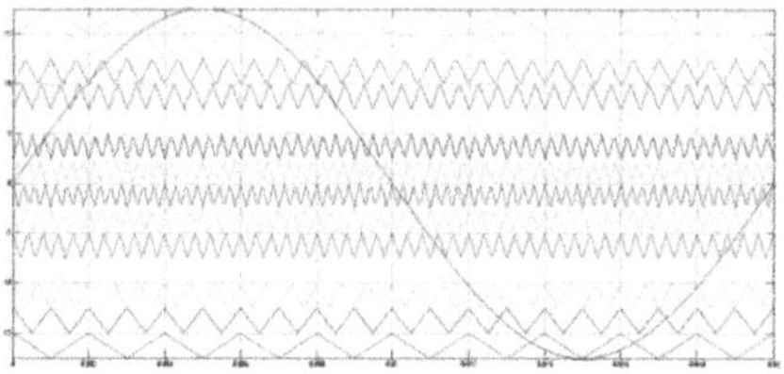

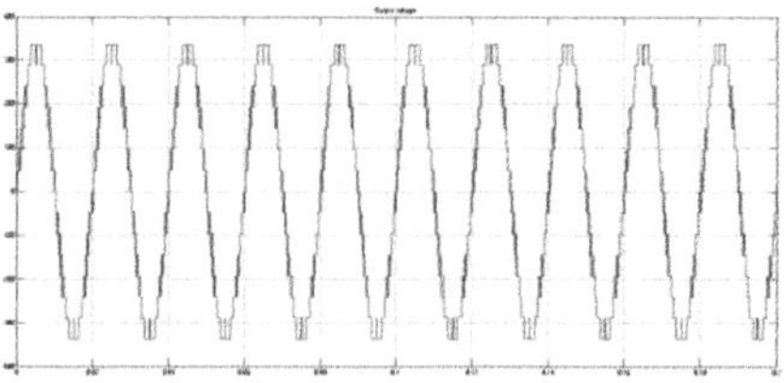

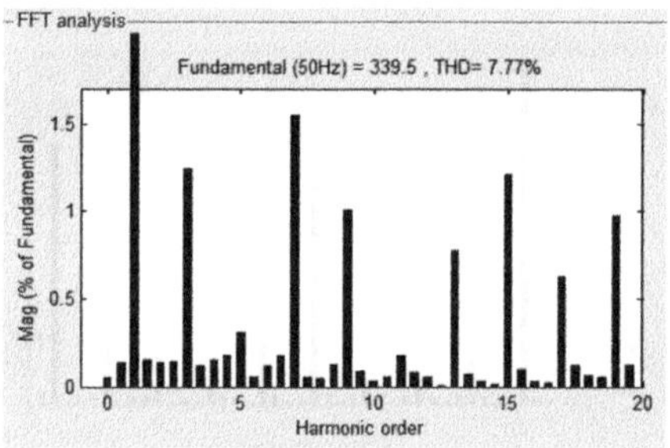

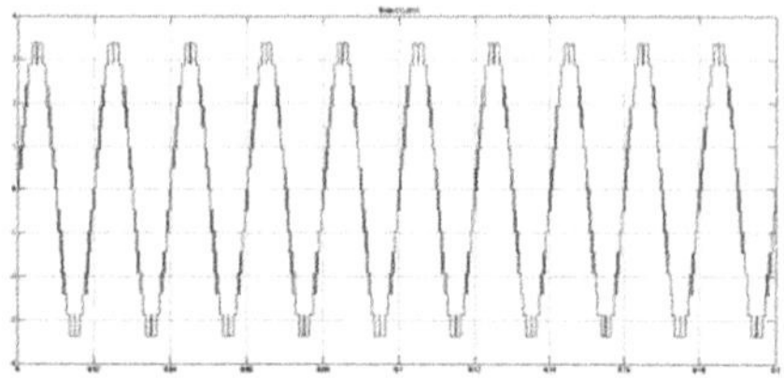

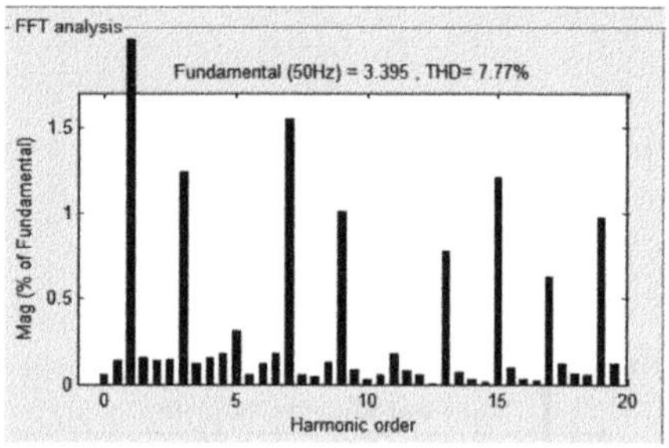

THD: V,I: 7.77, 4.65

Alternate Phase Opposition Disposition

Three level:

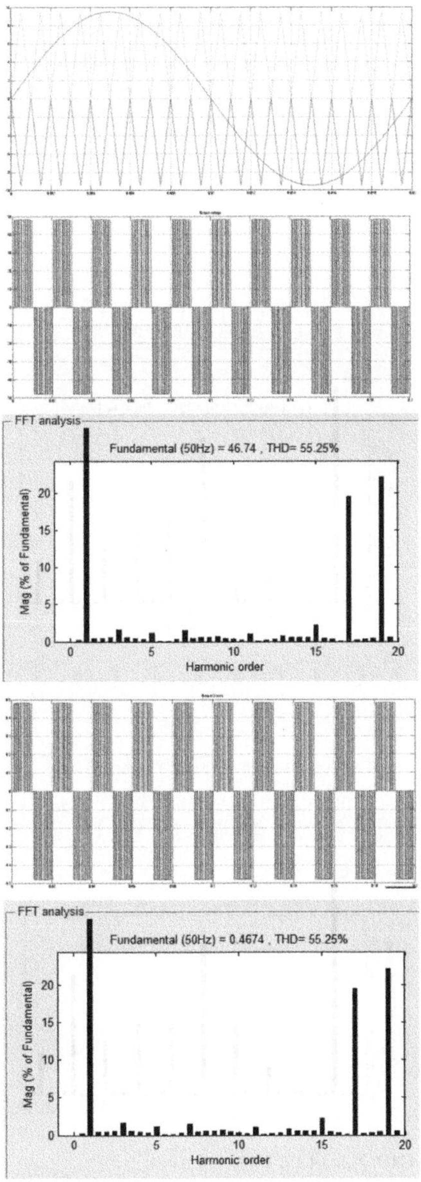

THD: V,I: 55.25, 38.95

Five level

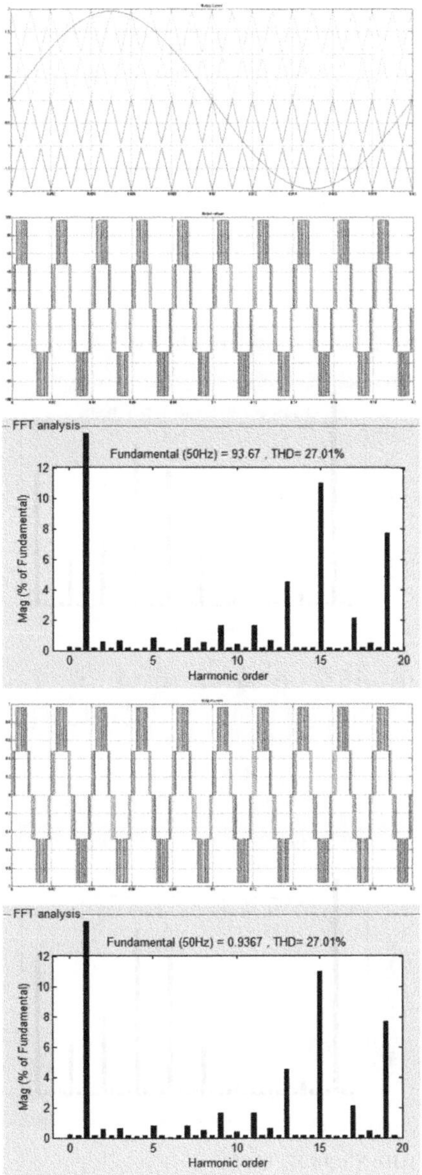

THD: V,I: 27.01,18.92

Seven level

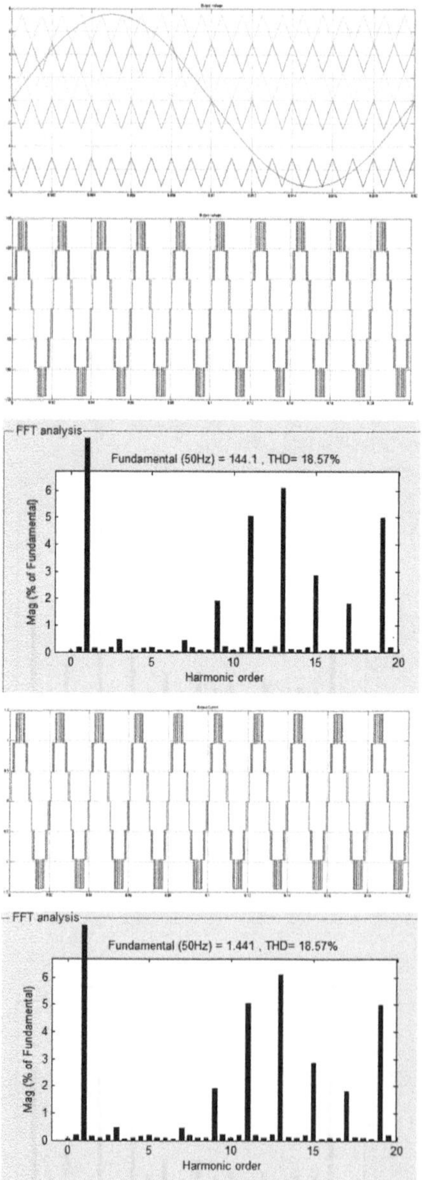

THD: V,I: 18.57, 13.55

Nine level

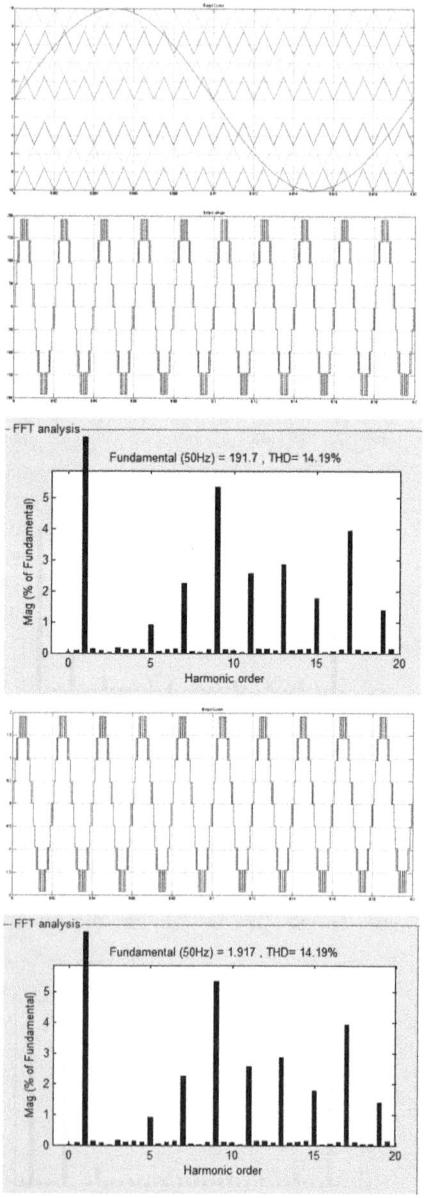

THD: V,I: 14.19, 10.45

Eleven Level

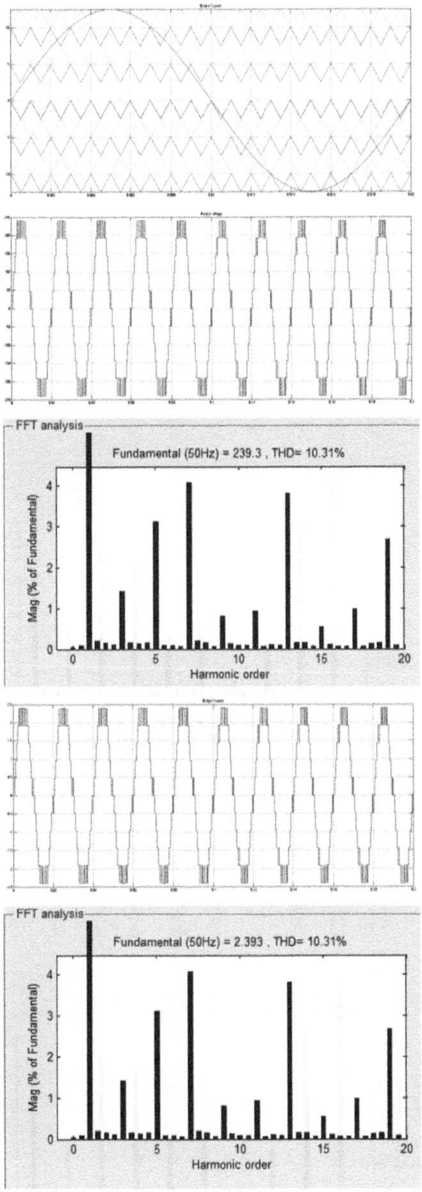

THD: V,I: 10.31,7.82

Thirteen level

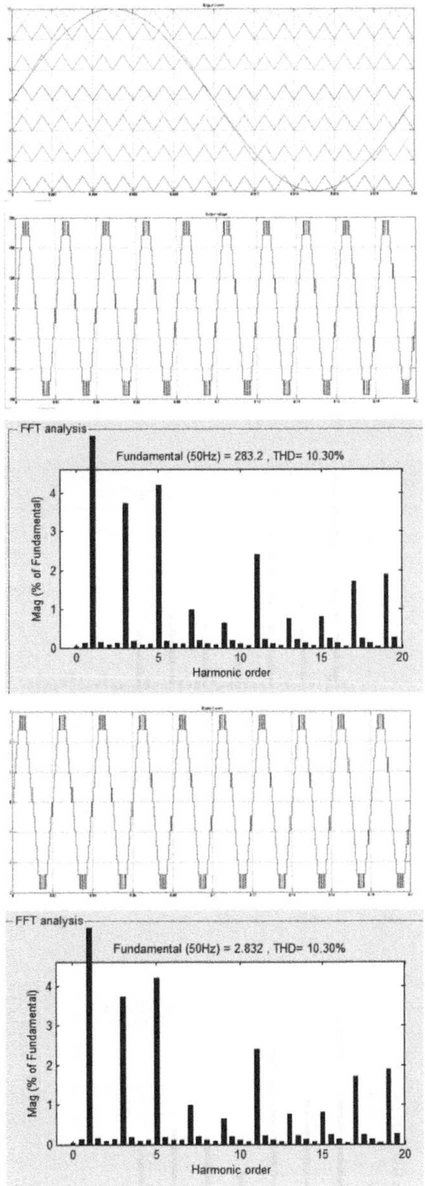

THD: V,I: 10.30, 8.03

Variable Amplitude

Fifteen level

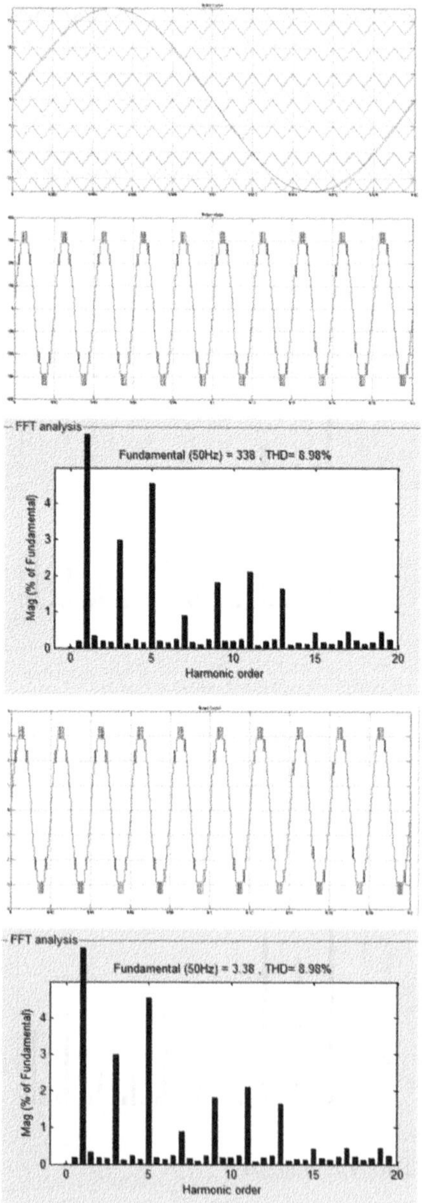

THD: V,I: 8.98,7.23

Five level

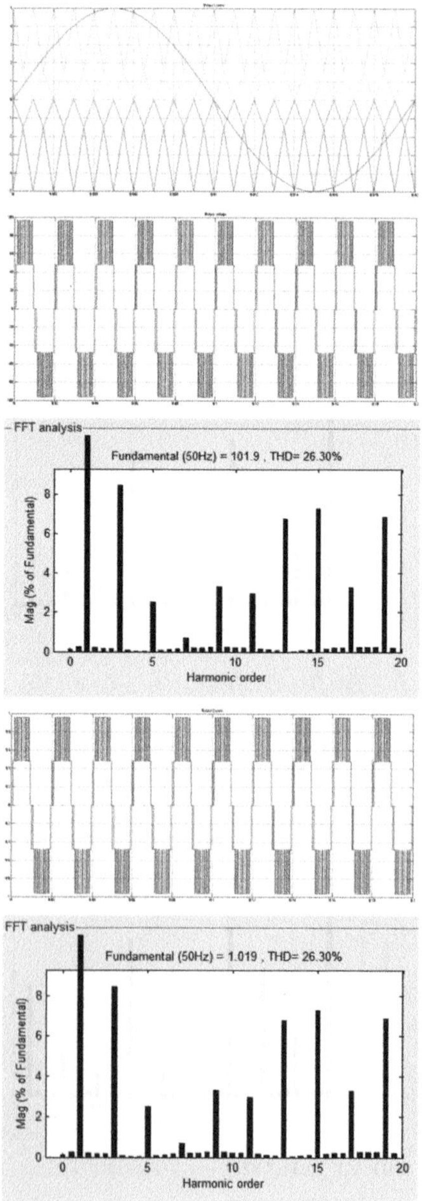

THD: V,I: 26.30, 19.74

Seven level

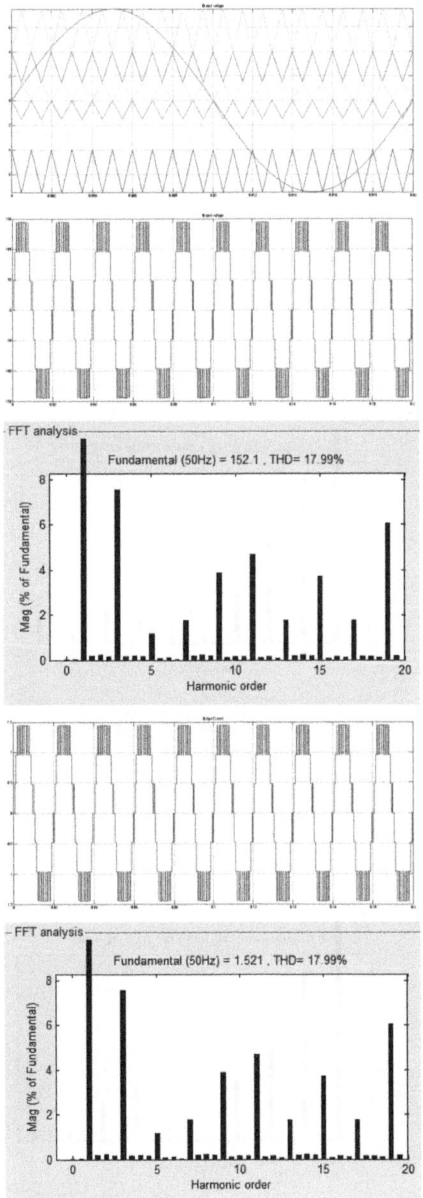

THD: V,I: 17.99, 13.85

Nine level

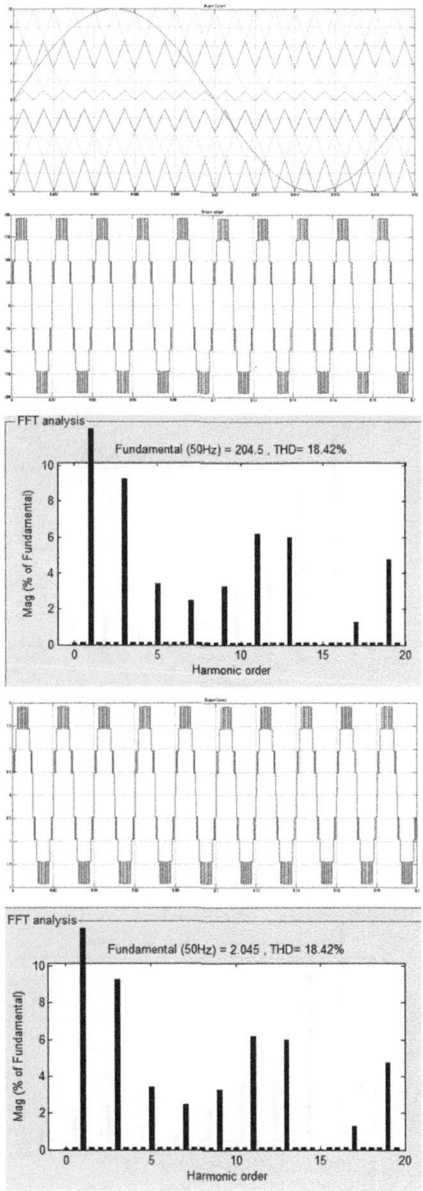

THD: V,I: 18.42, 15.31

Eleven level

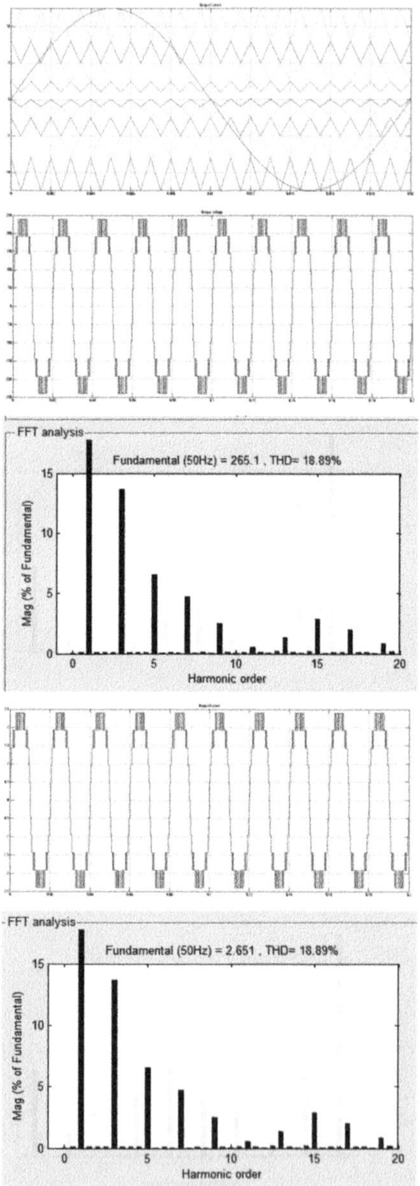

THD: 18.89,17.3

Thirteen level

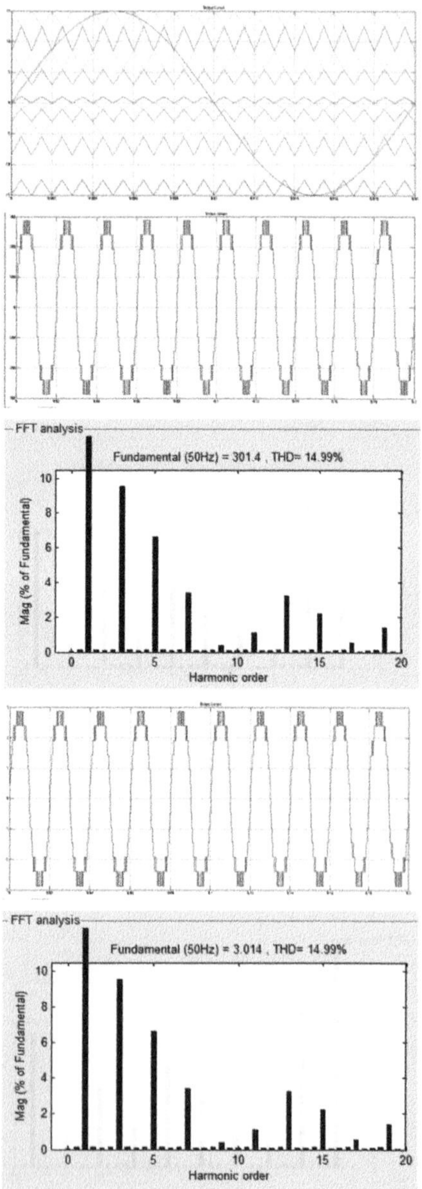

THD: V,I: 14.99, 13.46

Variable Time

Fifteen level

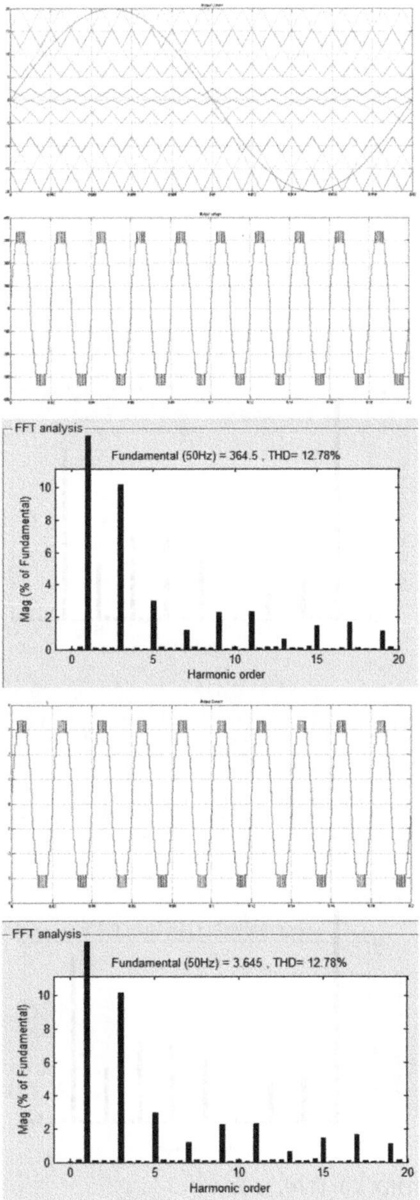

THD: V,I: 12.78, 11.70

Five level

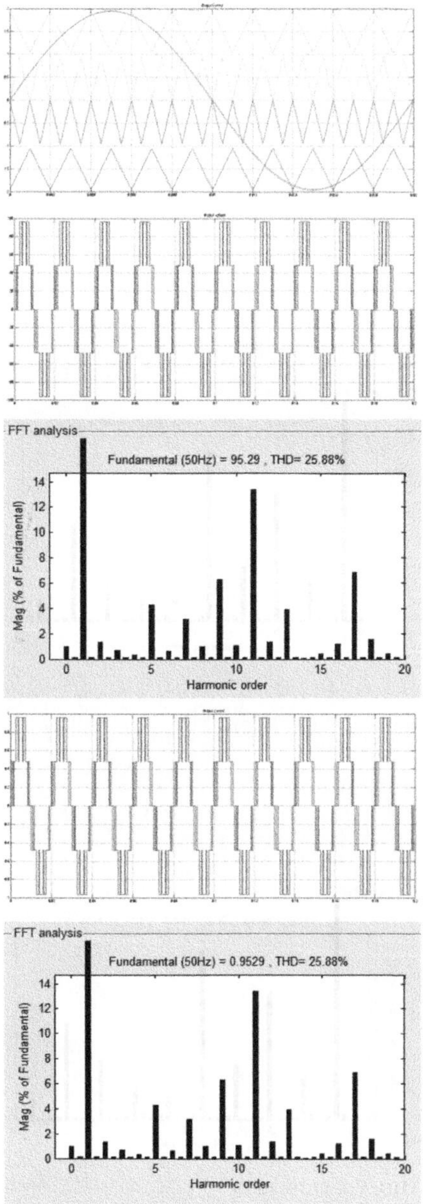

THD: V,I: 26.88, 20.20

Seven level

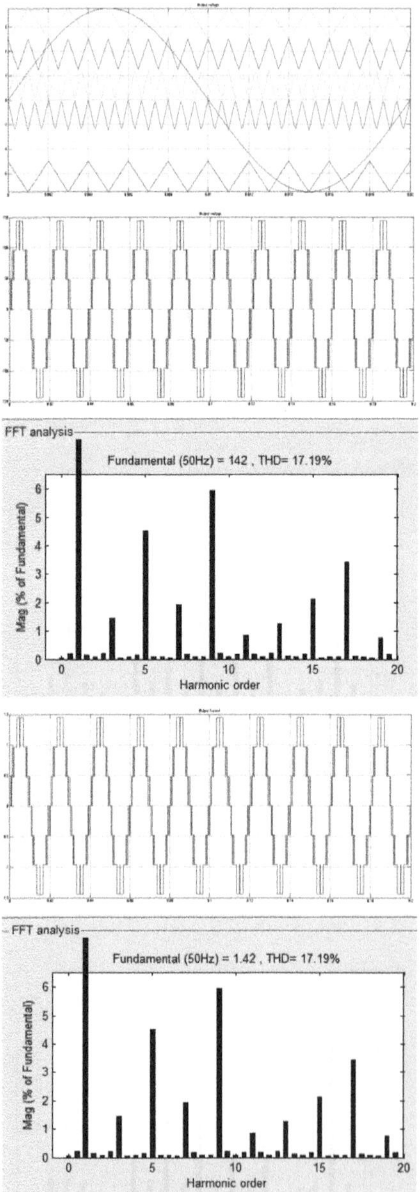

THD: V,I: 17.19, 12.79

Nine level

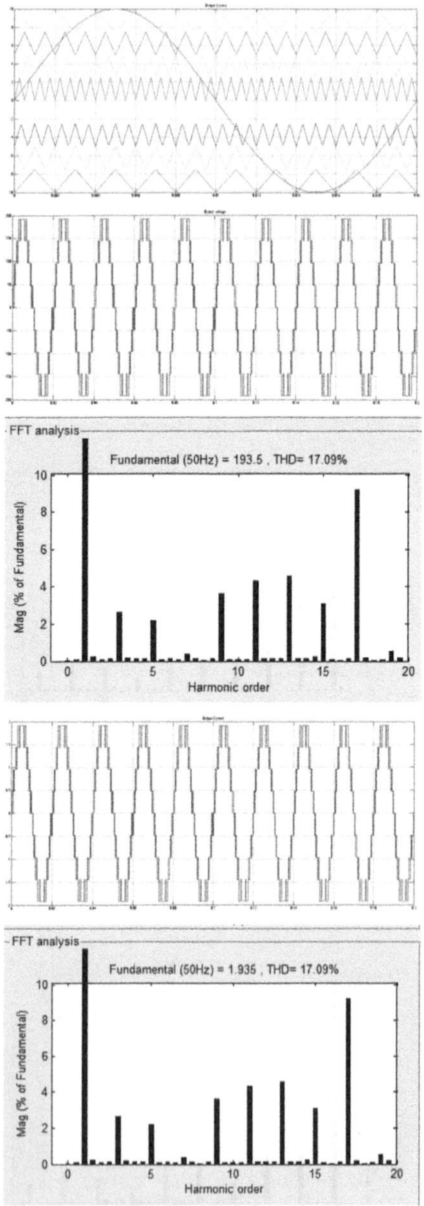

THD: V,I: 17.09, 13.11

Eleven level

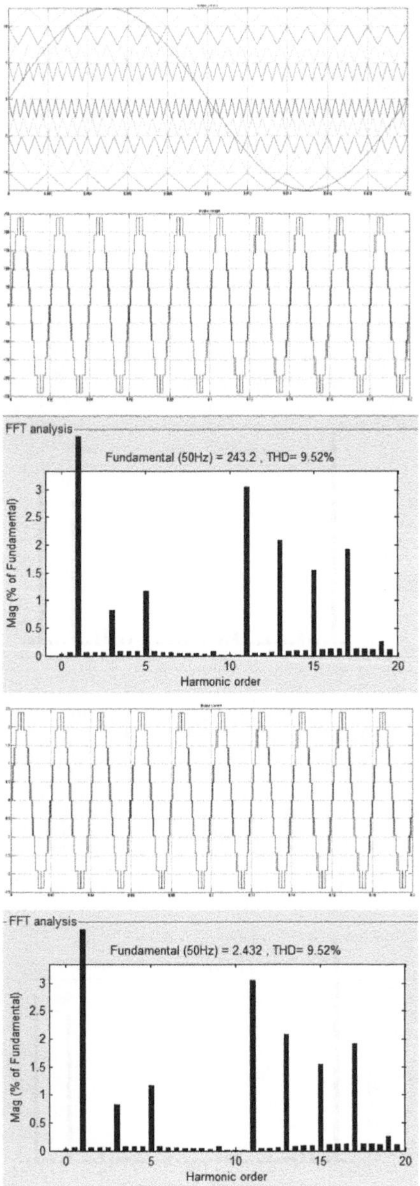

THD: V,I: 9.52,6.03

Thirteen level

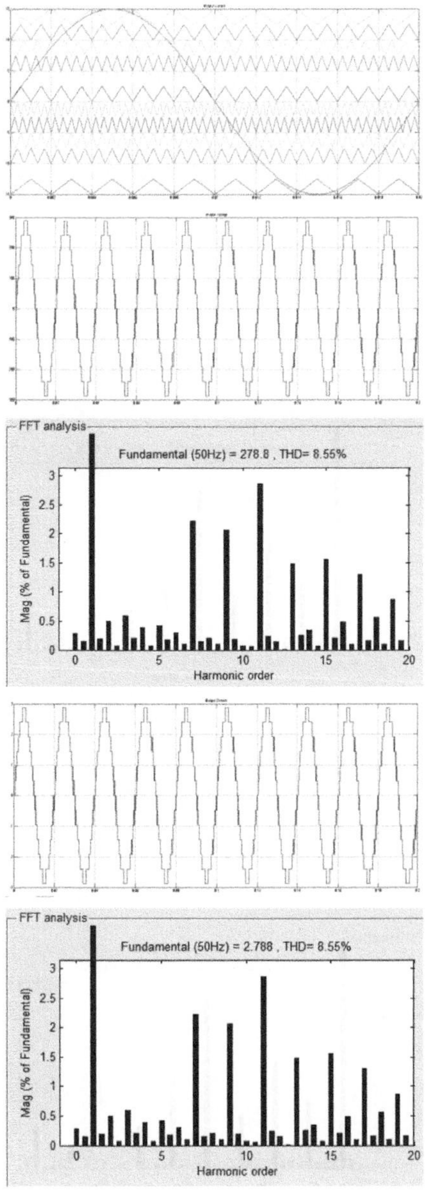

THD: V,I: 8.55, 6.08

Fifteen level | Three level

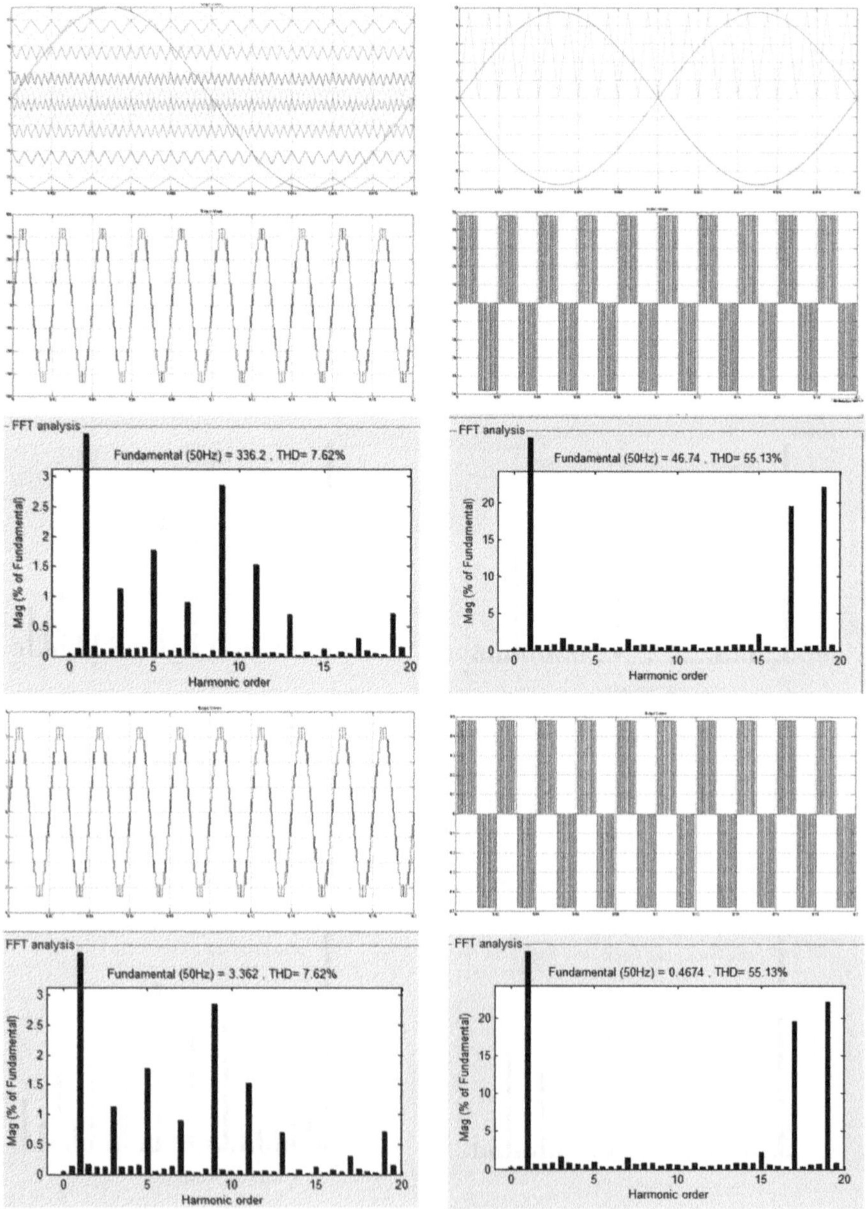

THD: V,I: 7.62, 5.12

THD: V,I: 55.13, 38.92

The clear understanding of the output voltage levels is given below:

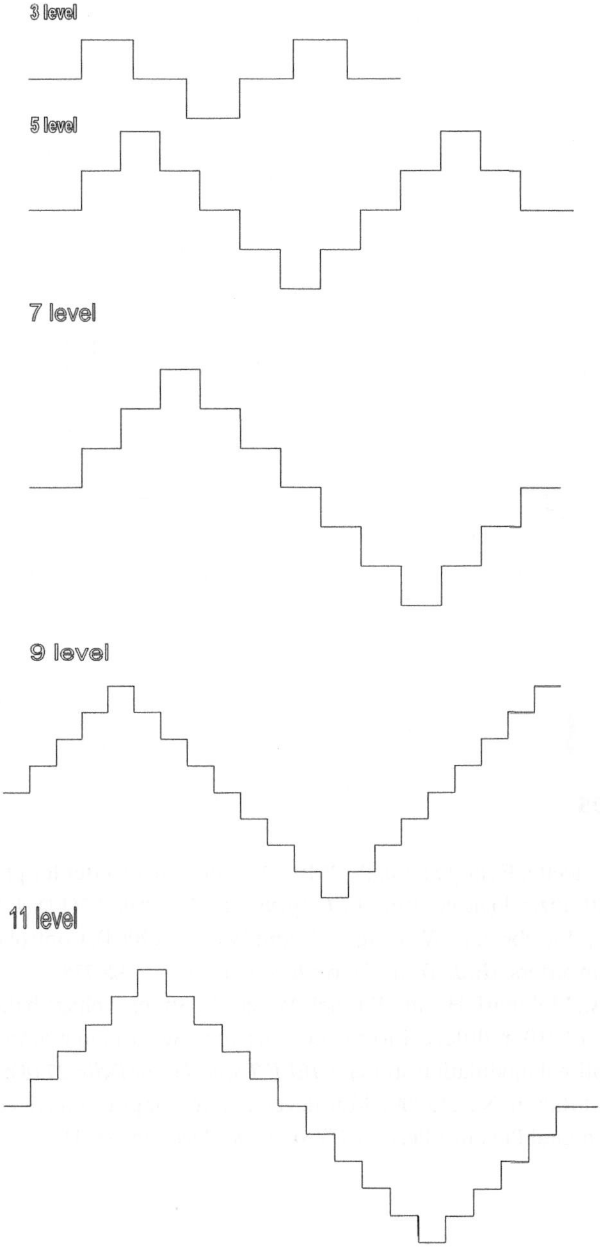

13 level

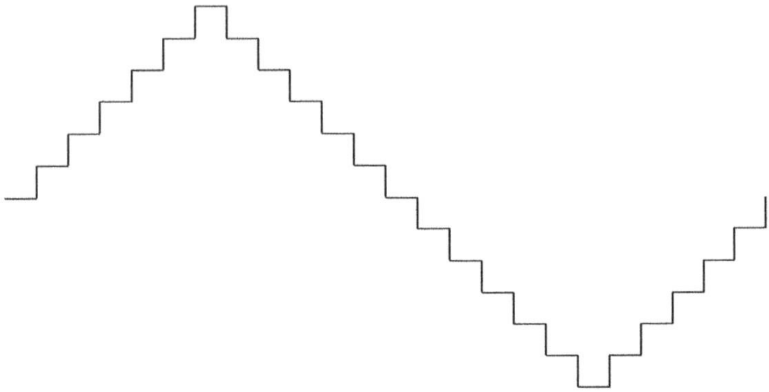

15 level

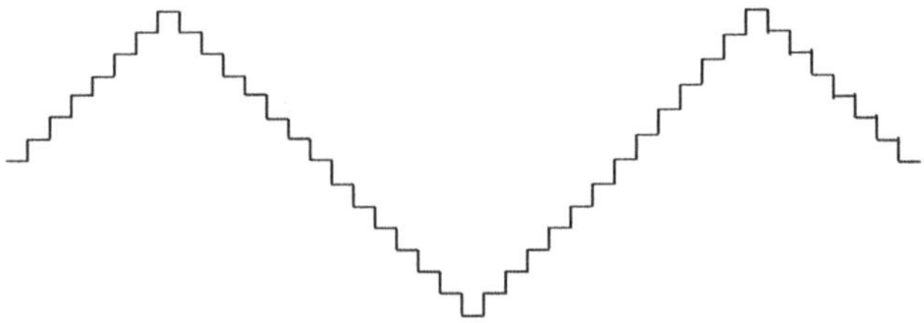

References

Cecati, C., Ciancetta, F., and Siano, P. (2010). A multilevel inverter for photovoltaic systems with fuzzy logic control. *IEEE Trans. Ind. Electron.* 57 (12): 4115–4125.

Corzine, K.A., Wielebski, M.W., Peng, F.Z., and Wang, J. (2004). Control of cascaded multilevel inverters. *IEEE Trans. Power Electron.* 19 (3): 732–738.

Ghazanfari, A., Mokhtari, H., and Firouzi, M. (2012). Simple voltage balancing approach for CHB multilevel inverter considering low harmonic content based on a hybrid optimal modulation strategy. *IEEE Trans. Power Deliv.* 27 (4): 2150–2158.

Selvaraj, J. and Rahim, N.A. (2009). Multilevel inverter for grid connected PV system employing digital PI controller. *IEEE Trans. Ind. Electron.* 56 (1): 149–158.

5

PWM For Multilevel Inverters

5.1 Introduction

Multilevel inverters (MLIs) are advised in contrast to conventional inverters because their synthesized multilayer outputs are of higher quality, which reduces the need for filters and the size of the entire system. Power quality is made possible by several pulse width modulation (PWM) approaches that regulate the MLI's switching sequence. A solar photovoltaic (PV) system, a direct current (DC) windmill, a battery, a fuel cell, etc., are examples of input sources that can provide the desired amount of power. The same hardware structure used in the distributed generating system makes it easy to combine the functions of power generation and power quality enhancement.

Power quality refers to any power issue that manifests as variations in voltage, current, or frequency that cause client equipment to malfunction or fail. The quality of electricity is becoming more and more important for a number of reasons. They are as follows: (i) the electrical supply is becoming more and more important to modern culture. Industrial consumers are greatly impacted financially by (i) minor power outages; (ii) the introduction of new power electronic equipment; (iii) the deregulated environment, which lowers system maintenance and investments and, consequently, system margins; and (iv) the emergence of distributed power generation. The main issues related to the quality of power are interference, malfunction, additional losses, human safety concerns, damage to sensitive equipment, low utilization, and low power factor.

A collection of values for factors including service continuity, voltage magnitude variation, transient voltages and currents, and harmonic content in AC power waveforms can be used to characterize the quality of electrical power. Among these, harmonics are essential to every part of the electrical system. Sinusoidal voltages or currents with frequencies that are integer multiples of the fundamental frequency at which the supply system is intended to function are known as harmonics. The nonlinear properties of devices and loads on the power system are the source of harmonic distortion.

Transformer feeder overheating, unintentional circuit breaker trips, fuse blowing, particularly on distribution feeder laterals, equipment malfunction (sensitive equipment), increased kVA demand and the need to oversize, total power factor and its reduction, electric energy waste, and flickering lights are all issues brought on by harmonics.

According to Equation (5.1), total harmonic distortion (THD) is a measurement of the effective value of the harmonic components of a distorted waveform. Either voltage or current can be used to calculate this index. "How close the waveform is to pure sine" is measured by this metric.

$$\mathrm{THD} = \frac{\sqrt{\sum_{h>1}^{h_{\max}} X_h^2}}{X_1} \tag{5.1}$$

where X_1 is the fundamental component, and X_h is the root mean square (RMS) value of the harmonic component h of the quantity X. Up to the harmonic $h_{\max}$, which is usually 20, the THD is computed. Either the peak values or the RMS values can be used to determine the THD. Equation (5.2) illustrates the relationship between the waveforms' true RMS value (XRMS) and THD.

$$X_{\mathrm{RMS}} = X_1 + \sqrt{1 + \mathrm{THD}^2} \tag{5.2}$$

The THD can be expressed as in Equation (5.3) if there are no harmonics above $h_{\max}$ and the waveform is periodic with a fundamental wave period.

$$\mathrm{THD} = \sqrt{\left(\frac{X_{\mathrm{RMS}}}{X_1}\right)^2 - 1} \tag{5.3}$$

Power quality issues can be mitigated in two ways: either from the utility side or from the customer side. The first strategy is load conditioning, which guarantees that the apparatus is less susceptible to power fluctuations and permits functioning even in cases of severe voltage distortion. Installing line conditioning systems that reduce or eliminate power system disruptions is the second option.

Active power filters provide a flexible and adaptable solution to voltage quality issues. Series active power filters function as a programmable voltage source, whereas shunt active power filters function as a controllable current source. Using voltage source PWM inverters and a DC bus with a reactive component like a capacitor is the preferred method for implementing both schemes. One or more of the tasks necessary to improve power quality and compensate power systems can be carried out by active power filters.

The power rating and response time also affect how well they operate. However, utility-side solutions or line conditioning systems will be crucial in enhancing the intrinsic supply quality due to the power sector's restructuring and the tendency toward dispersed and distributed generation. Power quality is defined from the utility's point of view as the voltage characteristic that impacts the customer's supersensitive equipment.

5.2 Types of Multilevel Inverters

An alternative for high-power medium-voltage applications is a multilevel power converter arrangement. Despite their widespread use in industrial settings, traditional PWM inverters have the following disadvantages:

- An extremely high carrier frequency is required. The carrier frequency should be higher than 1 kHz if the output waveform has a frequency of 50 Hz.
- There is a very high pulse height. The output voltage of a typical PWM waveform (not multistage PWM) has a wide leaping range, and all of the pulse heights are equal to the DC link voltage. Large dv/dt and significant electromagnetic interference (EMI) are the results of this.
- The pulse width would be quite narrow when the output voltage was low.
- PWM generates a lot of harmonics, which results in low THD. The switching devices suffer significant switching power losses under extremely demanding switching conditions.
- The devices and control circuitry used to generate switching pulses are expensive and complicated.

By accumulating the output voltage in horizontal levels, a multilevel power converter solves the following problems with traditional inverters:

- The switching devices' switching frequencies are modest, matching or slightly exceeding the frequency of the output signal.
- The pulse heights are rather low. The pulse heights of an "m" level inverter with output amplitude V_m are either V_m/m or a tiny multiple of it. It typically results in ignorable EMI and low dv/dt.
- very pulse has an appropriate pulse width that is comparable to the output signal.
- Multilevel converters generate less THD and are unable to produce many harmonics.
- Provides a smooth switching environment with minimal switching power losses in the switching devices.
- The gadgets are inexpensive, and the control circuitry is straightforward.

The three-level converters are the source of the phrase "multilevel." To generate a high voltage at the output, the power switches' commutation aggregates these many DC sources. High-voltage capability, minimal switching losses, improved electromagnetic compatibility, and superior power quality are all benefits of this multilevel technique. Neutral point clamped (NPC) or diode-clamped multilevel inverter (DCMLI), flying capacitor multilevel inverter (FCMLI), and cascaded multilevel inverter (CMLI) are the three types of MLI structures.

Baker and Bannister, who invented the Cascaded H-Bridge (CHB) in 1975, are credited with creating the first multilevel converter. To provide a staircase AC output voltage, this cascaded inverter was initially defined using a structure that joins independently sourced DC full-bridge cells in series. In 1981, Nabae, Takahashi, and Akagi introduced a diode-clamped architecture that divides the DC bus voltage using a bank of series capacitors.

When the diode-clamped inverter was initially employed in a three-level inverter, the mid-voltage level was designated as the neutral point. This type of inverter was also known as the NPC inverter. Without the need for exact voltage matching, the NPC inverter efficiently doubles the device voltage level. The flying capacitor (FC) architecture, which employs floating capacitors to clamp voltage levels, was patented by Meynard and Foch in 1992.

Ogasawara introduced a typical current source inverter (CSI) in 1992; however, instead of increasing voltage levels, he increased current levels. Voltage source inverters (VSI) have great dynamic performance, but current source topologies have restricted dynamic performance since they employ huge DC chokes as DC links. Despite being developed earlier, the cascade inverter was not widely used until the mid-1990s. The cascade inverter has attracted a lot of attention because of the rising need for medium-voltage, high-power inverters.

"Stages" and "levels" are the two terms that make up how MLIs operate. One "stage" refers to a single H-bridge inverter. It generates a square-shaped output voltage during conduction in both positive and negative half cycles, which are referred to as two "levels." A "single-stage three-level" inverter is produced by adding the common zero potential as an extra level. By adding more inverter stages, it is possible to reach any number of inverter output voltage levels.

The quantity of switching devices and the control circuits that go with them should be taken into account while raising the levels. The power circuit of a 15-level diode-clamped and FC-type MLI with a solar PV array at its input is shown in Figure 5.1 (a) and (b). For applications with a single input source, these two MLI designs work well. In these situations, the solar PV array's design necessitates a high-power rating or a boost converter circuit to increase the voltage in order to achieve the higher levels.

5.2.1 Diode-Clamped Multilevel Inverter (DCMLI)

The 15-level DCMLI power circuit is depicted in Figure 5.1(a). It is also known as an NPC multilevel inverter (NPCMLI), in which the desired voltage rating

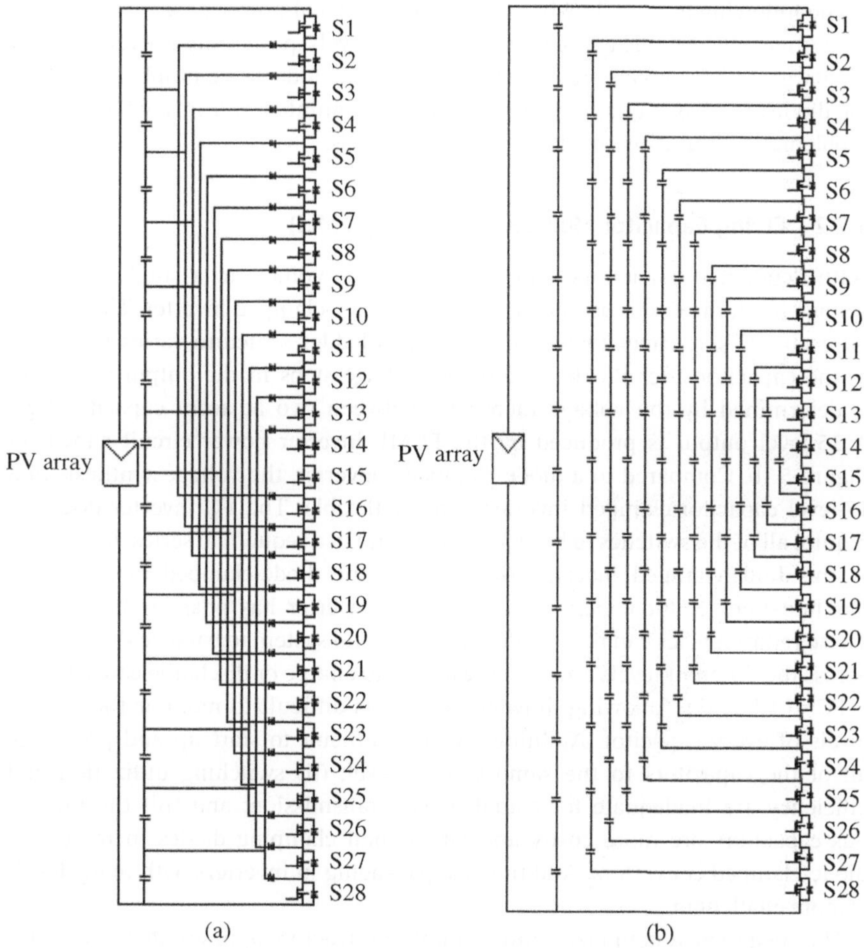

Figure 5.1 (a) The power circuits for 15-level DCMLI and (b) 15-level FCMLI, respectively.

and output levels are achieved by connecting the switching devices in series. One high-frequency capacitor or two more diodes are used to clamp the inner voltage points. A voltage level of $V_{dc}/(m-1)$ must be blocked by the switching devices of an m-level inverter. For varying inner voltage levels, the clamping diodes require varied voltage ratings. Seven bulk capacitors connected in series divide the DC bus voltage (PV array) into 15 levels in the power circuit. The neutral point is the center point of each of the two capacitors. The diodes are the main elements that set this circuit apart from a traditional inverter. The switch voltage is clamped by these diodes to half of the DC bus voltage.

The following are the drawbacks of DCMLI: (i) real-power flow is challenging for a single inverter due to the intermediate DC levels' propensity to overcharge or discharge without precise monitoring and control; and (ii) the number of clamping diodes needed is quadratically related to the number of levels, which can be problematic for units with a large number of levels.

5.2.2 Flying Capacitor Multilevel Inverter (FCMLI)

With dependent capacitors clamping the device voltage to a single capacitor voltage level, the circuit is known as the capacitor-clamped inverter. The voltage on each DC side capacitor in this topology's ladder structure varies from that of the capacitor after it. The size of the voltage steps in the output waveform is determined by the voltage increment between two adjacent capacitor legs. A 15-level output is produced by the FCML inverter power circuit shown in Figure 5.1b. Compared to a diode-clamped converter, the voltage synthesis in a 15-level capacitor-clamped inverter is more flexible. The FC inverter does not require all of the switches to be ON (conducting) in a sequential series, in contrast to the diode-clamped inverter. Additionally, the diode-clamped inverter only has line–line redundancies, whereas the FC inverter has phase redundancies. These redundancies can be included in the control system to balance the voltages across the different levels and give the option to charge or discharge capacitors.

FCMLI has the following drawbacks: (i) it is difficult to monitor the voltage levels of every capacitor. Additionally, it is difficult to start up and precharge all of the capacitors to the same voltage level; (ii) switching utilization and efficiency are inadequate for actual power transmission; and (iii) the numerous capacitors are more costly and larger than clamping diodes in multilevel diode-clamped converters. Additionally, packaging in inverters with many levels is more challenging.

The suggested system takes into account the structure appropriate for solar PV with the help of power quality enhancement in order to solve these shortcomings of both NPCMLI and FCMLI.

5.2.3 Cascaded H-Bridge Multilevel Inverter (CHBMLI)

Series H-bridge (single-phase complete bridge) inverter units make up a CMLI. This MLI's general purpose is to synthesize a desired voltage from independent DC sources (SDCs), which can come from solar cells, fuel cells, or batteries. The voltages produced by various H-bridges are added to create the final phase voltage. The CMLI is referred to as symmetrical if the DC link voltages of the HBs are the same. Nonetheless, the DC link voltages of HBs may differ, in which case the circuit may be referred to as asymmetrical. Asymmetrical inverters are

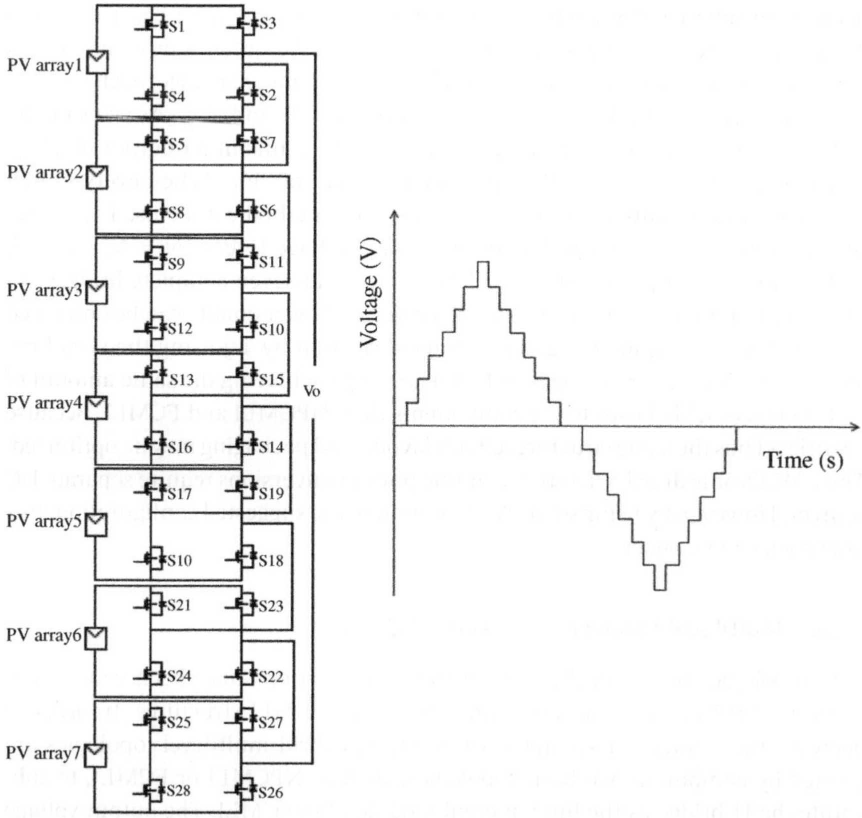

Figure 5.2 The output voltage waveform of a solar-fed cascaded 15-level inverter.

strongly advised because solar PV voltages are dependent on environmental conditions. A solar-fed cascaded 15-level inverter's power circuit and output voltage waveform are depicted in Figure 5.2. Various level inverters have their AC terminal voltages connected in series. The CMLI does not need voltage clamping diodes or voltage balancing capacitors, in contrast to the NPCMLI or FCMLI.

Because of its modular and straightforward structure, the cascade configuration, one of the different MLI topologies, has been used for medium-voltage and high-voltage renewable energy systems like solar PV. Because different DC sources are required, cascaded inverters are perfect for linking renewable energy sources to an AC grid. This is true for applications like fuel cells and PVs. The output filter design can be made much simpler by using a greater number of voltage levels to efficiently reduce the harmonic content of the staircase output.

The phase voltage waveform of a 15-level cascaded inverter with seven PV array inputs is also displayed in Figure 5.2. The seven inverter outputs are added together to create the phase voltage, which is expressed as van = va1 + va2 + va3 + va4 + va5 + va6 + va7. By connecting the PV array source to the AC output side using various combinations of the four switches in each individual inverter stage, each inverter level can provide three distinct voltage outputs: $+V_{dc}$, 0, and $-V_{dc}$. As an example, at the inverter's first stage, turning S1 and S4 ON results in an output of $+V_{dc}$, whereas turning S2 and S3 ON results in an output of $-V_{dc}$. When every switch is turned off, the output is 0. Likewise, the same method can be used to obtain the AC output at each level. The output phase voltage level is $m = N_s + 1$ if N_s is the number of input PV sources. Seven SDCs and seven complete bridges are therefore required for a 15-level cascaded inverter. Power quality can be enhanced by reducing the output voltage's harmonic distortion by adjusting the switching angles at various inverter stages. When it comes to achieving the same amount of voltage levels, CMLI uses fewer components than NPCMLI and FCMLI. Because every level has the same structure, circuit layout, and packaging may be optimized. The CMLI's only drawback is that genuine power conversions require separate DC sources. However, by using solar PV at its input in the suggested configuration, this drawback can be offset.

5.2.4 Multilevel Inverter with Reduced Order

New topologies based on the current multilevel notions have been categorized as hybrid MLI in addition to the three fundamental MLI structures. In order to decrease the number of separated DC sources, hybrid multilevel topologies are created by combining two basic topologies, such as NPCMLI or FCMLI, to substitute the H-bridge as the fundamental module of the CMLI. The output voltage waveforms with lower harmonic content are created by the asymmetric hybrid MLIs. This benefit is obtained by employing different voltage levels in various modules, which can provide an output voltage waveform with a greater number of levels. Additionally, it keeps the number of switching devices and sources from rising while maintaining the THD value.

The efficiency and THD compensating features of an inverter can be enhanced by operating each power module of a hybrid MLI at a unique DC voltage and switching frequency. However, because of the switching devices of the higher voltage modules, traditional PWM techniques, which provide a switching frequency at fundamental frequency (50 Hz), are inappropriate for the hybrid inverters. Additionally, they would only need to run at high frequencies during specific instances of inversion. Hybrid modulation techniques that switch low-power cells at high frequency and higher-power cells at low frequency can be employed to accomplish this control scheme.

For high-power and medium-voltage industrial drives, hybrid MLIs have great promise. Without adding more power devices, asymmetrical multilayer inverters aid in reducing the output voltage's harmonic components. In an attempt to maximize the system's overall power processing, hybrid MLI topologies employ several DC voltages in the supply.

5.2.5 Multilevel Inverter Comparison

One benefit of NPCMLI is that the three phases share DC input capacitors. Its primary drawback is the excessive number of diodes utilized for clamping, which also complicates the physical layout's implementation. Higher levels of the NPCMLI structure are possible, but they are less appealing due to increased losses and an unequal distribution of these losses in the inner and outer devices. The clamping diodes, which must be linked in series to block the higher voltages, increase conduction losses and generate reverse recovery currents during commutation, which further impacts the other devices' switching losses.

Voltage synthesis redundancies allow the clamping capacitor voltages in an FCMLI to be balanced in a few cycles. Its primary drawback is the large number of capacitors required for clamping, even if these capacitors are less expensive than the DCMLI's clamping diodes. Additionally, the clamping capacitors get larger at low switching frequencies, which lowers the inverter's power density.

Clamping capacitors and diodes are not required in a CMLI. Moreover, it is possible to employ a low switching frequency. Because every cell has the same structure and there are no additional FCs or clamping diodes, modularized circuit layout and packaging are feasible. The utilization of separate DC sources for each cell is its primary drawback. Because of this, the CMLI can be applied to fuel cell-based power generation systems and solar PV.

The three MLIs use the same number of semiconductor switches per phase when their power circuits are examined. Nevertheless, the NPCMLI is more expensive than the other two structures since it employs $(m-1)\,(m-2)$ clamping diodes. The NPCMLI requires two $(m-1)+(m-1)\,(m-2)$ heat sinks, while the FCMLI and CMLI require two $(m-1)$. The CMLI has the lowest implementation cost since it employs the fewest semiconductor devices to provide an m-level inverter output. The CHB converter also has the benefit of allowing the inverter to continue operating at a lower power level if any device in the H-bridges fails.

5.2.6 Multilevel Inverters' Utilization

The commercial and industrial sectors are paying more attention to multilevel converters as the best options for converting electrical power in medium- and high-power applications. High-voltage DC (HVDC) lines, electrical drives, and flexible AC transmission system (FACTS) equipment are just a few of the numerous uses for MLIs. Compressors, extruders, pumps, fans, grinding mills, rolling mills, conveyors, crushers, blast furnace blowers, gas turbine starters, mixers, mine hoists, reactive power compensation, marine propulsion, hydro pumped storage, wind energy conversion, and railway traction are just a few of the many devices that are currently powered by MLIs.

5.3 Switching Sequences for Multilevel Inverter

The process of giving the inverter switches the proper switching signals is known as modulation. With a basic component that is usually sinusoidal in steady state and has variable amplitude, frequency, and phase, modulation approaches aim to create a stepped switching waveform that most nearly mimics an arbitrary reference signal. Since the modulation technique is intended to be used in high-power converters, the main figures of merit desired are high power quality and low switching frequency.

The struggle between these two requirements is one of the primary challenges in multilayer converter technology. Figure 5.3 illustrates how various modulation control systems are categorized. Techniques that use high switching frequencies feature many commutations for the power semiconductors in a single period of the fundamental output voltage. When analyzing the performance in terms of improving power quality in a CMLI fed by solar electricity, selecting the right switching mechanisms is crucial.

The traditional carrier-based sinusoidal PWM (SPWM), which employs the phase-shifting technique to lower the harmonics in the load voltage, is a widely used technology in industrial applications. Based on the traditional SPWM

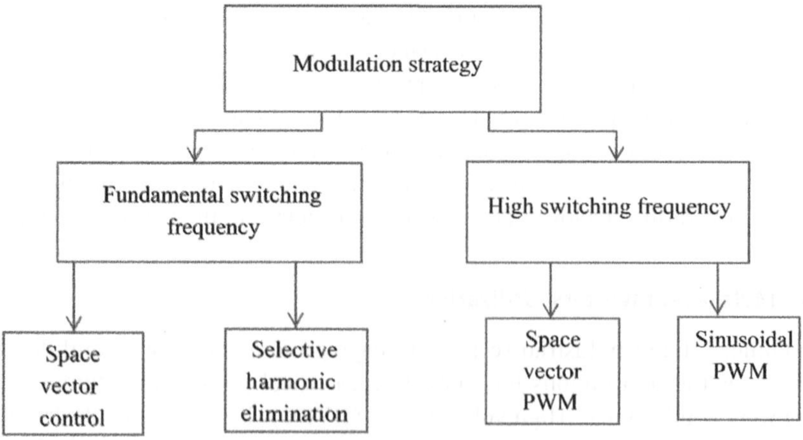

Figure 5.3 Classification of MLI modulation strategies.

with triangle carriers, a number of multicarrier methods can be employed to lessen the distortion in MLIs. Phase shifting of several carrier signals is used in some techniques, whereas carrier disposition is used in others.

The effective switching frequency of the load voltage equals N_s (where N_s is the number of inverter stages) times the switching frequency of each cell, as indicated by its carrier signal, which is a benefit of multilayer SPWM. This characteristic makes it possible to lower each cell's switching frequency, which lowers switching losses. Good DC link voltage usage, minimal current ripple, and very simple hardware implementation using a digital signal processor (DSP) are typical characteristics of space vector modulation (SVM). It is appropriate for high-voltage, high-power applications because of these qualities. Redundant switching states and the difficulty of choosing switching states both sharply rise with the number of levels.

Low switching frequency methods often produce a staircase waveform by performing one or two power semiconductor commutations in a single output voltage cycle. Space vector control (SVC) and multilevel selective harmonic elimination (SHE) are examples of this family. In SHE, high-frequency harmonic components can be easily eliminated by employing extra filter circuits, while the most important low-frequency harmonics are selected for elimination by carefully choosing angles among various level inverters. The primary drawback of this modulation approach is that it essentially offers a limited range of modulation index.

In contrast to the SVM principle, SVC operates at low switching frequencies and does not produce the mean value of the target load voltage in each switching period. Space vector techniques are generally quite useful for three-phase systems. Because there are many switching vectors in cascaded topologies, the use of SVM is typically restricted to a small number of levels.

Even though SHE was first launched in 1973, a significant number of tasks still require the use of clever techniques. In order to obtain high-precision output despite situations of imprecision, ambiguity, and partial truth, control approaches have recently been updated to become intelligence-based. Actually, every method is based on a natural occurrence and has its unique control algorithm. Neural networks have been inspired by the type of computation that occurs in the human brain since they are simplified representations of the biological nervous system. Generally speaking, it is an architecture modeled after the brain that consists of a highly interconnected network of many processing units called neurons.

It aims to demonstrate parallel distributed processing and is capable of huge parallelization. Neural networks pick up knowledge from examples. They can learn about it by being trained on well-known problem cases. The network can be effectively used to solve unknown or untrained instances of the issue once it has been somewhat trained.

Classical set theory is expanded upon by fuzzy logic. The goal of fuzzy logic representations, which are based on fuzzy set theory, is to depict how people use real-world information to represent and reason in the face of ambiguity. A fuzzy set can be mathematically defined by giving a value that represents the degree of membership in the fuzzy set to every potential human in the universe of discourse. The degree to which the person is comparable to or compatible with the idea represented by the fuzzy set is shown by this grade.

Based on the need that the solution meets the constraints, optimization approaches yield the optimal solution within the specified search space. These fall into two categories: derivative-based and derivative-free. Derivative-free optimization techniques include GA, PSO, and BO. A population of individual solutions is repeatedly altered by the GA. In order to create the offspring for the following generation, the genetic algorithm randomly chooses members of the present population to be parents at each stage. The population "evolves" toward an ideal solution over the course of several generations.

The sociological behavior of swarms, including schools of fish and flocks of birds, served as an inspiration for PSO. Particles are the individuals that make up the population. Every particle represents a potential solution to the optimization problem and attempts to fly through a multidimensional space in pursuit of the optimal position. BO is an optimization algorithm that finds the best solution by using honeybees' natural foraging behavior. The three stages of this algorithm correspond to the three different kinds of bees that are present in the colony. The following elements influence the output voltage waveform of the MLIs: (i) the topology employed, (ii) the application, (iii) the control algorithm, (iv) the filter size, and (v) the switching frequency selection.

5.4 Analog PWM for Multilevel Inverter

Unique modulation approaches, control, and protection features are needed to enhance the modular characteristics and performance of MLIs. Compared to two-level inverters, a higher level of complexity is brought about by a large number of power electronic devices and switching redundancies. By lowering the switching frequency (f_s), limiting the common mode voltage (CMV), and balancing the DC voltages, this complexity could be utilized to enhance the modulation techniques' capabilities.

The following are the modulation schemes of multilevel converters categorized by switching frequency: (i) High switching frequency: each inverter has multiple commutations per cycle, such as SVM and SPWM; (ii) Fundamental switching frequency: each inverter has only one commutation per cycle, such as SVC and SHE.

Numerous publications address basic frequency switching with equal step output voltage modulation control for a seven-level CMLI driven by a single DC power supply and capacitors. In order for the charging energy to be much higher than the discharging energy, the capacitor value.

Because there are so many switching vectors in cascaded topologies, the use of SVM is typically restricted to a small number of levels. Although SVC lowers switching losses, the fundamental component's magnitude error varies. While SVM is the most widely used method for three-phase applications, SPWM is a good option for single-phase inverters. Using the volt-second equivalency principle, the SVM calculates the effective action time of various switch states to determine the output voltage waveform. Its high-voltage utilization ratio is another significant benefit. However, the SVM technique's calculation complexity increases significantly as level numbers rise. As a result, the MLI with more than five levels cannot use it.

The fundamental voltage of the inverter can be regulated and the harmonics reduced by employing SPWM techniques. This technique creates gating signals for the switching devices by generating a carrier signal at the desired frequency and comparing it with the modulating voltage signal. The upper switch is turned on when the modulating signal is above the carrier, and the lower switch is turned on when it is below the carrier. Multiple carriers PWM with different sequence modifications are meant to enhance power quality in CMLI, in contrast to the single carrier utilized in the SPWM technique for the two-level inverters.

The output voltage harmonic content can be decreased by adjusting the phase and/or vertical location of multiple triangular carrier signals in multiple-carrier-based PWM. The frequency of the modulating signal defines the inverter output frequency, while the frequency of the triangle signal controls the switching frequency of the switches.

Typically, the multicarrier SPWM uses two methods, such as: (i) the carriers' vertical distribution and (ii) their horizontal distribution. The distribution of carriers along the vertical and horizontal axes is depicted in Figure 5.4(a, b), respectively.

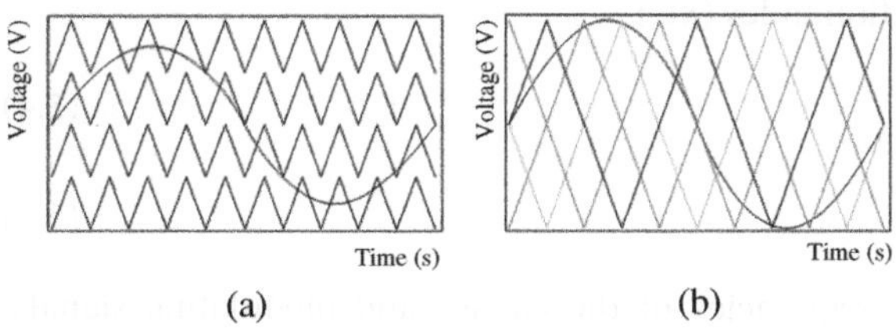

Figure 5.4 (a) Carriers distributed vertically and (b) horizontally.

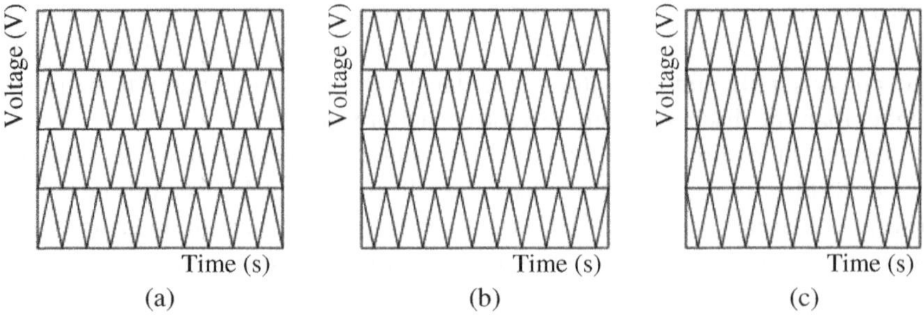

Figure 5.5 Waveforms of (a) PD, (b) POD, and (c) APOD.

5.4.1 Vertical Distribution of Carriers

The equivalent carrier frequency is not raised by modulation techniques based on the vertical dispersion of carriers. Phase opposition disposition (POD), alternate phase opposition disposition (APOD), and phase disposition (PD) are other classifications for this method. Compared to PS PWM, PD PWM offers a better output voltage harmonic profile and the capacity to balance voltage. When it comes to the differential mode of phase currents, POD is superior. The carrier arrangements for PD, POD, and APOD are depicted in Figure 5.5, which takes into account a five-level inverter in a typical scenario with an equal distribution of carriers and no changes to frequency or amplitude.

Phase disposition: In this instance, every carrier in every band is in phase.
Phase opposition disposition: This occurs when carriers are 180° out of phase with those below the reference zero point.
Alternate phase opposition disposition: This involves a 180-degree phase shift of carriers in nearby bands.

This level-shifted modulation technique requires (m–1) triangle carriers with the same frequency and peak-to-peak amplitude for an "m" level inverter. The bands that the (m–1) carriers inhabit are continuous due to their vertical dispersion. Equations (3.1) and (3.2) provide the frequency modulation index (m_f) and amplitude modulation index (m_a).

$$m_f = \frac{f_{cr}}{f_m} \tag{5.4}$$

$$m_a = \frac{2V_m}{V_{cr}(m-1)} \tag{5.5}$$

where f_{cr} and f_m are the frequencies of the carrier and modulating signals, respectively. V_{cr} is the peak amplitude of the modulating signal, and V_m is the peak amplitude of each carrier signal. The value of m_a is between 0 and 1, beyond which it is termed as the overmodulation region, which has to be avoided to achieve better results.

5.4.2 Carriers' Horizontal Distribution

The equivalent carrier frequency can be raised and the output voltage's harmonic content reduced by using modulation techniques based on the horizontal dispersion of carriers. Each pair of switches in this design has a carrier signal with a phase shift. Generally speaking, $(m-1)$ triangle carriers are needed for an MLI with m voltage levels. According to Equation (5.6), there is a phase shift (Δ_{cr}) between any two adjacent carrier signals, yet all of the triangular carriers in this approach have the same frequency and peak-to-peak amplitude. Therefore, Δ_{cr} equals 25.710 for a 15-level inverter.

$$\Delta_{cr} = \frac{360^\circ}{m-1} \tag{5.6}$$

Equation (5.6) provides the frequency modulation index (m_f) for horizontal carrier distribution; however, Equation (5.7) provides the amplitude modulation index (m_a).

$$m_a = \frac{V_m}{V_{cr}} \tag{5.7}$$

In order to minimize harmonic distortions, a solar-fed cascaded 15-level inverter is designed and implemented using a variety of modulation techniques. This covers both SPWM and multiple carrier PWM methods, including APOD, PD, and POD. Despite differences in the solar PV input, the carrier and reference configurations were modified to provide the optimal method for solar PV applications.

By lowering the system's harmonics, an appropriate modulation strategy with the selection of several factors, including modulation index, switching frequency, and signal arrangement, will undoubtedly improve the power quality. From single-stage 3-level inverters to 7-stage 15-level inverters, the following techniques are taken into account and examined for every multicarrier PWM topology:

a) Multiple carriers with sinusoidal reference
b) Multiple carriers with varying frequencies
c) Multiple carriers with varying amplitudes
d) Several carriers using a modified sinusoidal reference
e) Trapezoidal amalgamated reference (TAR) with many carriers
f) Carrier phase shift.
g) Modulations that are unipolar and bipolar

The investigation stems from the fact that carrier-based PWM and SVM are the most often utilized PWM techniques for CMLI when the output voltage levels exceed five. Because there are more switching states as the levels are raised, SVM becomes much more complex. The numerous carrier-based approaches are therefore taken into account. Since the THD provides an estimate of the amount of energy present in frequencies other than the fundamental component, it can be used to measure the impacts of the modulation scheme. The criteria listed below can be used to determine which fs, ma, and THD measurements are appropriate.

The switching frequency of the power semiconductors has a significant impact on PWM approaches, which are now employed to regulate contemporary static converters, such as high-power machine drives. The power switches' switching frequency is unrestricted in low-power applications, and a low switching frequency can boost the converter's high-quality efficiency. Low switching frequency techniques are also favored in high power applications because they reduce switching losses, whereas high switching frequency algorithms are better suited for high dynamic range applications because of their greater bandwidth and superior output power quality. fs is between 500 and 700 Hz for high-power applications. Getting very low THD with a low switching frequency is currently a difficulty. The carrier frequency and switching functions determine the output voltage's harmonic components. It is recommended to run the inverter with a modulation index between $0.5 \leq ma \leq 1.0$ in order to maximize the power transferred from PV arrays to the grid.

The total of the output voltages of "N_s" H-bridge cells with isolated PV array inputs is the inverter output phase voltage. By lowering the harmonics in the generated current, the voltage waveform lowers the input filtering effort. It's crucial to remember that while reducing voltage THD is ideal in some situations, it's not a cure-all in others. For instance, the minimal current THD is preferred in applications like motor drives. Limiting each order harmonic to specific maximum permitted levels is desirable in some high-power applications.

The harmonic distortion, which is anticipated to be as low as feasible in many MLI applications, is measured by the THD of the inverter's output voltage. The THD of the inverter's output voltage is considered a crucial indicator of power quality in single-phase MLI for the PV power supply system. Applying the minimization algorithm to the phase voltage is a typical practice in THD minimization of an MLI output.

The line-to-line voltage may not always have the lowest THD when the phase voltage's THD is minimized. All odd-order harmonics, including the triplens, are present in phase voltage; their THD minimization provides the ideal condition for the phase voltage alone, but not for the line voltage. For a 15-level inverter powered by a 3 kWp solar PV plant, a single chip topology is suggested and put into practice to provide the corresponding changes of carrier signals above and below the zero reference.

5.5 Simulation Exercises

MATLAB/Simulink software is used to simulate each modulation strategy. The input for each of the CMLI's individual inverter stages is the simulated

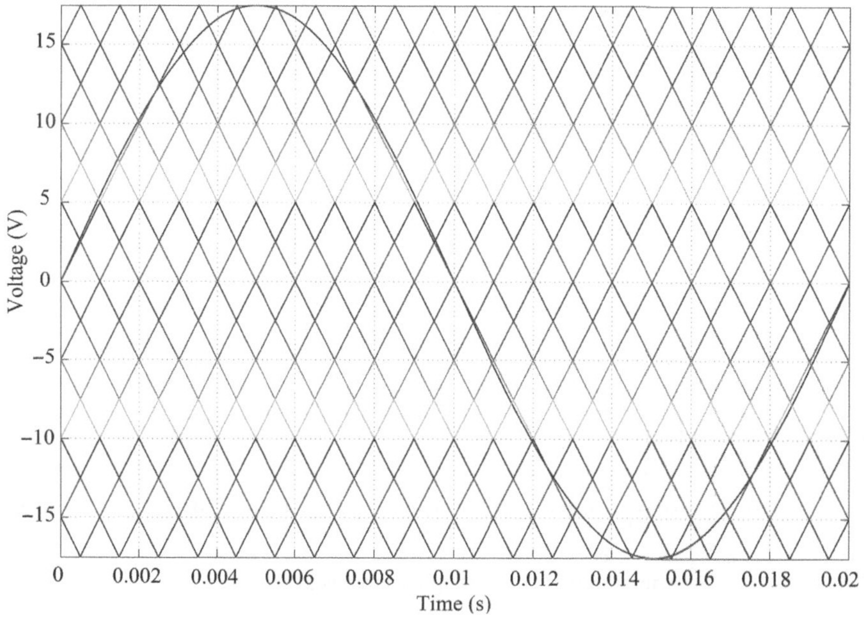

Figure 5.6 Normal mode carrier configuration based on APOD.

solar PV panel with an input of 48 V and 7 A. Amplitude modulation index $m_a = 1$, switching frequency $f_s = 1000$ Hz, inverter output frequency = 50 Hz, and frequency modulation index $mf = 20$ are the parameters utilized for simulations. For PWM inverters, a switching frequency of 1 kHz will be ideal for lowering losses. The load under consideration is RL, with values of $R = 100\ \Omega$ and $L = 10$ mH.

The many carrier signal arrangement modes for a 15-level inverter in APOD are depicted in Figures 5.6–5.8. To create the switching pulses for a 15-level inverter, 14 carrier signals are needed, 7 of which are positioned above the zero reference and the remaining seven below. Since neither the amplitude nor the frequency of the carrier signals is altered, Figure 5.6 depicts carrier arrangements with equal amplitude and frequency, which is regarded as the "normal" scenario. The APOD carrier arrangement with variable amplitude is depicted in Figure 5.7. The carriers are positioned so that the amplitude decreases closer to the zero reference and increases as it approaches the peak.

In a similar vein, Figure 5.8 depicts the carrier setup with variable frequency, where the frequency range from the zero reference begins at 500 Hz and ends at 3.5 kHz. The carrier configuration for APOD that modifies the sinusoidal reference signal is depicted in Figure 5.9.

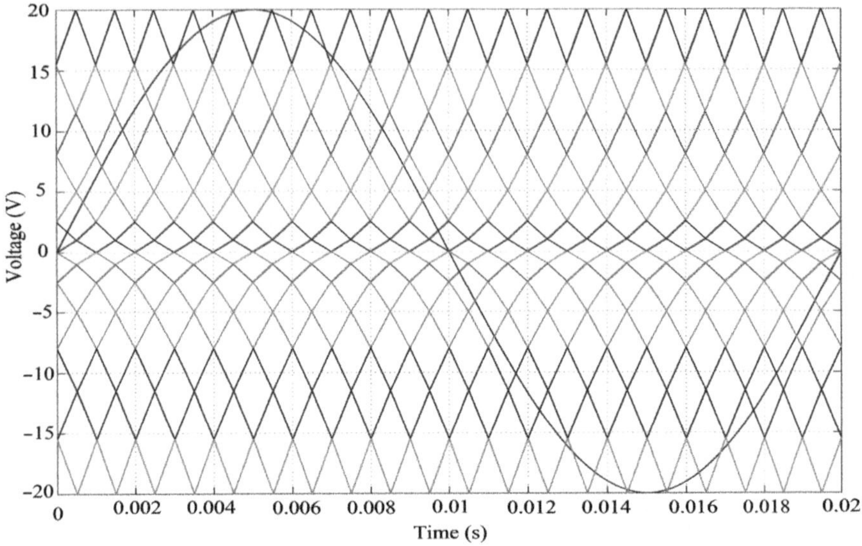

Figure 5.7 Variable amplitude carrier configuration based on APOD.

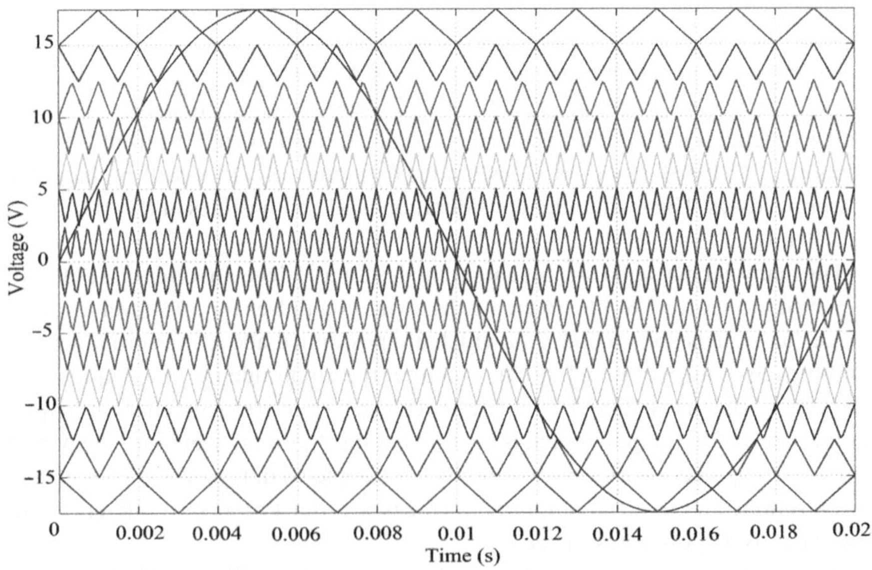

Figure 5.8 Variable frequency carrier configuration based on APOD.

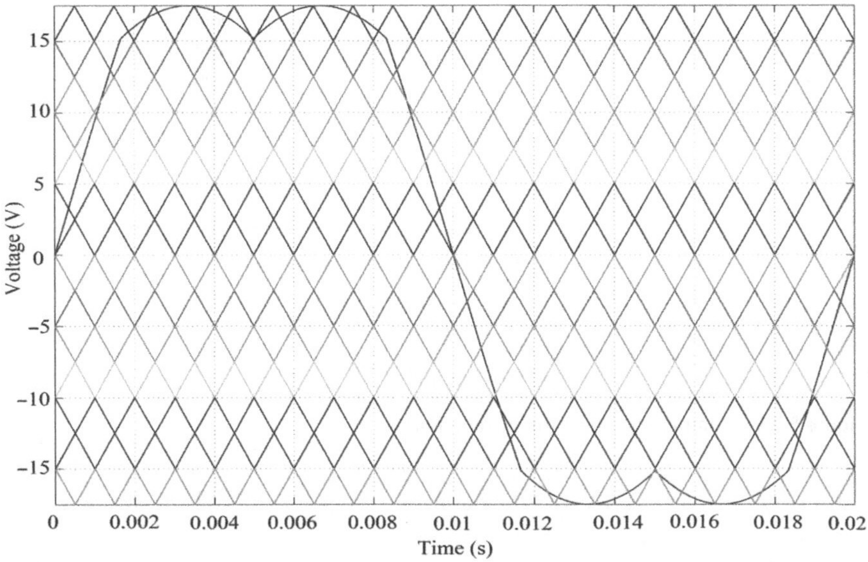

Figure 5.9 APOD-based carrier arrangement with modified reference.

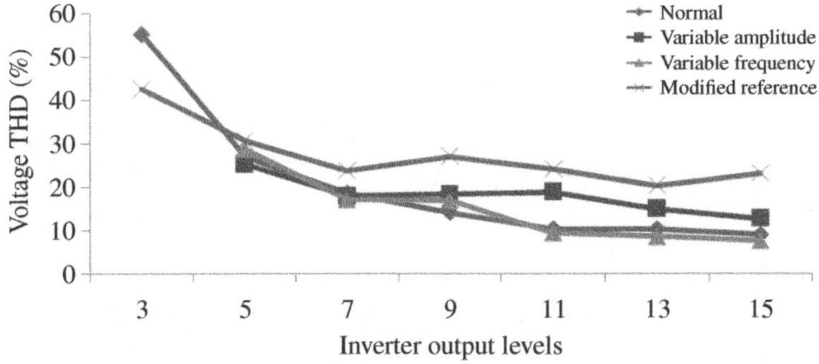

Figure 5.10 Comparison chart for voltage THD.

The voltage THD and current THD variations for the different levels of solar-fed CMLI with APOD are shown in Figures 5.10 and 5.11.

Figures 5.12–5.14 depict similar techniques implemented for APOD applied to POD.

The carrier arrangement with adjusted reference for POD is shown in Figure 5.15. The voltage THD and current THD variations for the different levels of solar-fed CMLI with POD are shown in Figures 5.16 and 5.17.

Figures 5.18–5.20 represent the similar techniques shown above for APOD; POD is applied to PD. The carrier configuration for PD with a modified modulating sinusoidal reference is shown in Figure 5.21.

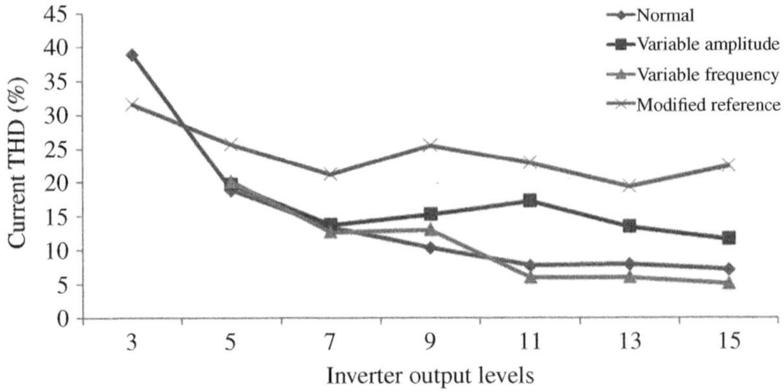

Figure 5.11 Comparison chart for current THD.

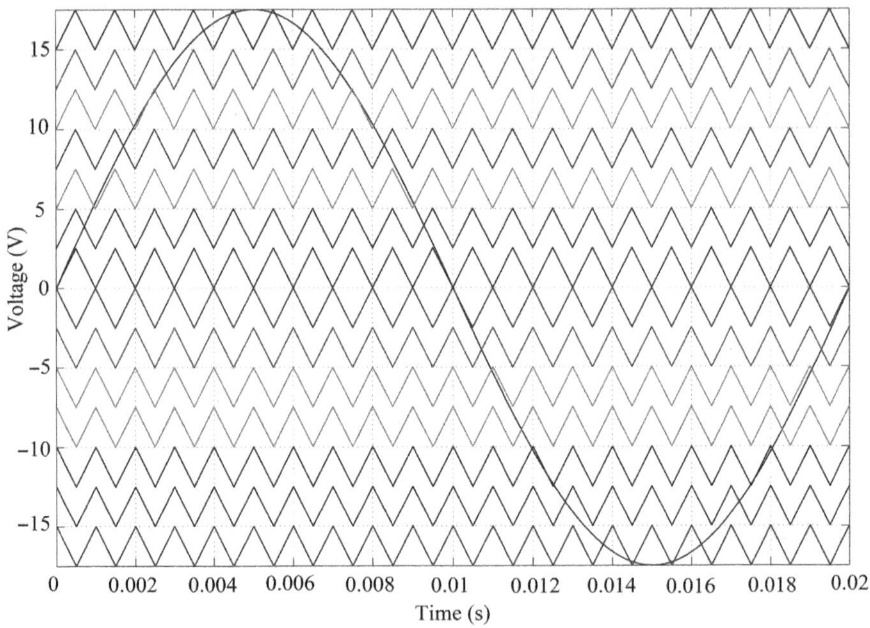

Figure 5.12 POD-based carrier configuration in standard mode.

The voltage THD and current THD variations for the different amounts of solar-fed CMLI with PD are displayed in Figures 5.22 and 5.23.

Figure 5.24 shows the PSC arrangement under the horizontal distribution scheme. This method comprises 14 carriers, which are phase shifted by an angle of 25.71^0 as per Equation (5.3).

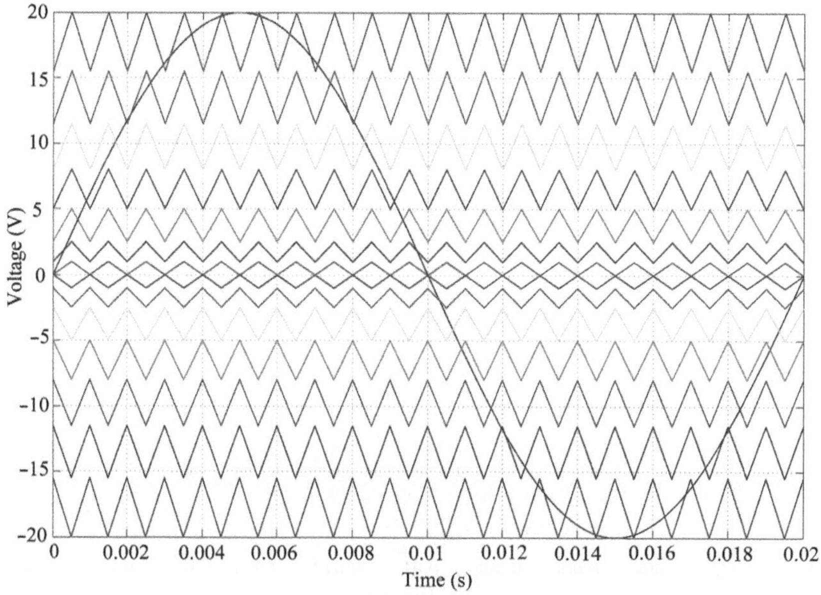

Figure 5.13 Variable amplitude carrier configuration based on POD.

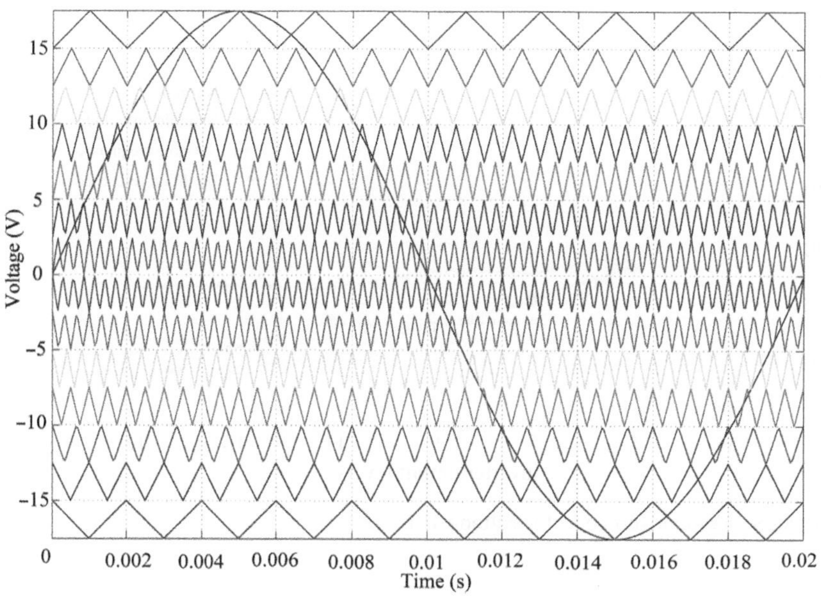

Figure 5.14 Variable frequency carrier configuration based on POD.

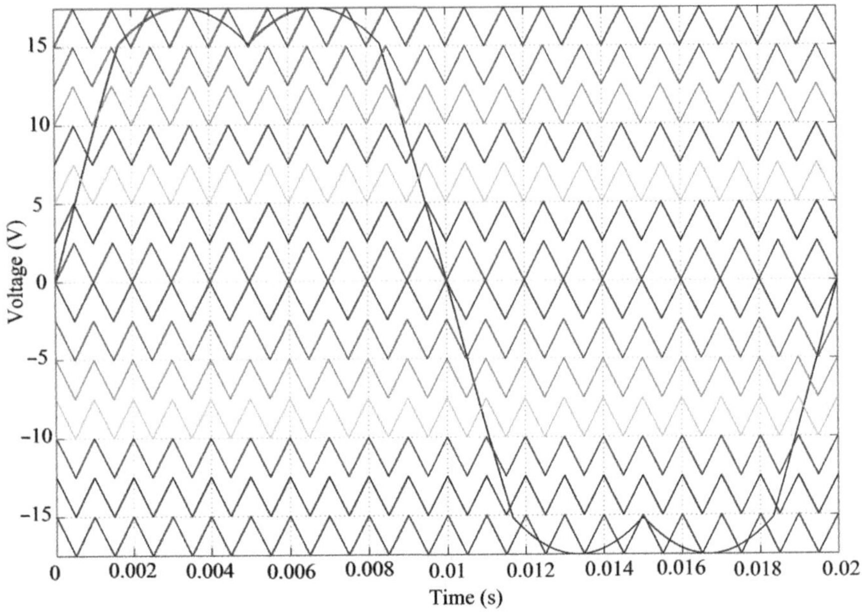

Figure 5.15 POD-based carrier configuration with updated reference.

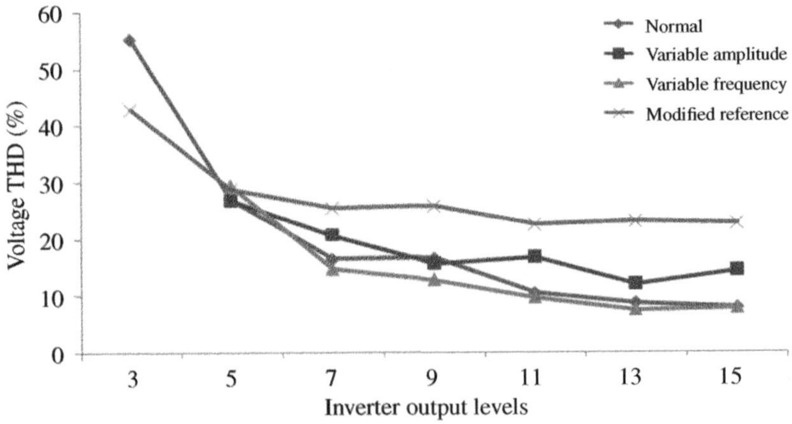

Figure 5.16 Comparison chart for voltage THD.

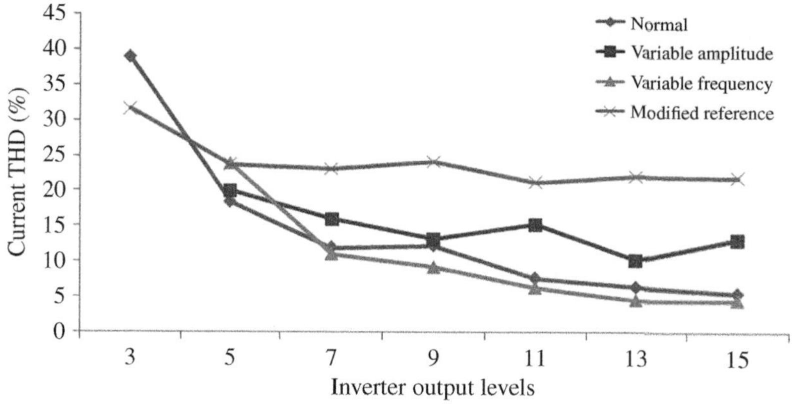

Figure 5.17 Comparison chart for current THD.

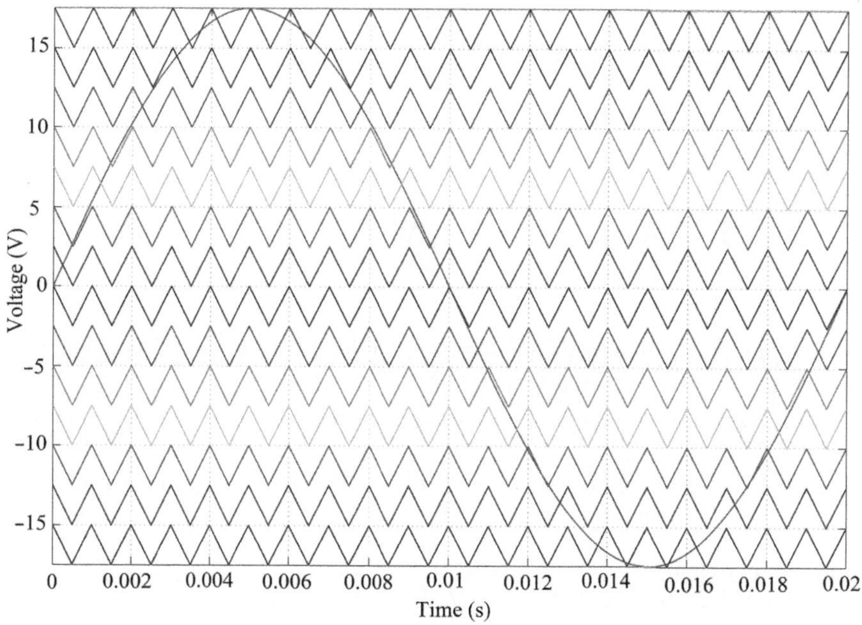

Figure 5.18 Normal mode carrier placement based on PD.

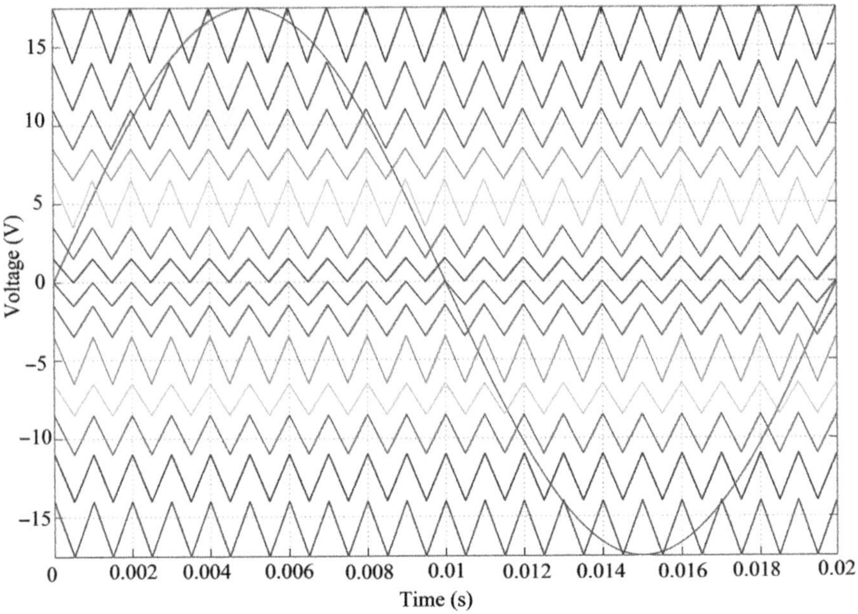

Figure 5.19 Variable amplitude carrier configuration based on PD.

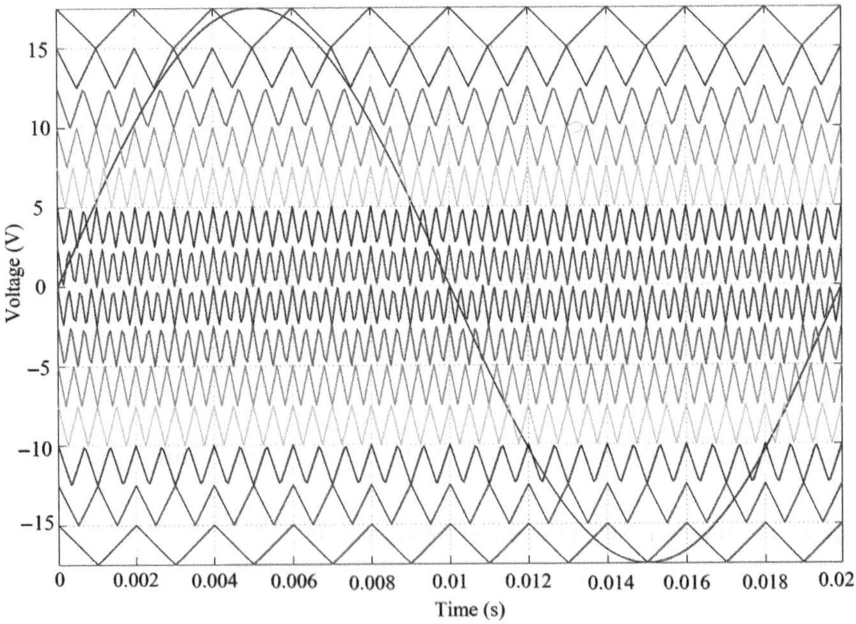

Figure 5.20 Variable frequency carrier combinations based on PD.

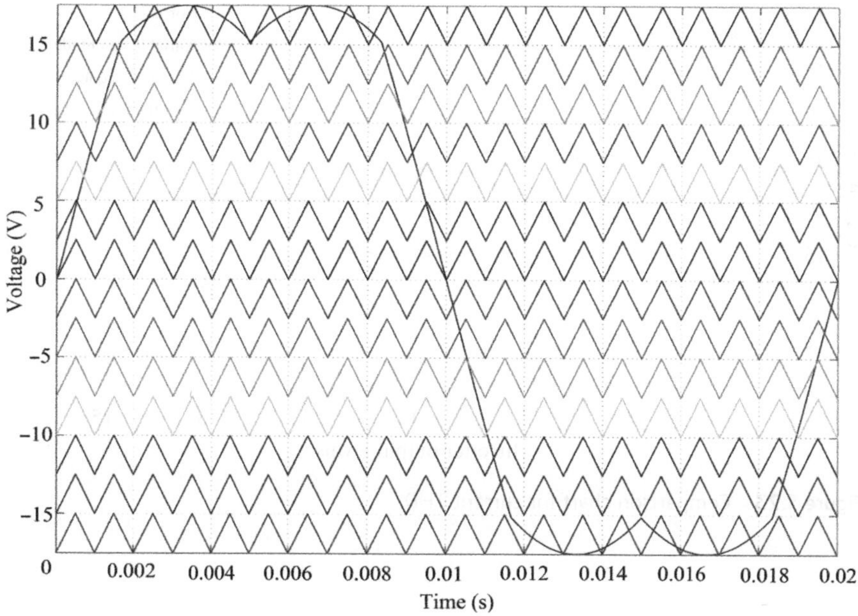

Figure 5.21 PD-based carrier configuration with altered reference.

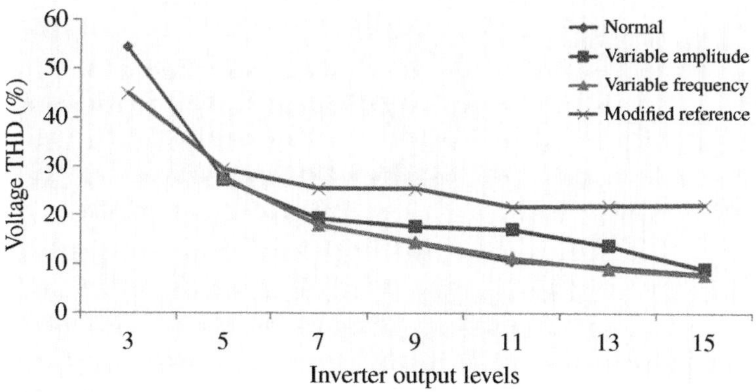

Figure 5.22 Comparison chart for voltage THD.

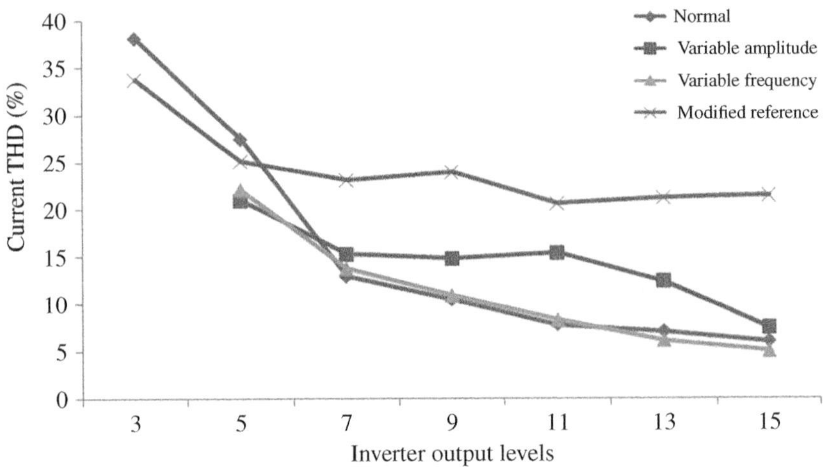

Figure 5.23 Comparison chart for current THD.

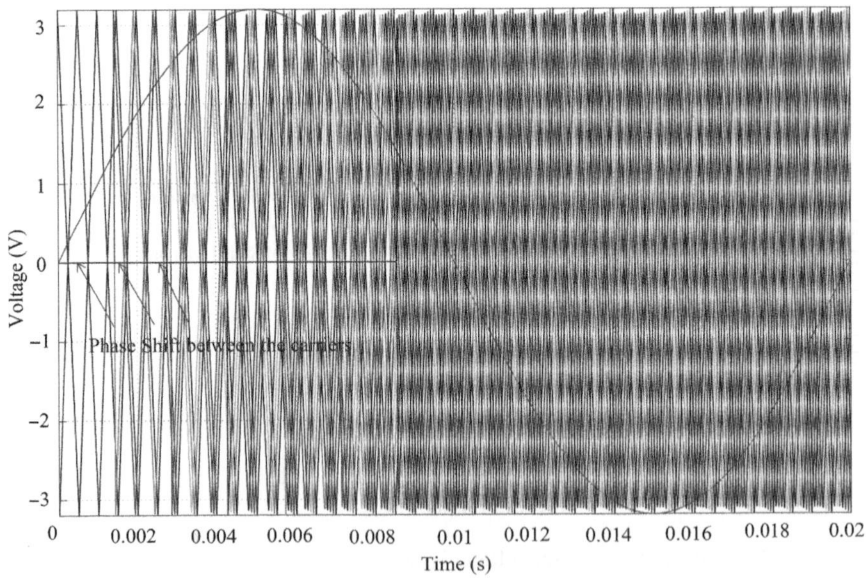

Figure 5.24 PSC arrangement for 15-level inverter.

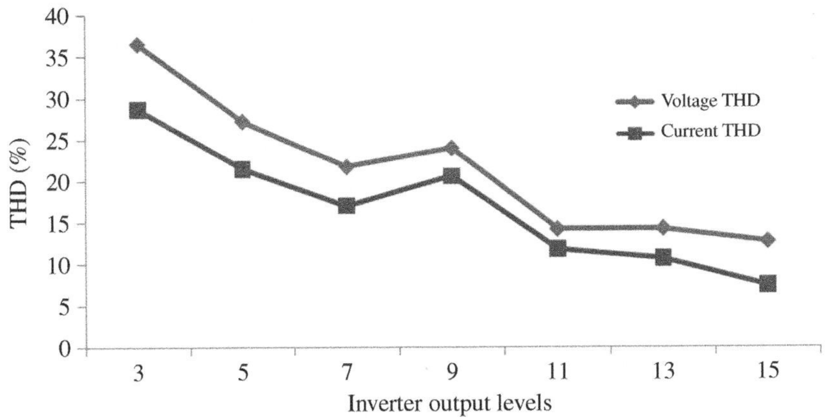

Figure 5.25 Comparison of THD for PSC modulation.

Figure 5.25 shows the variation of voltage and current THD for various inverter output voltage levels. It is also noted that PSC modulation cannot be used for DCMLI.

The listed modulation techniques are bipolar, meaning that carrier signal configurations are maintained above and below the zero reference. Unipolar modulation, on the other hand, requires half as many carriers as bipolar modulation. However, the drawback of unipolar PWM is that the system complexity is increased by the extra delay circuits that supply the pulses in a negative sequence.

By demonstrating the proper fluctuations in both carrier and reference signals, the following unipolar schemes are provided for comparison with the bipolar. The carrier arrangement for PSC modulation in unipolar is depicted in Figure 5.26, where the modulating signal is fixed at a bias of 3.2 V, and the carriers are positioned above the zero reference.

Figure 5.27 shows the methodology where the multiple sinusoidal carriers are used with TAR.

Figure 5.28 displays the THD comparison for the different unipolar modulation techniques.

According to the simulation and comparative research, THD decreases as the number of layers increases. When the inverter output level reaches 15, a minimum THD is reached. Further increases in levels can also be made, but as the design procedures are developed for the 15 levels, it is limited with this level. The lower THD values are found at APOD, PD, and POD in variable frequency modes as well as PSC in unipolar mode, when comparing the THD for the multiple carrier PWM techniques.

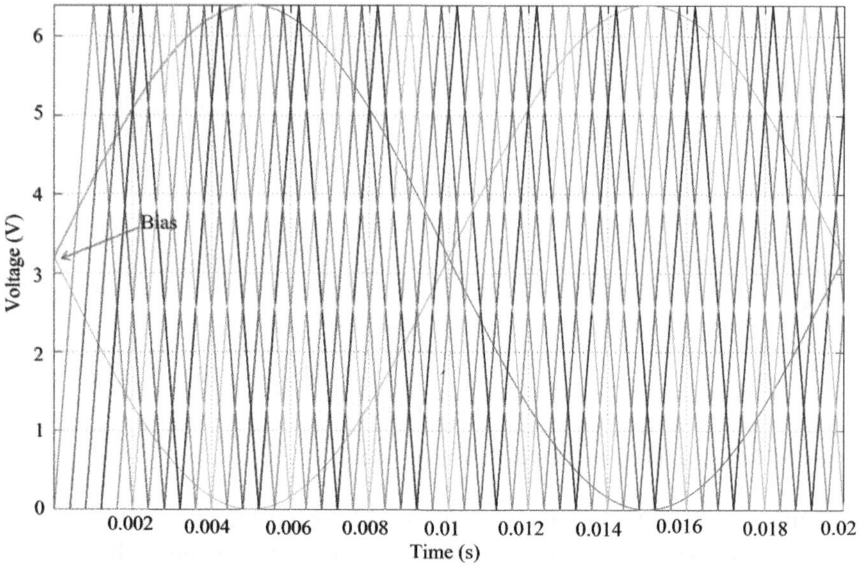

Figure 5.26 Unipolar PSC arrangement with dual reference.

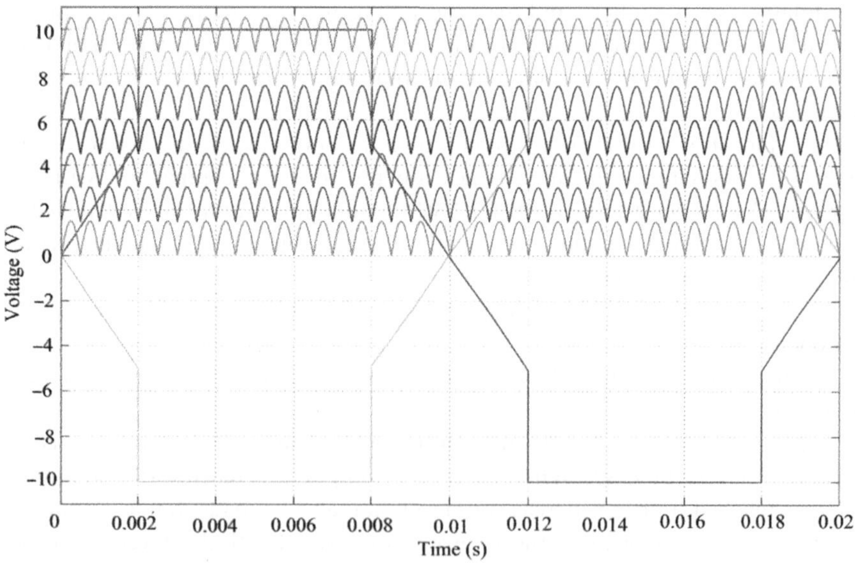

Figure 5.27 Sinusoidal carrier arrangement with TAR.

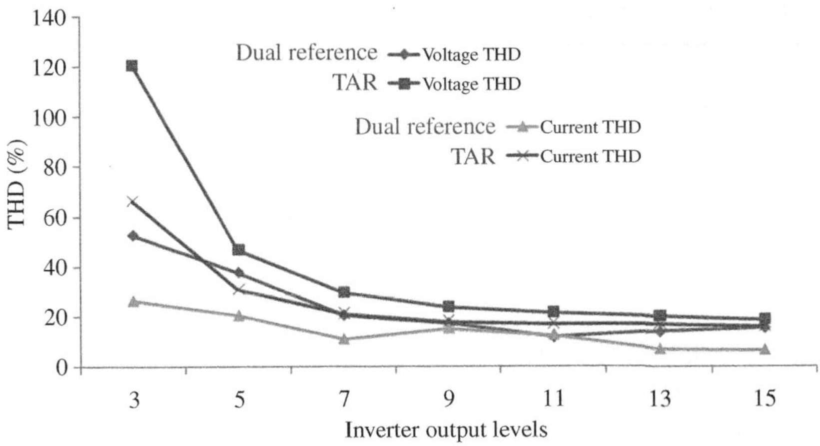

Figure 5.28 Comparison of THD for unipolar modulation.

5.6 Results of the Experiment

For the 7-stage, 15-level solar-fed CMLI, a 3 kWp solar PV power supply unit with multiple carrier PWM generation in a single chip is designed and built. A single-phase 230 V, 50 Hz AC system powers a large portion of home loads in the majority of nations. As a result, the system's output voltage is intended for a 230 V AC supply. This is accomplished by using a 48 V input supply for each inverter step in the system.

The ratings for each solar PV module are displayed in Table 5.1. Each module has a 115 W rated power and a voltage range of 16–21 volts (nominal 12 V) based on various working conditions, including light intensity. The single individual source for CMLI is a nominal voltage of 48 V obtained by connecting four solar panels in series.

Direct, freestanding, and grid-connected solar panels can all be connected to loads. The first one does not apply to application-oriented. The setup uses a stand-alone type that can be expanded to include grid-connected systems. A nominal DC bus of 48 volts is obtained by connecting four 12 V, 100 Ah battery packs in series. Through a regulated charging circuit, the PV unit charges these batteries. Since they generate the necessary $48 \times 7 = 336$ Vp (230VRMS) in compliance with the simulation model rating, seven sets of separate PV supply sources are used for the hardware implementation.

The hardware configuration specifications are shown in Table 5.2. The 3 kWp solar PV plant is depicted in Figure 5.29, and the full hardware configuration and matching output waveform of the suggested CMLI are shown in Figures 5.30 and 5.31. In this case, the individual inverter stage is indicated by the notation INV. The primary challenge for a transformerless PV inverter with high efficiency

Table 5.1 Solar PV panel specifications.

Parameter	Value
Model	Solectric 9000
P_{mpp}	115 Wp
V_{oc}	21.2 V
I_{sc}	7.4 A
V_{pm}	16.5 V
I_{pm}	6.95 A
Max system voltage	540 V
Tolerance at peak power	± 5%
Number of panels	28
Total power	3220 Wp

Table 5.2 Details of the experimental configuration.

Parameter	Rating
Charge controllers (CC)	
Make	Sukaam
V-I rating	48 V, 10 A
Number of CC	7
Battery bank	
Make and model	EXIDE 6LMS100L
Rating	12 V, 100 Ah
Number of batteries	28
Power circuit	
Semiconductor devices	MOSFET IRF 840
Number of devices	28
Control circuit	
Controller	DSP - TMS320F2812
Measuring instruments	
PQ analyzer	Yokogawa WT3000
Oscilloscope	Tektronix

Figure 5.29 3 kWp 28-module solar PV plant.

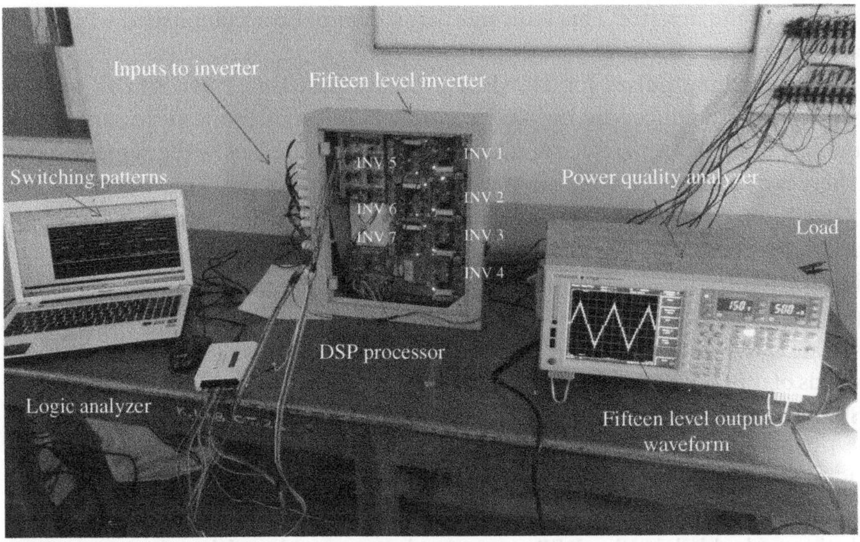

Figure 5.30 The suggested 15-level inverter's experimental configuration.

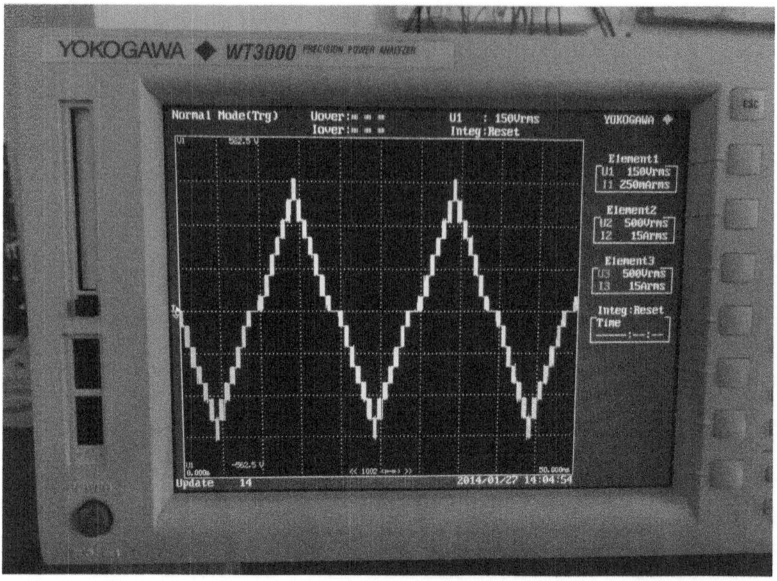

Figure 5.31 15-Level inverter output voltage waveform.

and dependability is that all switching devices must use MOSFETs in order to achieve high efficiency across a broad load range. MOSFETs are used in MLIs to offer a high-bandwidth sinusoidal output voltage at high efficiency or to lower the cost of the inverter. This indicates that it cannot be accomplished with linear amplifiers.

The switching pulses for MOSFET switches based on POD, APOD, and PD schemes are produced by the DSP controller. To produce the necessary pulses, the carrier signals are created and level-shifted above and below the zero reference. There is a changeover switch that will then select one of the three systems.

Additionally, each bridge contains a plug-in MiniDSP processor card built around a Texas Instruments TMS320F240 DSP chip and its own integrated input circuitry for conditioning and processing analog and digital signals. This processor has two expansion connectors (analog and I/O), a 150 MIPS operating speed, 18K words of on-chip RAM, 128K words of on-chip Flash memory, 64K words of off-chip SRAM memory, a 30 MHz clock, and a TI F28xx Code Composer Studio tools driver.

To guarantee that all necessary computations can be carried out in real time, the software designed for the DSP is divided into foreground (time-critical interrupt) and background jobs. The event manager (EV manager) peripheral device embedded in the DSP generates interrupts that drive these processes. Three general-purpose timers (GP timers 1, 2, and 3) in the EV management can be configured to provide the necessary triangular carrier for PWM pulse production. To produce PD, POD, and APOD signals, these carrier signals are level shifted above and below the zero reference in accordance with the program. The gating signals needed for the MOSFET switches are produced by comparing these signals with the sinusoidal reference.

The DSP can be redirected from the present background activity to carry out a PWM interrupt service routine (ISR) by configuring one of these timers to initiate a high-priority interrupt precisely once per ($1/2f_s$) interval, where f_s is the bridge switching frequency. The primary purpose of the ISR is to compute the new CHB switching times at each half-carrier period. After these times are computed, they are stored in registers so that, at the conclusion of the half-carrier period, they can be compared to the GP timer's counter value in preparation for the subsequent half-carrier period.

As is common with all sampled PWM systems, a half-carrier interval is thus added between the computation of the switching timings and the actual comparisons of these times with the counter value of the GP timer. Every time a state is traversed, the DSP also carries out other housekeeping duties, like looking for gate faults to confirm the power bridge is operating correctly. When a fault occurs, the DSP is placed into the "fault" mode with all of its PWM output channels disabled in order to safeguard the bridge module. Before moving to the "Stop" state and preparing for the power bridge to start up again, the DSP will remain in this mode until the fault has been resolved.

Figures 5.32–5.34 display the carrier configurations that the controller produced for the APOD, PD, and POD modulation schemes.

The harmonic analysis performed for each modulation strategy using the power quality analyzer (PQA) WT3000, which displays both voltage and current THD values, is shown in Figure 5.35. The WT3000 is a high-precision analyzer that can show 20 parameters in addition to the harmonic order magnitude and THD.

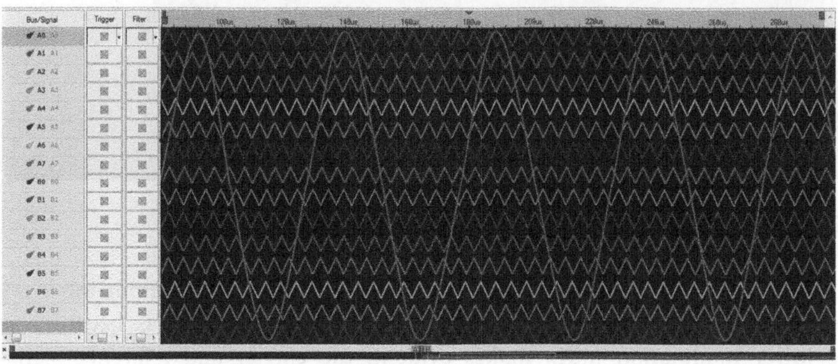

Figure 5.32 APOD-based carrier configuration.

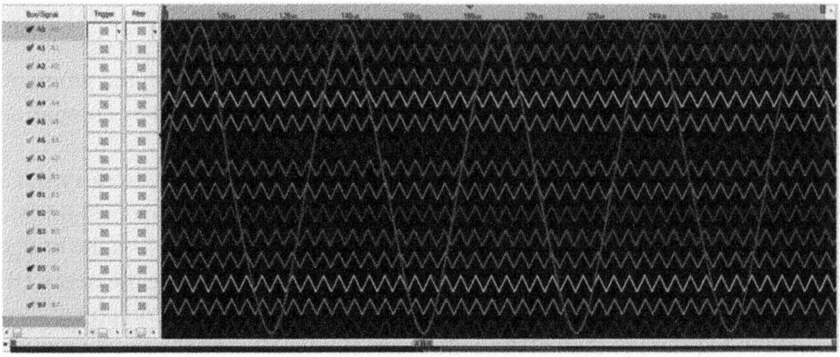

Figure 5.33 Carrier layout based on POD.

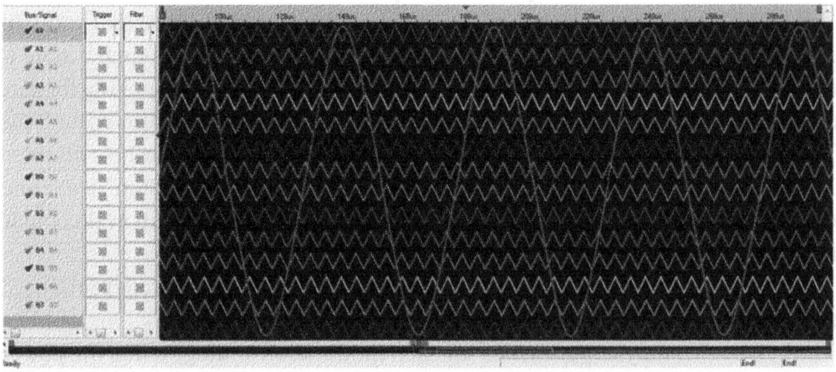

Figure 5.34 Carrier layout based on PD.

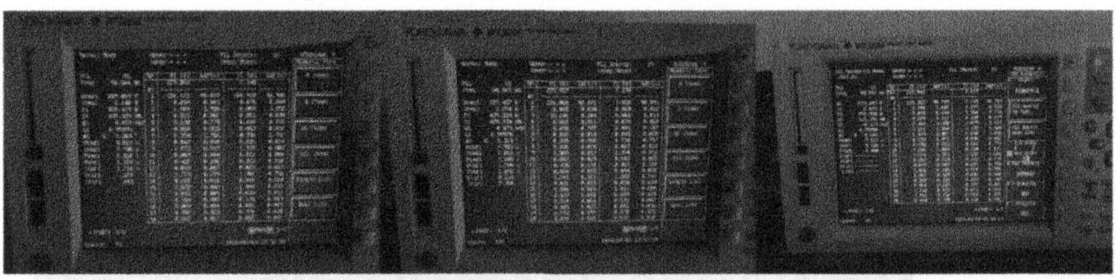

Figure 5.35 APOD, POD, and PD harmonic analysis.

According to the findings, the inverter produces a 15-level output for every modulation scheme; however, the resulting THD varies. This is because, as the lower modulation index makes evident, the carrier and reference signal compassion differs for each of the three approaches. A comparison of simulation

Table 5.3 Comparison of results from simulations and experiments.

		Simulation		Experimental	
S. no.	**Modulation**	V_{THD} (%)	I_{THD} (%)	V_{THD} (%)	I_{THD} (%)
1.	APOD	8.98	7.23	8.813	7.740
2.	POD	7.84	5.63	7.955	7.851
3.	PD	8.04	5.96	9.514	9.526

Table 5.4 A comparison with alternative approaches.

Authors	Levels	Methodology	THD (%)
Zhao et al. (2014)	5	PD	38.73
Malinowski et al. (2010)	5	PSC	30.2
	5	Bipolar	29.9
	7	Bipolar	21.8
Gupta et al. (2010)	5	Multiband	27.0
Zambra et al. (2010)	9	PD	8.29
Proposed	15	POD	7.9

and hardware results for a solar-fed cascaded 15-level inverter is presented in Table 5.3. The findings obtained for the current approaches using the resultant THD and the number of levels taken into consideration are displayed in Table 5.4.

5.7 Digital PWM

MLIs have several benefits, but their increasing component count is their main disadvantage. Even though an MLI uses switches with low voltage ratings, each switch needs a corresponding gate driver circuit. The system as a whole may become more costly and complicated as a result. Therefore, it is crucial to reduce the number of switches and gate driver circuits in practical implementation.

The MIL-HDBK-217F standard states that the number of components in a system has an inverse relationship with its reliability. Consequently, the converter's reliability declines as the number of switches rises. The power converter's efficiency declines along with its overall dependability. Voltage imbalance issues are introduced, and control complexity is increased by the large number of levels.

The CHB MLI needs the fewest total main components of the three MLI topologies. The flexibility to use various DC voltages on the individual H-bridge cells is one feature that distinguishes the cascaded H-bridge from other MLIs. Asymmetrical MLIs can be obtained as a result of splitting the power conversion. Asymmetric multilevel converters can be utilized to give a high number of output steps without adding more DC voltage sources. It is possible for the CHB to function as an asymmetric or symmetric converter. The DC voltage sources in asymmetric multilevel converters must be selected with varying values based on several techniques.

Applications based on solar PVs are a good fit for the asymmetric structure. On the other hand, the widespread use of higher-level inverters in real-world applications has been hindered by the large switch count and the challenge of balancing the voltage of the capacitors in the diode-clamped design.

Table 5.5 gives a basic comparison of the power component requirements per phase leg across different MLIs, where m is the number of output levels. Apart from the quantity of primary switching devices, FCMLI and DCMLI require clamping diodes and balancing capacitors, which are not necessary for CMLI. Additionally, CMLI requires half as many DC bus capacitors as other MLIs. It is evident that in three-phase applications, the DCMLI provides the benefit of having common DC input capacitors for the three phases. Its primary drawback is the excessive number of diodes utilized for clamping, which also complicates the physical layout's implementation.

Voltage synthesis redundancies in an FCMLI allow the clamping capacitor voltages to be balanced in a matter of cycles. Its primary drawback is the large number of capacitors required for clamping, even if these capacitors are less expensive than the DCMLI's clamping diodes. Additionally, the clamping capacitors get larger at low switching frequencies, which lowers the inverter's power density.

Clamping capacitors and diodes are not required in a CMLI. Moreover, it is possible to employ a low switching frequency. Because every cell has the

Table 5.5 Power component needs for different MLI structures are generally compared.

Inverter configuration	Diode-clamped	Flying–capacitor	Cascaded–inverter
Main switching devices	$2(m-1)$	$2(m-1)$	$2(m-1)$
Main diodes	$2(m-1)$	$2(m-1)$	$2(m-1)$
Clamping diodes	$(m-1)(m-2)$	0	0
DC bus capacitors	$(m-1)$	$(m-1)$	$(m-1)/2$
Balancing capacitors	0	$(m-1)(m-2)/2$	0

same structure and there are no additional FCs or clamping diodes, modularized circuit layout and packaging are feasible. The utilization of separate DC sources for each cell is its primary drawback. Because of this, PV and fuel cell-based power production systems can employ the CMLI.

The drawbacks of connecting MLIs in series include the need for switches with varying voltage ratings, which limits their use in high-power applications, and the requirement to alter the voltage polarity every half cycle. A thorough lookup table is necessary for the serial connection of a low-voltage conventional inverter and a high-voltage diode-clamped inverter.

While standard MLIs can produce up to nine levels with the same number of power supplies, the Mondal approach uses two numbers of two-level and three-level inverters that can only produce five output levels with four DC sources. Bidirectional switches that can block voltage and current in both directions are necessary for the various switch reduction techniques, and each bidirectional switch needs a gate drive circuit, which increases power loss. Additionally, the switching sequence is controlled by PWM approaches, and the several studies listed are only implemented with low-power DC sources.

Table 5.6, which was created by taking into account the different MLI architectures, lists the power components needed for a solar-fed 15-level inverter. Substituting the value of $m = 15$ in Table 5.5 yields the comparison list.

Even though CMLI has fewer switching components than other types, more reduction will undoubtedly improve the system's dependability. Therefore, 28 switching devices are needed for all MLI topologies in order to create a 15-level inverter. However, additional clamping diodes and balancing capacitors are not required in CMLI alone. Additionally, there are far fewer DC bus capacitors in CMLI. Achieving a high number of levels with fewer devices and sources is the methodology's goal.

It is also considered that the problems involved in increasing levels include control complexity and voltage imbalance. In order to minimize the number of semiconductor switches in a solar-fed CMLI, a control circuit design with

Table 5.6 Comparison of a 15-level inverter's power component requirements.

Inverter configuration	Diode-clamped	Flying-capacitors	Cascaded-inverter
Main switching devices	28	28	28
Main diodes	28	28	28
Clamping diodes	182	0	0
DC bus capacitors	14	14	7
Balancing Capacitors	0	91	0

less complexity and voltage imbalance is suggested. The design incorporates topologies based on "binary," "trinary," and "modified multilevel converter" (MMC) that are appropriate for different solar PV input sources.

The output voltage levels in binary mode are $2N_s$+1−1, where N_s is the number of separate inverter stages. Digital logic functions, such as counters, flip-flops, and logic gates, do this. $3N_s$ levels are reached in trinary mode using a matching look-up table. By using suitable switching sequences, MMC aims to deliver comparable output voltage levels in both control and power circuits. Therefore, 12 switches are required in binary mode, and 28 switches are required in the conventional way to achieve a 15-level inverter. While just seven switches are used in MMC to get 15 levels at the inverter output, twenty-seven levels can be obtained in trinary mode using the same twelve switches.

These three approaches have the advantage of lowering THD by raising the levels. In order to accomplish more levels with fewer switching devices without the use of transformers, filter components, bidirectional switches, front-end rectifiers, and comprehensive lookup tables, an experimental examination is conducted for each design.

5.8 Binary Mode

In the traditional method, the necessary gating signals for the inverter switches are obtained by comparing reference and carrier signals using PWM techniques. Equation (5.8) indicates how many output voltage levels were produced using this method. (8) where N_s represents the different inverter stages and m represents the output voltage levels. Equation (5.8) provides the number of switches (l) needed to reach m levels.

$$l = 2(m - 1) \tag{5.8}$$

As in the case of CMLI, 28 switches are needed to build a 15-level inverter with seven separate inverter stages. In NPC or DCMLI, 182 clamping diodes are required in addition to the 28 switches. In the instance of FCMLI, 15-level output requires 28 switching devices, 91 balancing capacitors, and 14 DC bus capacitors. Increasing the number of levels will thus lessen harmonic distortion, improving both system dependability and power quality. Equation (5.9) indicates how many levels can be attained in binary mode operation for a given set of inverter stages.

$$m = 2^{N_s+1} - 1 \tag{5.9}$$

Therefore, just three inverter stages with twelve switches are needed to achieve a 15-level output. This is accomplished by implementing a switching circuit for the solar-fed CMLI with a control strategy that includes digital logic functions. Different solar PV input sources supply power to the three inverter stages.

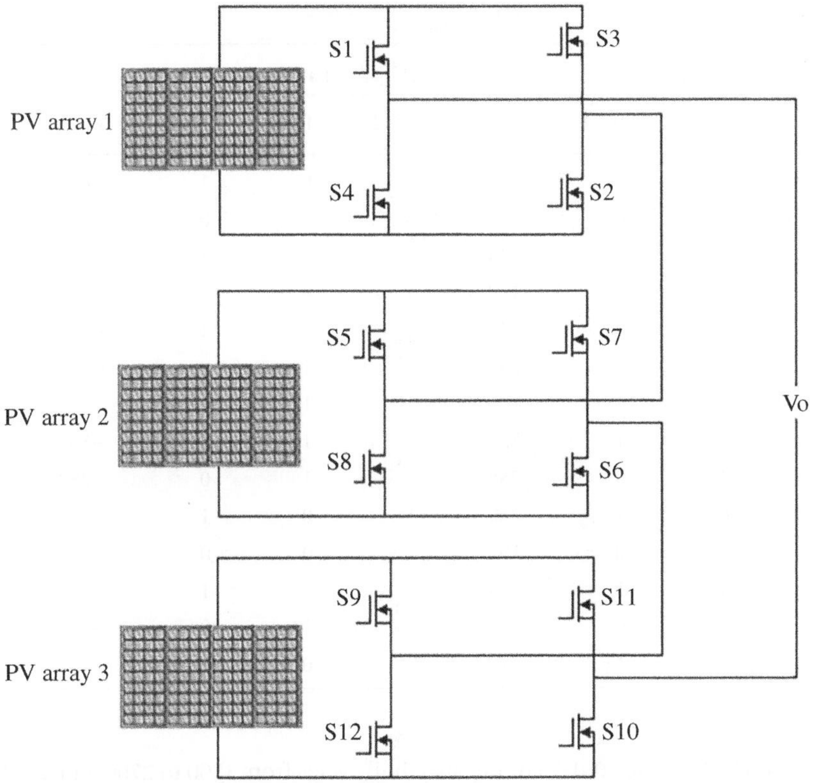

Figure 5.36 Three-stage CMLI power circuit.

The solar-fed three-stage CMLI power circuit shown in Figure 5.36 can produce seven levels using a traditional manner, but it can produce 15 levels using a digital switching strategy. Binary counters can achieve an output voltage in the range of $2N_s$ by scaling the input voltages to the power of 2. To meet the necessary condition specified in Equation (5.9) at the output level, an incremental and descent operator is necessary.

The switches S1, S5, and S9 in the circuit seen in Figure 5.36 are ON during the positive half cycle, whereas S3, S7, and S11 are ON during the negative half cycle. The inverter's output frequency is 50 Hz since the conduction period for each half cycle is fixed at 10 ms.

The switching pattern to achieve the appropriate voltage levels is provided by the sequence displayed in Table 5.7. The three inverter stages and the output voltage, Vo, are specified by INV1, INV2, and INV3. The switches' "ON" and "OFF" states are indicated by the numbers "0" and "1." The conduction period from 00 to 900 is represented by the sequence "001" to "111" in the positive half cycle, and from

Table 5.7 Three-stage, 15-level inverter's ON/OFF states.

Positive half cycle (10 ms)				Negative half cycle (10 ms)			
INV3	INV2	INV1	V_o	INV3	INV2	INV1	V_o
0	0	1	48	0	0	1	−48
0	1	0	96	0	1	0	-96
0	1	1	144	0	1	1	−144
1	0	0	192	1	0	0	−192
1	0	1	240	1	0	1	−240
1	1	0	288	1	1	0	−288
1	1	1	336	1	1	1	−336
1	1	1	336	1	1	1	−336
1	1	0	288	1	1	0	−288
1	0	1	240	1	0	1	−240
1	0	0	192	1	0	0	−192
0	1	1	144	0	1	1	−144
0	1	0	96	0	1	0	-96
0	0	1	48	0	0	1	−48

900 to 1800 by "111" to "001." For negative half cycles from 1800 to 2700 and 2700 to 3600, a similar condition applies.

To obtain the 15-level output, the switches in the CMLI are designed to function according to the switching sequence indicated in the truth table. 48 V is regarded as the input voltage for the first inverter stage. The remaining stages are then scaled to the power of two, with $48 \times 2 = 96$ V and $96 \times 2 = 192$ V.

According to the order shown in Table 5.7, the output voltage with a magnitude of 48 V square pulse arrives in the positive half cycle under the first condition [0 0 1]. In a similar vein, [0 1 0] yields an output voltage of magnitude $48\text{ V} + 48\text{ V} = 96$ V, and the next sequences follow suit to achieve the necessary 336 V at [1 1 1] until the first half of the positive cycle.

The bits must advance from 00 to 450 in order to do this operation. Reversing the above sequence from [1 1 1] to [0 0 1] completes the second half of the positive cycle, causing the bits to travel in the opposite direction from 450 to 900. Once more, the procedure is repeated in the same way during the negative half cycle.

The block diagram for the binary mode implementation is shown in Figure 5.37. The order in which the inverter's power switches work is similar to a counter's truth table. Therefore, N_s+2-bit counters are needed to create a CMLI with firing sequence constraints. $N_s + 2$ $(3 + 2 = 5)$ bit counters, which function as an up

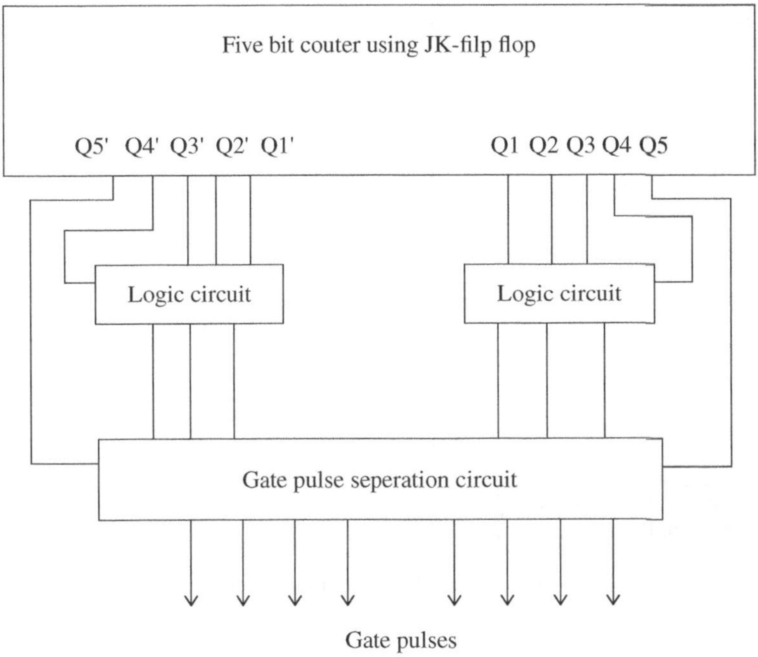

Figure 5.37 The suggested logic circuit's block diagram.

counter, are needed for the three-stage inverter. During the first half of the positive half cycle, the combinational logic circuit causes the three bits (Q1, Q2, and Q3) to travel forward; during the second half of the positive half cycle, they move in the opposite manner. Both halves of the negative half cycle repeat the same situation. The bits (Q1, Q2, and Q3) are modified by Q4 and Q4′, where Q4′ is used to regulate the incremental operation in the first half of positive or negative half cycles, and Q4 is used for the decrement operation in the latter half of positive or negative half cycles. In addition to controlling other bits in the circuit, the pulse separation block uses bits Q5′ and Q5 to separate the pulses needed for both positive and negative half cycles. Figure 5.38 displays the methodology's detailed arrangement, where "n" stands for the number of separate inverter stages.

Assuming that the input voltages are scaled to the power of two, let $V_{PV1} = V$, $V_{PV2} = 2V$, and $V_{PV3} = 4V$ represent the input sources of the three inverters. In relation to the voltage rating indicated in Table 5.7, the process is explained as follows for reaching the 15 levels:

Level 1: $V_{PV1} = 0$, $V_{PV2} = 0$, $V_{PV3} = 0$
Level 2: $V_{PV1} = V$, $V_{PV2} = 0$, $V_{PV3} = 0$
Level 3: $V_{PV1} = 0$, $V_{PV2} = 2\text{V}$, $V_{PV3} = 0$
Level 4: $V_{PV1} = V$, $V_{PV2} = 2\text{V}$, $V_{PV3} = 0$

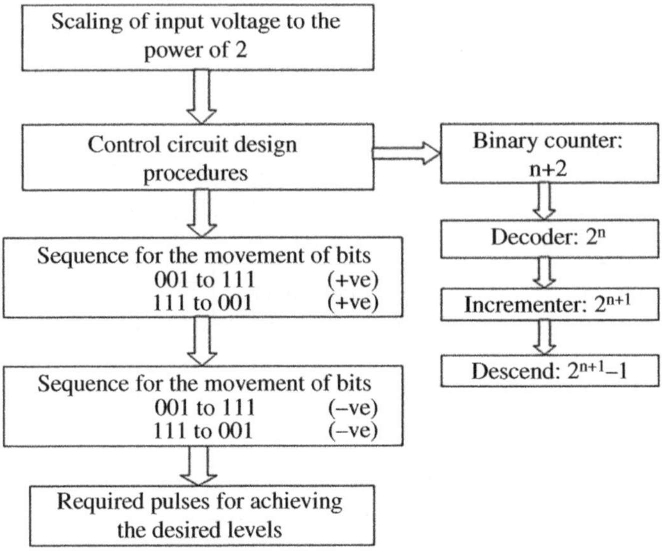

Figure 5.38 Flow diagram for the suggested logic circuit function.

Level 5: $V_{PV1} = 0$, $V_{PV2} = 0$, $V_{PV3} = 4\,V$
Level 6: $V_{PV1} = V$, $V_{PV2} = 0$, $V_{PV3} = 4\,V$
Level 7: $V_{PV1} = 0$, $V_{PV2} = 2\,V$, $V_{PV3} = 4\,V$
Level 8: $V_{PV1} = V$, $V_{PV2} = 2\,V$, $V_{PV3} = 4\,V$
Level 9: $V_{PV1} = -V$, $V_{PV2} = 0$, $V_{PV3} = 0$
Level 10: $V_{PV1} = 0$, $V_{PV2} = -2\,V$, $V_{PV3} = 0$
Level 11: $V_{PV1} = -V$, $V_{PV2} = -2\,V$, $V_{PV3} = 0$
Level 12: $V_{PV1} = 0$, $V_{PV2} = 0$, $V_{PV3} = -4\,V$
Level 13: $V_{PV1} = -V$, $V_{PV2} = 0$, $V_{PV3} = -4\,V$
Level 14: $V_{PV1} = 0$, $V_{PV2} = -2\,V$, $V_{PV3} = -4\,V$
Level 15: $V_{PV1} = -V$, $V_{PV2} = -2\,V$, $V_{PV3} = -4\,V$

The output waveform of a 15-level inverter is depicted in Figure 5.39, where level 1 is reached at the origin and levels 2 through 8 show the seven voltage steps reached in the positive half cycle, denoted as "7 V." The negative half cycle from level 9 to level 15 exhibits a similar conduction, with the final seven voltage steps tending to "−7 V." Because there are fewer commutations in the inverter with a higher voltage level (4 V), switching losses are decreased.

By simply increasing the counters in accordance with the number of inverter stages and control logic functions, this technology can reach any number of levels. The suggested methodology considers the example $V_{PV1} = 48\,V$, $V_{PV2} = 96\,V$, and $V_{PV3} = 192\,V$ to generate 15 levels with an output voltage of 336 V_p, rather than $V_{PV1} = V$, $V_{PV2} = 2\,V$, and $V_{PV3} = 4\,V$ as shown in Figure 5.39.

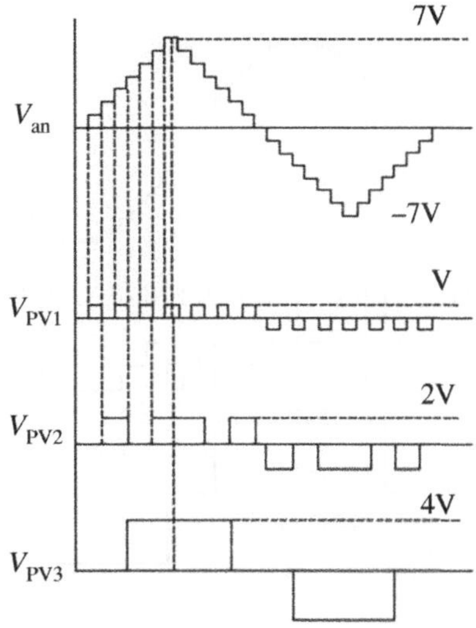

Figure 5.39 Binary mode switching sequence for a 15-level inverter.

5.9 Trinary Mode

Equation (5.10) provides the number of levels that can be attained in trinary mode operating for the specified set of inverter stages.

$$m = 3^{N_s} \tag{5.10}$$

The switching process for a three-stage, twenty-seven-level inverter is depicted in Figure 5.40. Twenty-seven levels can be achieved for the power circuit depicted in Figure 5.36 using just 12 switches and three inverter stages. The switching sequence for switches S1 through S12 to obtain a 27-level output is shown in Table 5.8.

Instead of using a digital logic function in binary mode, an embedded controller is suggested to accomplish this without the need for transformers or complex algorithms. The input voltages are scaled to the power of three in the trinary method. Only the positive half cycle of the sequence in Table 5.8 yields thirteen levels; in the negative half cycle, the same sequence is rotated by 900 degrees to provide the remaining 13 levels. The intended 27 levels will be attained by incorporating level zero.

An independent power supply with a high-frequency connection, a chopper circuit, a transformer with six secondary windings with a 9:3 turn ratio, and a diode bridge rectifier are all needed for a 3-stage, 27-level asymmetric inverter. The transformer's square wave operation necessitates a particular design, and its existence raises costs, space, weight, and related losses.

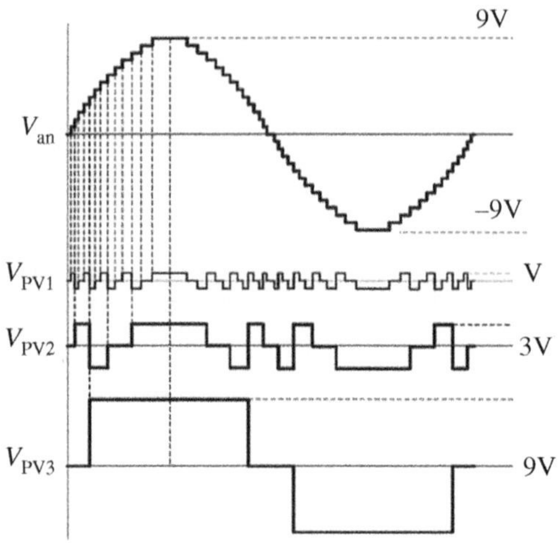

Figure 5.40 Sequence of switching for a 27-level inverter.

Table 5.8 ON/OFF states for a 3-stage, 27-level inverter.

Level	S1	S2	S3	S4	S5	S6	S7	S8	S9	S10	S11	S12
1	1	0	0	1	1	1	0	0	1	1	0	0
2	0	1	1	0	1	0	0	1	1	1	0	0
3	1	1	0	0	1	0	0	1	1	1	0	0
4	1	0	0	1	1	0	0	1	1	1	0	0
5	0	1	1	0	0	1	1	0	1	0	0	1
6	1	1	0	0	0	1	1	0	1	0	0	1
7	1	0	0	1	0	1	1	0	1	0	0	1
8	0	1	1	0	1	1	0	0	1	0	0	1
9	1	1	0	0	1	1	0	0	1	0	0	1
10	1	0	0	1	1	1	0	0	1	0	0	1
11	0	1	1	0	1	0	0	1	1	0	0	1
12	1	1	0	0	1	0	0	1	1	0	0	1
13	1	0	0	1	1	0	0	1	1	0	0	1

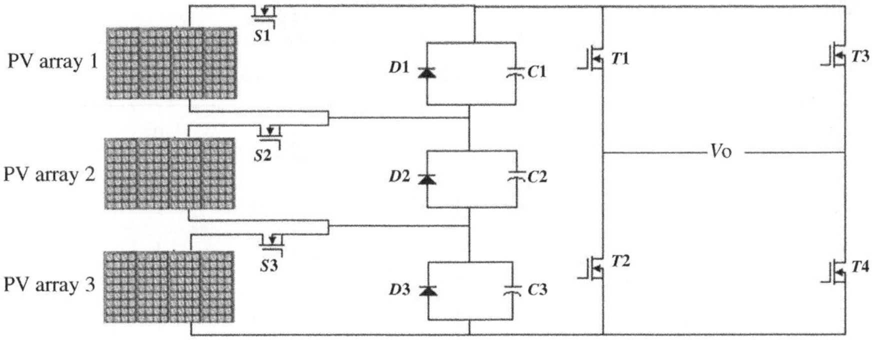

Figure 5.41 Single-stage 15-level inverter power circuit.

5.10 Modified Multilevel Converter

Only the CMLI control circuit is modified in the two aforementioned ways to reach 15 and 27 levels with three inverter stages. In order to achieve 15 levels with just 7 switches, this method modifies both the control circuit and, primarily, the power circuit. The circuit architecture for the MMC technique, where input scaling is not required, is shown in Figure 5.41. Both symmetrical and asymmetrical CMLI can use this topology. In addition to providing a conduction path from the source to the load, the addition of a diode and capacitor (1nF) aids in normalizing the output within the specified interval.

The switching process for obtaining the 15-level output is illustrated in Table 5.9. The input switches (S1, S2, and S3) are controlled to obtain the necessary output voltage by adding the voltages in steps; however, the inverter circuit (T1, T2, T3, and T4) is in the ON condition at all levels according to the sequence.

Let V1, V2, and V3 be the inputs for the PV array. Level 1 involves the inverter receiving only V1 as input while V2 and V3 are turned off. In a similar manner, the input voltages' ON/OFF states are controlled to reach the 15 levels. The seven-level output is the goal of the switching table provided; the remaining seven levels will be achieved by reversing the sequence. Table 5.10 compares the number of switches needed for each of the three suggested approaches.

5.11 Simulation Exercises

MATLAB/Simulink is used for simulations. Solar PV panels, which are scaled to the powers of two and three to increase the output levels, are regarded as the CMLI's input source. By connecting four 12 V modules in series, each panel is

Table 5.9 MMC switching sequence for a 15-level inverter.

Level	S1	S2	S3	T1	T2	T3	T4	Vo
0	0	0	0	0	0	0	0	0
1	1	0	0	1	0	0	1	48
2	0	1	0	1	0	0	1	96
3	1	1	0	1	0	0	1	144
4	0	0	1	1	0	0	1	192
5	1	0	1	1	0	0	1	240
6	0	1	1	1	0	0	1	288
7	1	1	1	1	0	0	1	336

Table 5.10 Comparison of the switches needed for each mode.

Number of stages	Number of switches	Number of levels			
		Conventional	Binary	Trinary	MMC
1	4	3	3	3	15 Achieved with only 7 switches
2	8	5	7	9	
3	12	7	15	27	
4	16	9	31	81	
5	20	11	63	243	
6	24	13	127	729	
7	28	15	255	2187	

rated at 48 V. Additional series connections between two or three PV panels result in inputs of 96 V and 192 V, respectively. The input voltages fall between 48, 96, and 196 volts. The simulation is carried out in "binary" and "trinary" modes for the circuit depicted in Figure 5.36. Digital switching is employed in "binary" mode, while embedded controllers are used in "trinary" mode in line with the truth table.

5.11.1 Conventional Approach

The seven-level output voltage waveform and associated THD analysis from the three-stage inverter for the power circuit shown in Figure 5.36 are displayed in Figures 5.42 and 5.43. The gating signals for the inverter switches in this simulation are produced using traditional PWM.

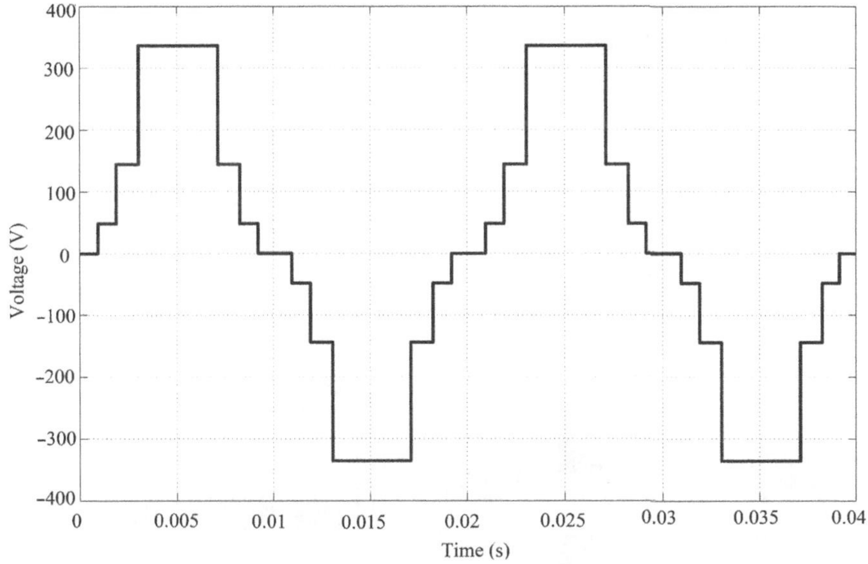

Figure 5.42 Three-stage, seven-level inverter's output voltage waveform.

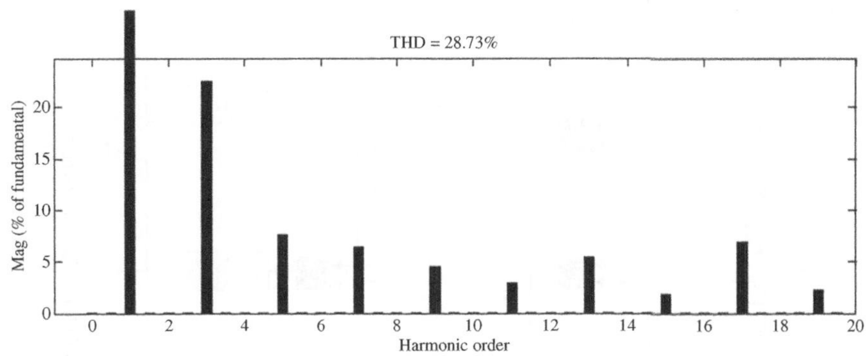

Figure 5.43 FFT analysis of a seven-stage, three-stage inverter.

5.11.2 Binary Mode

To reach the output voltage of 326 Vp in "binary" mode, the input to the three inverter stages must be scaled in powers of two as 48 V, 96 V, and 192 V. The suggested simulation block of the control circuit, which includes counters, a combination circuit, and a pulse separation circuit, is depicted in Figure 5.44. The five-bit asynchronous counter receives the clock signal as input to move bits that are appropriate for both positive and negative half cycles based on the truth

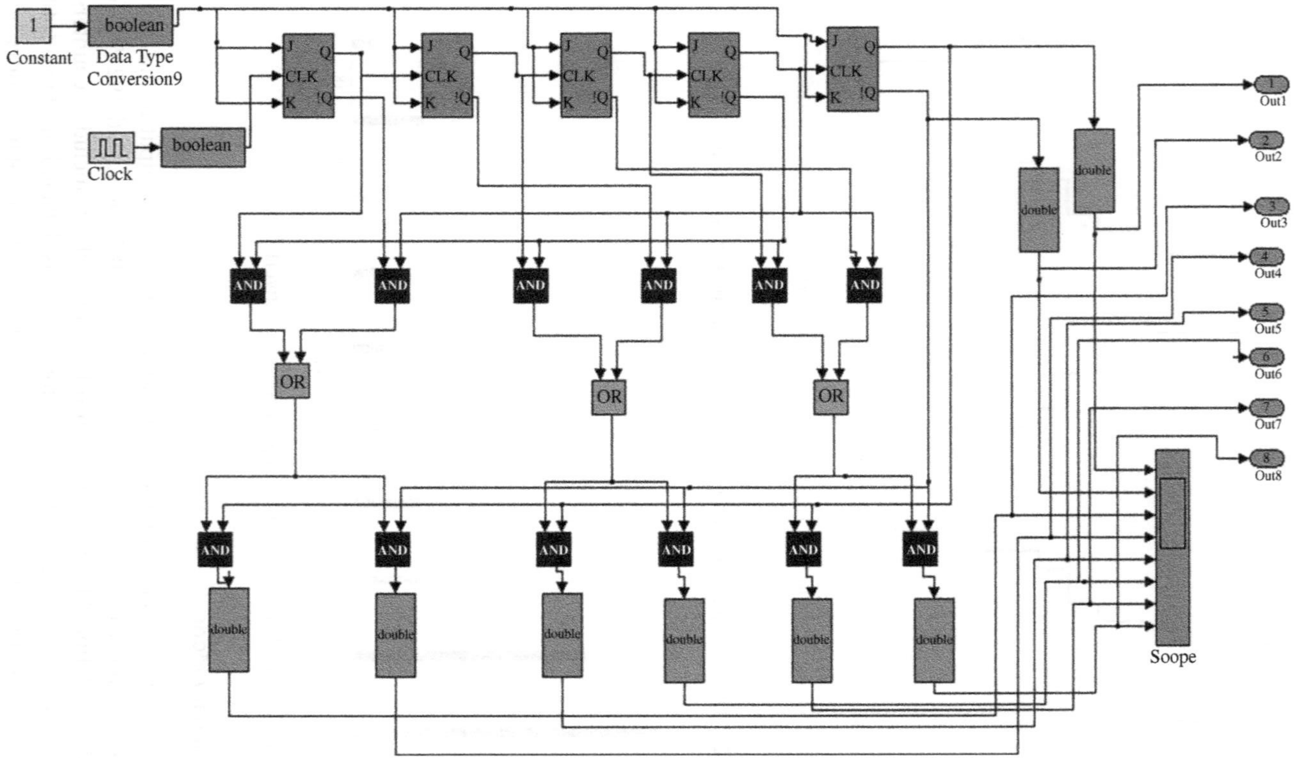

Figure 5.44 Three-stage, 15-level CMLI switching circuit.

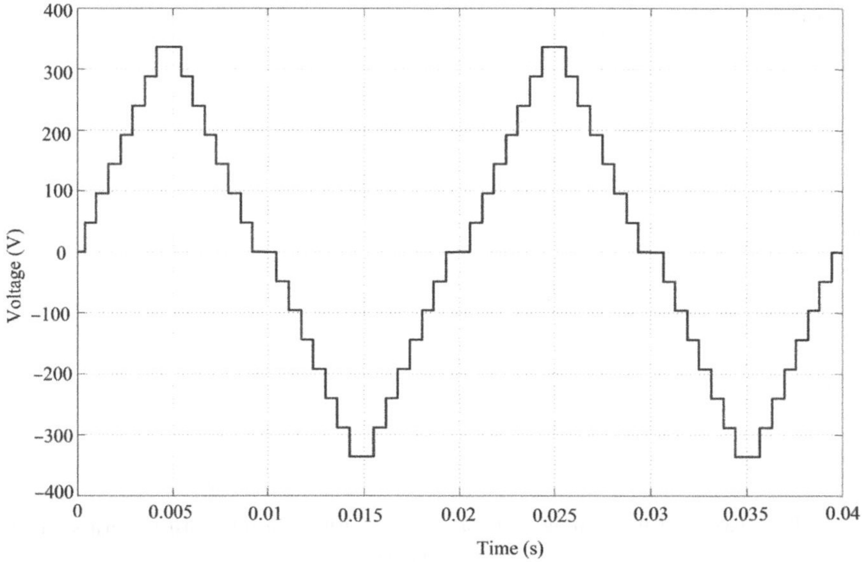

Figure 5.45 Waveform of 3-stage, 15-level output voltage.

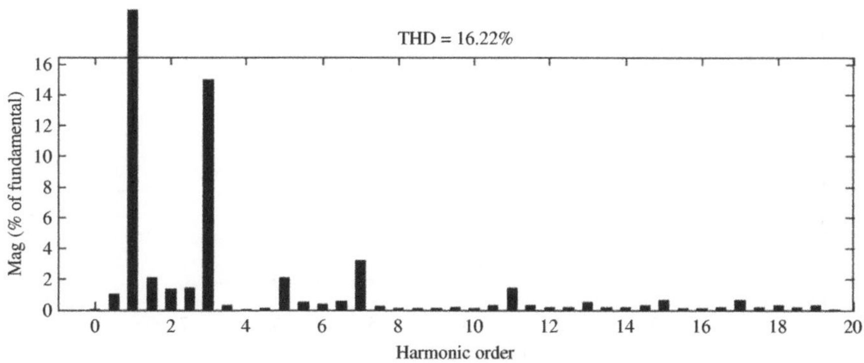

Figure 5.46 FFT analysis of a 15-level inverter with 3 stages.

table. The pulses are divided and sent to the inverter switches at the logic functions block. The 15-level output voltage waveform produced by a three-stage inverter is displayed in Figures 5.45 and 5.46 provides the appropriate FFT analysis.

The simulation is run in digital mode from a single-stage to a seven-stage CMLI. An extra counter, AND gates with multiples of four (four AND gates are required for a single stage, and eight AND gates are required for two stages), and OR gates are used in the suggested control circuit architecture to increase the number of levels. The 255-level output waveform produced by a seven-stage CMLI is displayed in Figure 5.47. Only 15 levels can be produced with the traditional

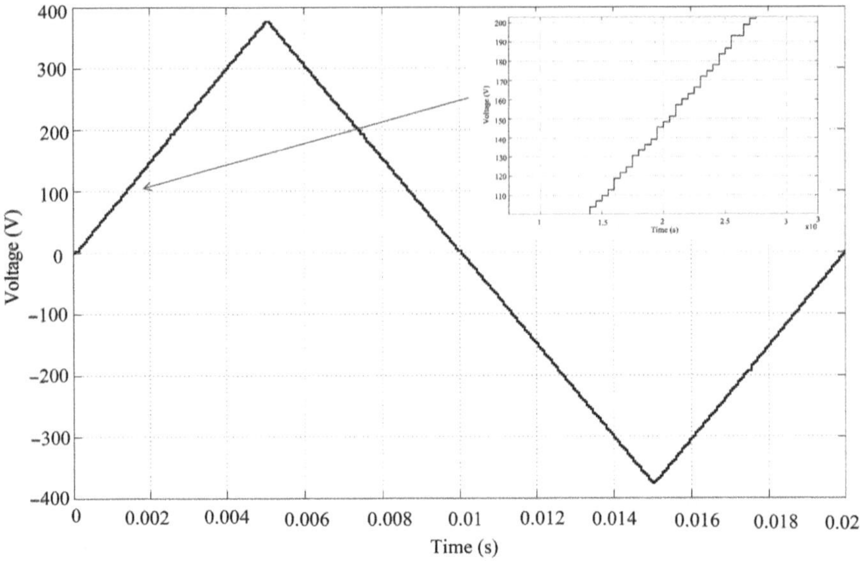

Figure 5.47 Seven-stage output waveform at 255 levels.

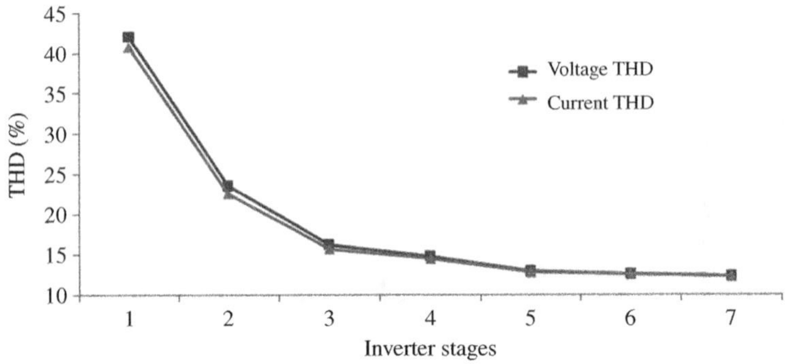

Figure 5.48 THD variation in binary mode.

seven-stage CMLI. A traditional inverter needs 508 switching devices and 127 inverter stages to reach a 255-level output. The fluctuation of THD measured for different levels with binary mode topology is displayed in Figure 5.48.

5.11.3 Trinary Mode

To reach the output voltage of 312 Vp in "trinary" mode, the inputs to the three inverter stages must be scaled in the power of three as 24 V, 72 V, and 216 V.

Even more than 1000 levels can be reached in this architecture thanks to the usage of an embedded controller, which can further minimize complexity. The truth table is integrated into the controller, which uses a repeating sequence block to produce the appropriate sequences to turn semiconductor switches on and off.

Any number of layers can be reached here as well with the right lookup table formulation. The output voltage waveform of a 3-stage, twenty-seven-level inverter is displayed in Figures 5.49 and 5.50 provides the appropriate harmonic spectrum. Table 5.11 displays the THD variance for different phases as determined

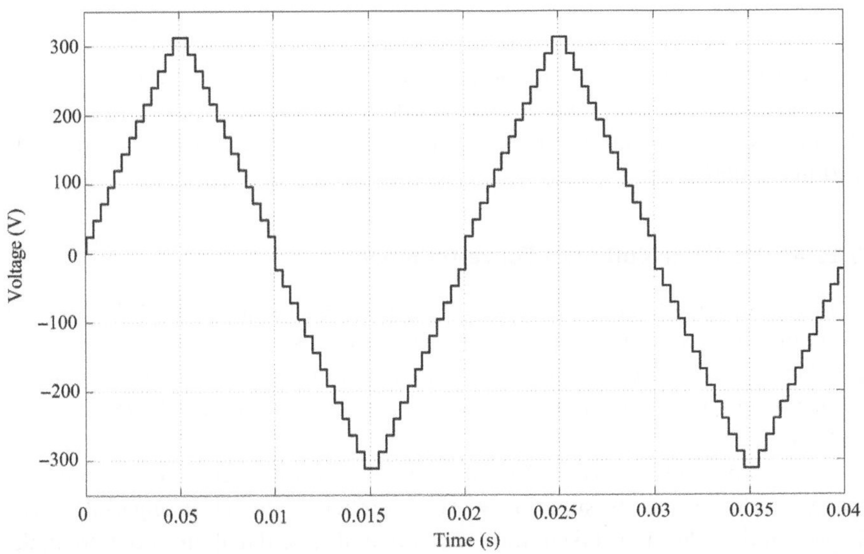

Figure 5.49 Three-stage 27-level output voltage inverter.

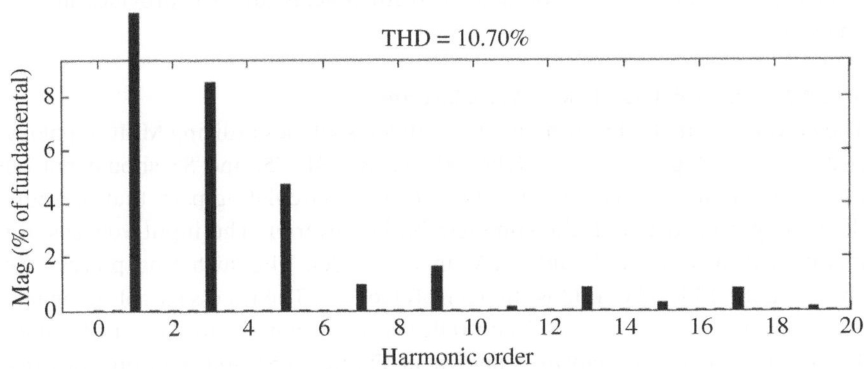

Figure 5.50 FFT analysis for a 3-stage 27-level inverter.

Table 5.11 Comparison of THD for the solar PV-fed CMLI.

Number of stages	Number of levels	V_{THD} (%)	I_{THD} (%)
3	27	10.70	7.71
4	81	11.37	8.66
5	243	11.77	9.00
6	729	11.76	11.96
7	2187	12.02	11.95

by the trinary mode. The values of harmonic distortions are found to summarily grow when the number of levels exceeds 100. Therefore, in order to demonstrate the necessary improvement in power quality, the increase in levels can be set within the limit.

5.11.4 Modified Multilevel Converter Mode

Both the power circuit and the control circuit are altered in the MMC in order to attain the necessary output levels. The gating signals in this method are provided by the embedded controller based on the switching sequence and the step-by-step addition of input voltages. The solar PV's input voltages are 48 V, 96 V, and 192 V.

Unlike binary or trinary modes, MMC mode cannot be compared to inverter stages to determine its superiority. The number of switches engaged in the 7-level and 15-level is taken into account and calculated in order to make the comparison.

The output voltage waveform for the circuit shown in Figure 5.40 is displayed in Figure 5.51, and Table 5.9 contains the truth table. Figure 5.52 provides an FFT analysis.

5.11.4.1 Ten Switches and a 15-Level Inverter

The power circuit of a 15-level inverter with ten switches utilizing MMC topology is depicted in Figure 5.53. Switches S1, S2, S3, S4, S5, and S6 should not be turned on simultaneously because this creates a circulating path that prevents the supply from reaching the complete bridge inverter. The input voltages are maintained at 48 V, 96 V, and 192 V, in that order. The switching process for obtaining the 15-level output is shown in Table 5.12. The necessary gating signals are produced by an embedded controller in accordance with the truth table. The output voltage waveform is shown in Figures 5.54 and 5.55 provides the related FFT analysis.

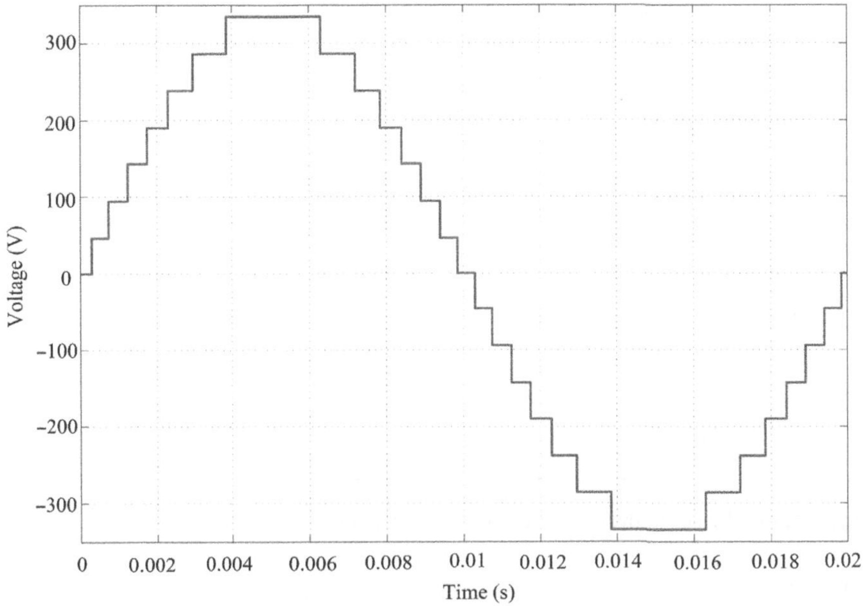

Figure 5.51 15-Level output waveform in MMC mode.

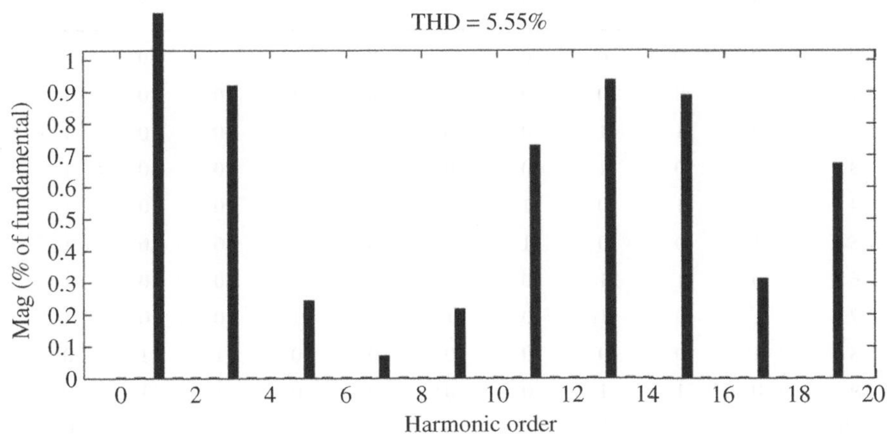

Figure 5.52 FFT analysis of an MMC-mode 15-level inverter.

5.11.4.2 Seven-Level Inverter with 10 Switches

For comparison, a seven-level inverter is used, which uses seven switches to produce the necessary levels, but a standard inverter needs twelve switches. The power circuit is depicted in Figure 5.56, and the switching sequence is provided

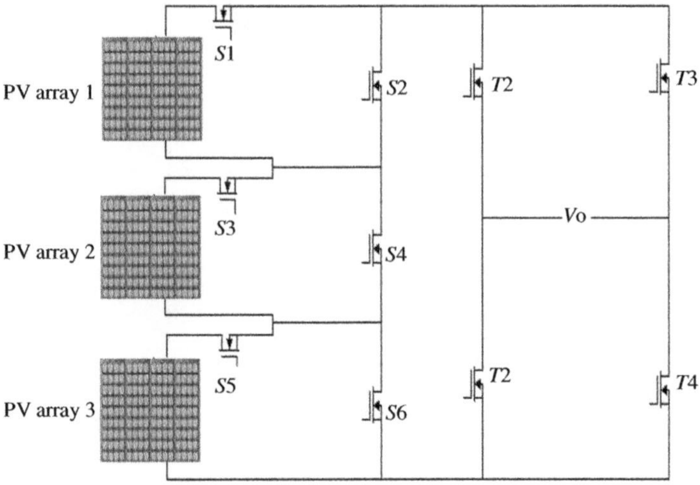

Figure 5.53 15-Level inverter circuit with 10 switches.

Table 5.12 Ten switches and a 15-level inverter's switching sequence.

Level	S1	S2	S3	S4	S5	S6	T1	T2	T3	T4
0	0	0	0	0	0	0	1	0	0	1
1	1	0	0	0	0	0	1	0	0	1
2	0	1	1	0	0	0	1	0	0	1
3	1	0	1	0	0	0	1	0	0	1
4	0	1	0	1	1	0	1	0	0	1
5	1	0	0	1	1	0	1	0	0	1
6	0	1	1	0	1	0	1	0	0	1
7	1	0	1	0	1	0	1	0	0	1
8	1	0	0	0	0	0	0	1	1	0
9	0	1	1	0	0	0	0	1	1	0
10	1	0	1	0	0	0	0	1	1	0
11	0	1	0	1	1	0	0	1	1	0
12	1	0	0	1	1	0	0	1	1	0
13	0	1	1	0	1	0	0	1	1	0
14	1	0	1	0	1	0	0	1	1	0

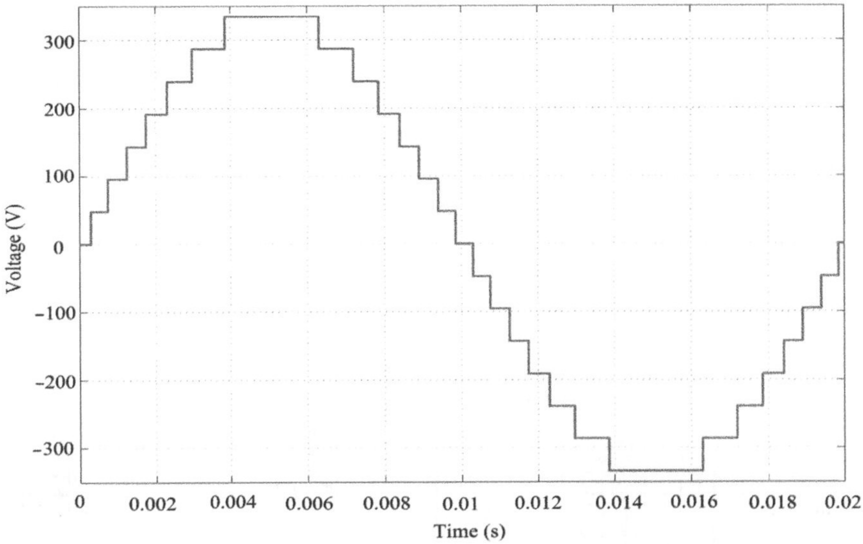

Figure 5.54 15 Waveforms of the level output from ten switches.

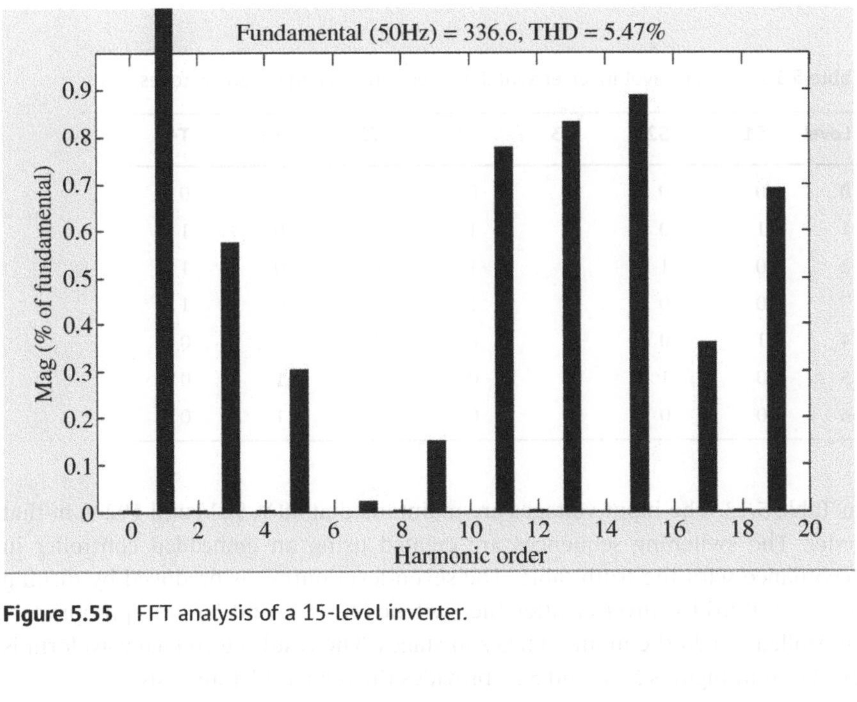

Figure 5.55 FFT analysis of a 15-level inverter.

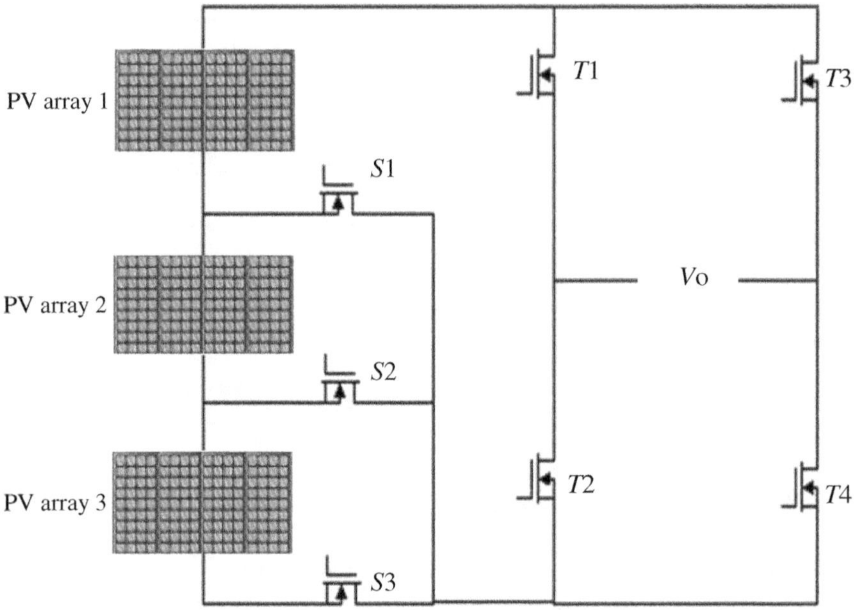

Figure 5.56 Seven switches and a seven-level inverter circuit.

Table 5.13 Seven-level inverter switching sequence using seven switches.

Level	S1	S2	S3	T1	T2	T3	T4
0	0	0	0	0	1	1	0
1	1	0	0	1	0	0	1
2	0	1	0	1	0	0	1
3	0	0	1	1	0	0	1
4	1	0	0	0	1	1	0
5	0	1	0	0	1	1	0
6	0	0	1	0	1	1	0

in Table 5.13. The input voltages are maintained at 48 V, 96 V, and 192 V, in that order. The switching sequences are created using an embedded controller in accordance with the truth table. The seven-level output is produced by turning on the full-bridge inverter after the switches S1, S2, and S3 are appropriately controlled to add the input voltages in stages. The resulting voltage waveform is displayed in Figures 5.57, and 5.58 provides the related FFT analysis.

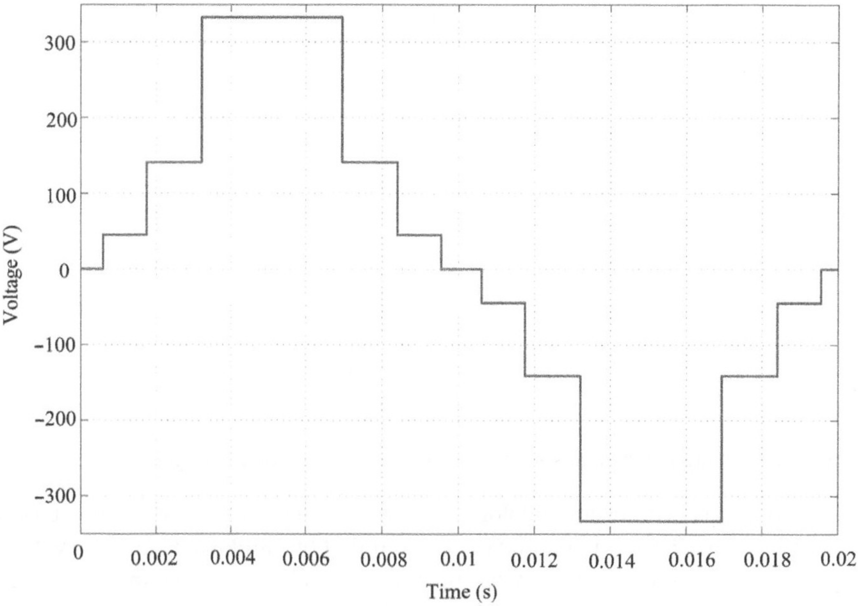

Figure 5.57 The output voltage waveform of a seven-level staircase.

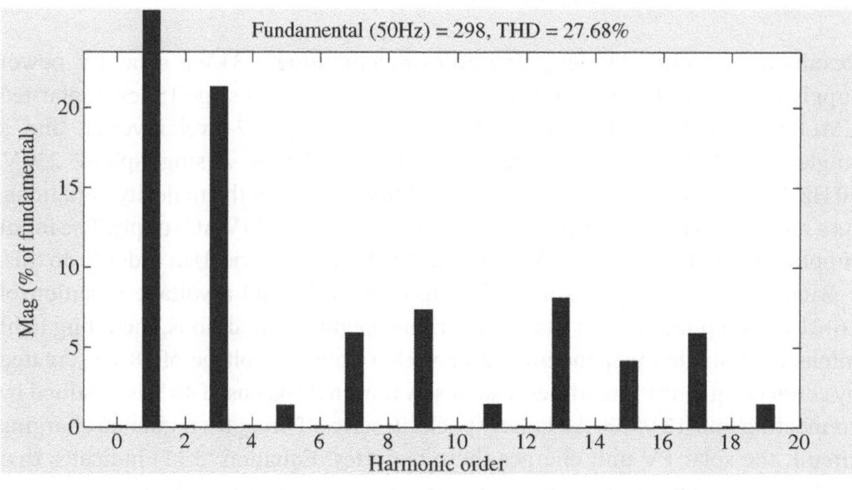

Figure 5.58 FFT analysis of output voltage waveform.

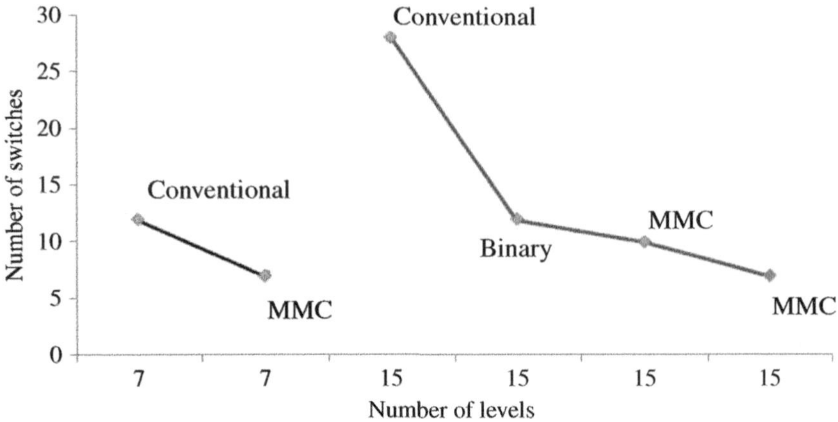

Figure 5.59 Graph of comparison for several switch reduction topologies.

The comparison of several topologies for solar PV-fed 7- and 15-level inverters is shown in Figure 5.59. It makes it very evident that by employing suitable switching sequences, the number of switching devices will be decreased, increasing the levels. According to the FFT study, the THD decreases as the number of levels increases.

5.12 Experimental Results

Results are acquired by designing and implementing a 3 kWp solar PV power supply unit for each of the three topologies separately. A 3-stage 15-level solar-fed CMLI with a digital logic control circuit, a 3-stage 27-level inverter, and a single-stage 15-level inverter are among the topologies. A single-phase 230 V, 50 Hz AC system powers a large portion of home loads in the majority of nations. As a result, the system's output voltage is intended for a 230 V AC supply. The input supplies are in the range of 48 V, 96 V, and 192 V (asymmetrical) in order to do this.

Each solar PV panel has a rated output of 115W and a voltage variation of 16–21 volts (nominal 12 V) based on various operating conditions, including light intensity. A single independent source with a nominal voltage of 48 V is created by connecting four PV modules in series. A nominal DC bus of 48 V is obtained by connecting four 12 V, 100 Ah battery packs in series. Through a regulated charging circuit, the solar PV unit charges these batteries. Equation (5.11) indicates that the hardware implementation uses seven sets of separate solar PV supply sources.

$$48 + (48 \times 2) + ((48 \times 2) \times 2) = 336V_p \tag{5.11}$$

Twelve MOSFET-IRF 840s make up the power circuit switches. Counter CD4520, IC 555, a combinational circuit with logic gates, and a buffer IC make

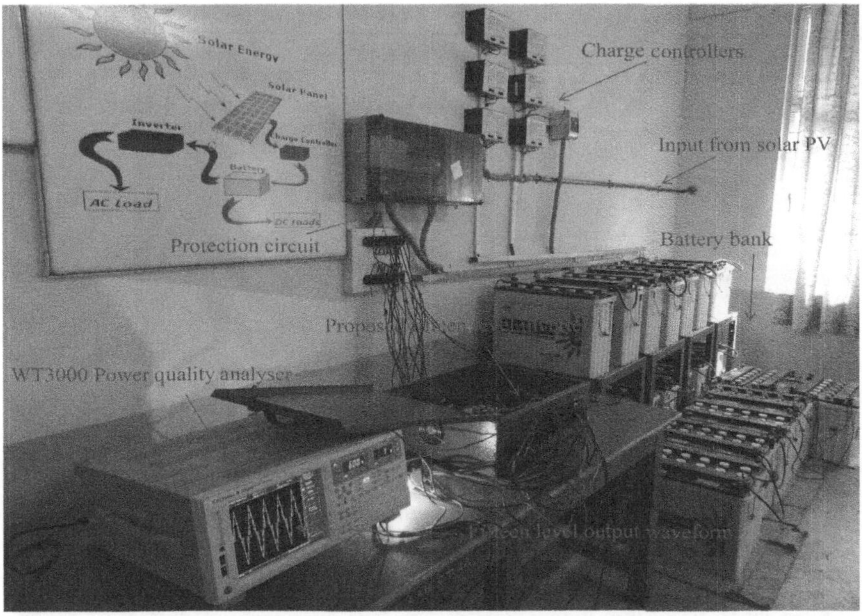

Figure 5.60 Solar-fed 15-level inverter experimental setup.

up the control circuit, which is an essential component of the system. The logic is programmed on an ATMEGA 16AVR single-chip microcontroller to further decrease complexity. The switching sequence specified in the truth table is used to generate the switching signals. A dead period of 1μs is inserted to ensure that the bits flow forward and backward in the proper directions without any mismatch.

Figure 5.60 depicts the entire experimental setup, including the battery bank, charge controller, suggested inverter circuit, and protective switchgear components. The connection link from the 3 kWp solar plant is indicated by the input from the solar PV. It is then linked to the charge controller, which uses three modes – bulk, absorption, and float stages – to monitor the battery's charging and discharging.

To prevent excess current from entering the inverter circuits, the protection circuit is made up of switchgear parts such as contactors and fuses. In order to achieve binary and trinary modes of operation, this block also incorporates the scaling of input voltages in powers of two and three. Fifteen and twenty-seven levels can be achieved from the same circuit while maintaining the same other component blocks by loading the binary and trinary switching sequence.

The integration of the suggested inverters with the solar PV system is depicted in Figure 5.60. The circuit connections in binary and trinary mode with 12 switches are shown in Figure 5.61. The three-stage inverter circuit's inputs are thought to

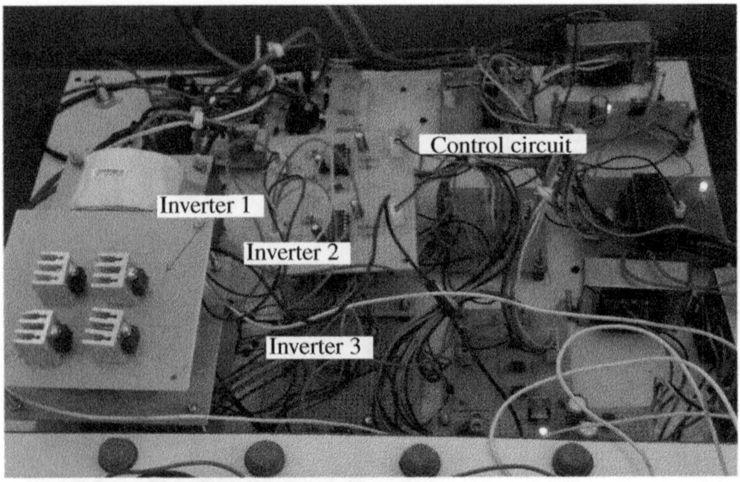

Figure 5.61 Three-stage inverter circuit experiment setup.

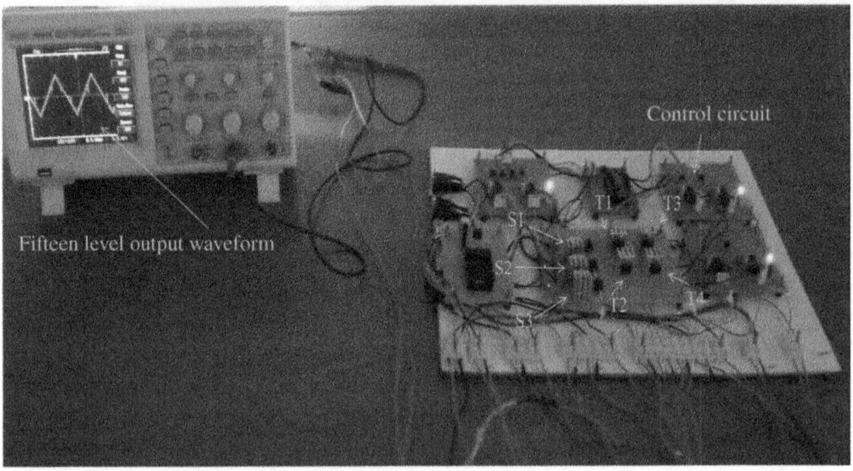

Figure 5.62 A 15-level inverter circuit with seven switches was set up experimentally.

be 48 V, 96 V, and 192 V (binary) and 24 V, 72 V, and 216 V (trinary). Because each battery has a 12 V, 100 Ah rating, connecting them in series causes the necessary input voltage scaling. The circuit for a 15-level inverter with seven switches and inputs of 48 V, 96 V, and 192 V, respectively, is shown in Figure 5.62.

The output voltage waveforms from binary and trinary modes are displayed in Figures 5.63 and 5.64, with 15 and 27 levels, respectively, according to the FFT analysis seen with the PQA WT3000. The binary and trinary modes can

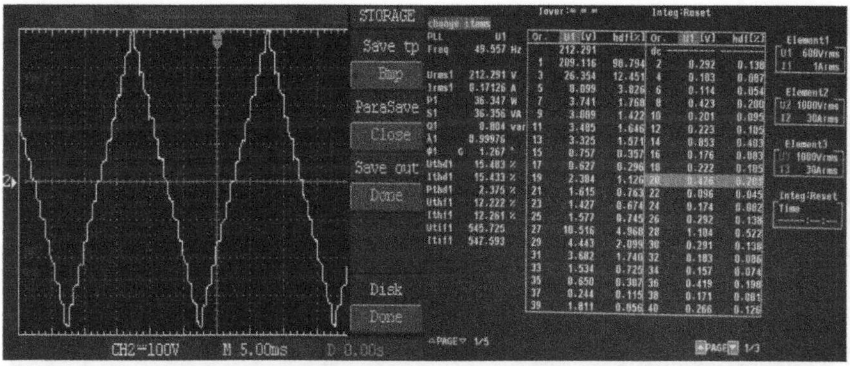

Figure 5.63 Waveform and FFT analysis at 3 stages and 15 levels.

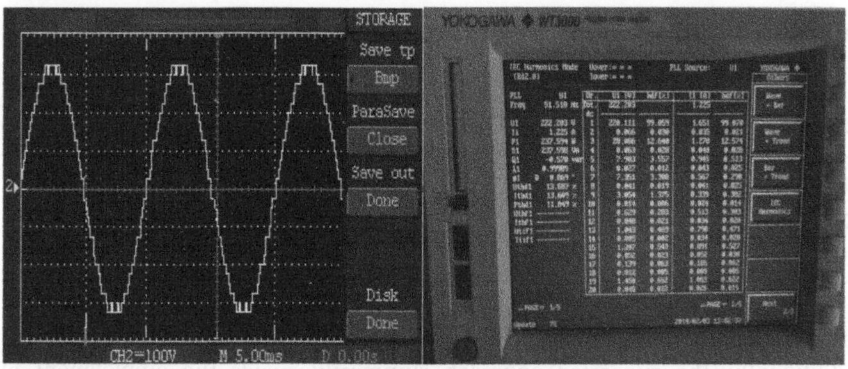

Figure 5.64 FFT analysis and a 3-stage, 27-level waveform.

be obtained by utilizing the identical circuit depicted in Figure 4.26 and merely altering the switching logic functions. The output voltage waveform and FFT analysis for a single-stage, 15-level inverter with just seven switches are displayed in Figure 5.65.

Table 5.14 compares the outcomes of the three MLI topologies. According to the findings, the single-stage 15-level inverter meets IEEE standard 519-1992 and offers a lower THD value. The harmonic distortion limits at the point of common coupling (PCC) or metering point with the utility are listed in this standard. For individual harmonics, the voltage distortion limitations are 3%, and for THD, they are 5%.

The number of devices taken into account in reaching the highest levels was the basis for the analysis. Numerous studies using low-power circuits have been documented, but none have been suggested for solar PV applications with the appropriate implementation. The studies mostly focus on reducing switches

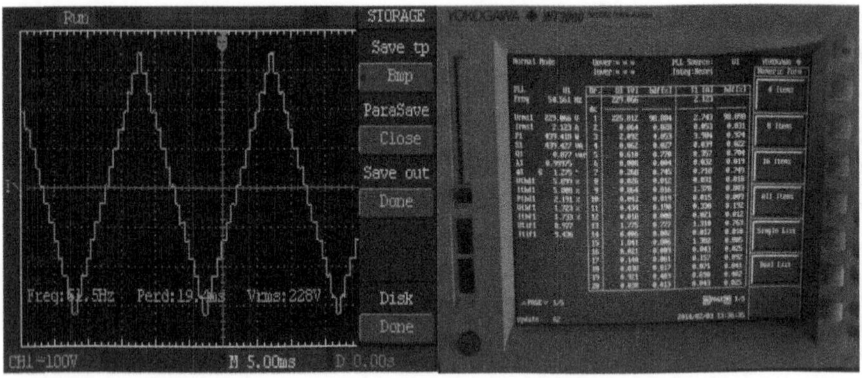

Figure 5.65 Single-stage 15-level waveform and FFT analysis.

Table 5.14 Comparison of results between the three topologies.

	Topology		Simulation		Experimental	
S. no.	**Stages**	**Level**	V_{THD} **(%)**	I_{THD} **(%)**	V_{THD} **(%)**	I_{THD} **(%)**
1.	3	15	16.22	15.76	15.483	15.433
2.	3	27	10.7	7.71	13.687	13.607
3.	1	15	5.55	3.5	5.899	5.809

rather than improving power quality by taking harmonic distortions into account. Additionally, the suggested approach summarily avoids the employment of output voltage filters, transformers, and comprehensive look-up tables, all of which contribute to the system's overall complexity.

Summary

The following has been illustrated in this chapter: for the decrease of THD in a solar-fed CMLI, multiple carrier PWM approaches are very appropriate. Variations are made in both carrier and reference signals in the multiple carriers, and the performance of solar-fed inverters with three to 15 levels is examined. According to the findings, the POD approach offers the least amount of THD when compared to its alternatives. Every approach taken into consideration for comparison deals with low-power systems and uses a DC power supply as its input source.

A 3 kWp solar PV system with a DSP-based controller for producing the switching signals to the MLI is the subject of an experimental inquiry.

In comparison to the other approaches, the harmonic profile is improved, and THD is decreased when employing the recommended POD method for solar PV applications.

The chapter also covered how to achieve the 15-level output with fewer switches. For the solar PV systems, three distinct plans are created and put into practice experimentally. Since the input sources must be scaled to the power of two or three, previous studies have documented the usage of input DC sources, which are implemented for low-power applications with peak voltages of 150 V, 15 V, 60 V, 10 V, etc. The system's advantages include its low cost, straightforward design, ability to achieve m levels with the addition of a counter, lack of a lookup table, and decreased computational complexity without the need for boost converters, output filters, and transformers, as well as fewer semiconductor switches and its gate drive circuits, which lower switching losses. The system is intended for a 230 Vac supply with a peak voltage of 324 V.

References

Gupta, R., Ghosh, A., and Avinash, J. (2010). Multiband hysteresis modulation and switching characterization for sliding-mode-controlled cascaded multilevel inverter. *IEEE Trans. Ind. Electron.* 57 (7): 2344–2353.

Malinowski, M., Gopakumar, K., Rodriguez, J., and Pérez, M.A. (2010). A survey on cascaded multilevel inverters. *IEEE Trans. Ind. Electron.* 57 (7): 2197–2206.

Zambra, D.A.B., Rech, C., and Pinheiro, J.R. (2010). Comparison of neutral-point-clamped, symmetrical and hybrid asymmetrical multilevel inverters. *IEEE Trans. Ind. Electron.* 57 (7): 2297–2306.

Zhao, J., He, X., and Zhao, R. (2014). A novel PWM control method for hybrid-clamped multilevel inverters. *IEEE Trans. Ind. Electron.* 57 (7): 2365–2373.

6

PWM Generation with the Aid of Intelligent Techniques

6.1 Introduction

Multilevel inverters (MLIs) have numerous structures and benefits. Unfortunately, the low THD of most current inverters prevents them from producing good output waveforms. This is due to the improper arrangement of each level's switching angle. The switching angles must be carefully arranged to get the lowest THD to obtain acceptable power quality. The integral multiples of the fundamental frequency are known as harmonics, and they emerge as the distortion of the voltage or current waveform.

Inverter output harmonics can cause microprocessor-based equipment to malfunction, neutral conductors, transformers, or induction motors to overheat, power factor correction capacitors to deteriorate or fail, breakers and relays to operate erratically, and strong magnetic fields to be present near transformers and switchgear. Several modulation techniques are provided to remove unwanted harmonics in MLI with equal DC sources. Nevertheless, the lower-order harmonic cannot be totally eliminated using PWM-based techniques. Additionally, a more effective way to degenerate the harmonics is needed for approaches related to uneven DC sources, like solar PV.

By selecting the proper switching angles, selective harmonic elimination (SHE) suppresses the output voltage of the inverter's particular order harmonics. Finding the arithmetic solution of nonlinear transcendental equations that naturally have numerous solutions and involve trigonometric components is a major problem with this approach.

Numerous iterative techniques can be used to solve the set of nonlinear equations. The dependency on an initial estimate and divergence concerns for a high number of inverter levels are the downsides of iterative approaches such as Newton–Raphson.

The resulting theory can only determine three switching angles for uneven DC voltage and up to six switching angles for equal DC voltage. Moreover, it is hard and time-consuming, which demands fresh expression when the voltage level or input DC voltage is altered. Step modulation determines the switching angles in real time, but this approach is not extended for unequal DC sources, which are most necessary for solar-supplied CMLI.

6.2 Intelligent Techniques

When the inequality of DC sources is applied, the transcendental equation set to be solved becomes asymmetrical, requiring the solution of a set of high-degree equations that are beyond the capability of current computer algebra software tools. The optimal harmonic stepped waveform (OHSW) strategy for a solar-fed 15-level inverter is combined with a variety of intelligent techniques, including artificial neural networks (ANN) and optimization methods (genetic algorithm (GA), particle swarm optimization (PSO), and bees optimization [BO]).

The ANN-based approach reduces or removes lower-order harmonics while preserving a stable fundamental by solving the harmonic equations and getting switching angles. Datasets with varying input voltages and switching angles are trained using an ANN. To decrease harmonic distortions, the solar PV system is coupled with the training network.

In a similar manner, the optimal switching angles needed for the inverter switches are obtained by formulating the objective function for the suggested approach and performing appropriate operations with regard to GA, PSO, and BO. The ideal switching angles in OHSW are found by taking into account the waveform's symmetry. To demonstrate the efficacy of the suggested system, an experimental setup for the method that produces the least THD is performed for a 3 kWp solar PV plant.

Equation (6.1) uses a Fourier series to explain the staircase output of MLI with nonequal input sources.

$$V(\omega t) = \sum_{n=1,3,5}^{\infty} \frac{4V_{PV}}{n\pi}[k_1 \cos(n\alpha_1) + k_2 \cos(n\alpha_2) + .. + k_m \cos(n\alpha_m)] \sin(n\omega t) \tag{6.1}$$

where $k_i V_{PV}$ is the ith DC voltage, V_{PV} is the nominal DC voltage from solar PV, n is the order of harmonics, m is the inverter output levels, and the switching angles α_1–α_m must match the demand indicated below.

$$\alpha_1 \le \alpha_2 \le \alpha_3 \le \ldots \le \alpha_m \le \frac{\pi}{2}c$$

The MLI can eliminate $(m-1)$ harmonics. Six harmonic orders can be reduced or removed for a seven-stage inverter. Equation (6.2) provides the formula controlling the 15-level inverter's SHE. Finding the switching angles α_1–α_7, as shown in Figure 6.1, is the goal here.

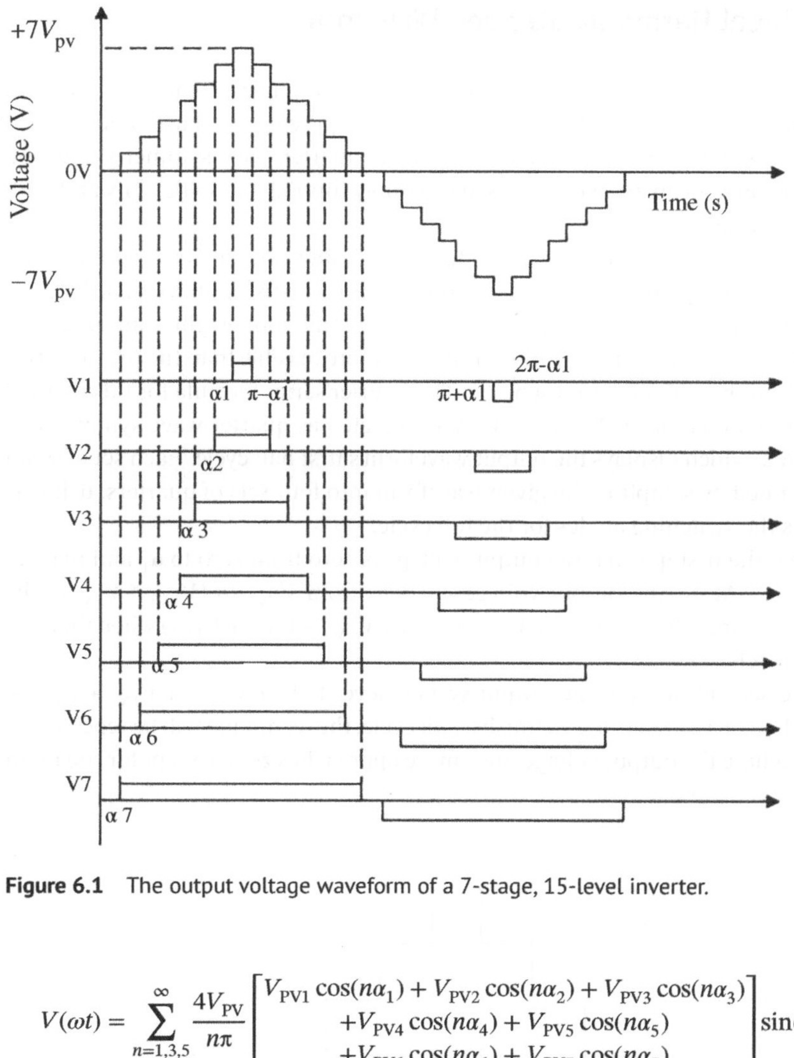

Figure 6.1 The output voltage waveform of a 7-stage, 15-level inverter.

$$V(\omega t) = \sum_{n=1,3,5}^{\infty} \frac{4V_{PV}}{n\pi} \begin{bmatrix} V_{PV1}\cos(n\alpha_1) + V_{PV2}\cos(n\alpha_2) + V_{PV3}\cos(n\alpha_3) \\ +V_{PV4}\cos(n\alpha_4) + V_{PV5}\cos(n\alpha_5) \\ +V_{PV6}\cos(n\alpha_6) + V_{PV7}\cos(n\alpha_7) \end{bmatrix} \sin(n\omega t) \tag{6.2}$$

Batteries in CMLI-based energy storage systems also have variable voltages due to their states of charge (SoC), while standard PV cells in CMLI-based PV systems have operating DC voltages between 15 and 35 V. Therefore, it is also vital to remember that the variable voltages that reach the inverter stages would result in uncontrolled fundamentals with increased magnitudes of low-order harmonics. Triple harmonics do not need to be eliminated for three-phase power system applications because the line-to-line voltage eliminates them right away.

6.3 Ideal Harmonic Stepped Waveform

The MLI can determine the appropriate switching angles to produce the output voltage waveform with a smaller THD by applying the harmonic stepped waveform technique, which makes use of the quarter-wave symmetry concept. Switching losses can also be decreased by simply flipping the switch ON and OFF once each cycle.

The low-order harmonics can be decreased using one of three optimization procedures. They are as follows: (i) evenly spaced steps optimize step heights; (ii) equally spaced steps optimize step spaces; and (iii) both heights and spaces are maximized. The third optimization strategy is more feasible than the previous two methods in the variable input scheme. The Fourier series formula for this method is given by Equation (6.2). The OSHW is depicted as quarter-wave symmetry in Figure 6.2, which displays the output waveform's first half cycle. Each set of α can be generated by sampling the given waveform into four sets of quarters, using α_1 to α_{28} as the switching angles for the full cycle.

During the first quarter, the output voltage is zero from zero to α_1 and reaches V_{PV1} at α_1. At α_2, the output voltage changes from V_{PV1} to $(V_{PV1} + V_{PV2})$. The voltage becomes $(V_{PV1} + V_{PV2} + V_{PV3} + V_{PV4} + V_{PV5} + V_{PV6} + V_{PV7})$ when this procedure reaches/2.

In the second quarter, the output voltage level falls to $(V_{PV1} + V_{PV2} + V_{PV3} + V_{PV4} + V_{PV5} + V_{PV6})$ at $-\alpha_7$. To finish a half cycle, the technique will be repeated up to $-\alpha_1$, where the output voltage once more approaches zero. Except for the input

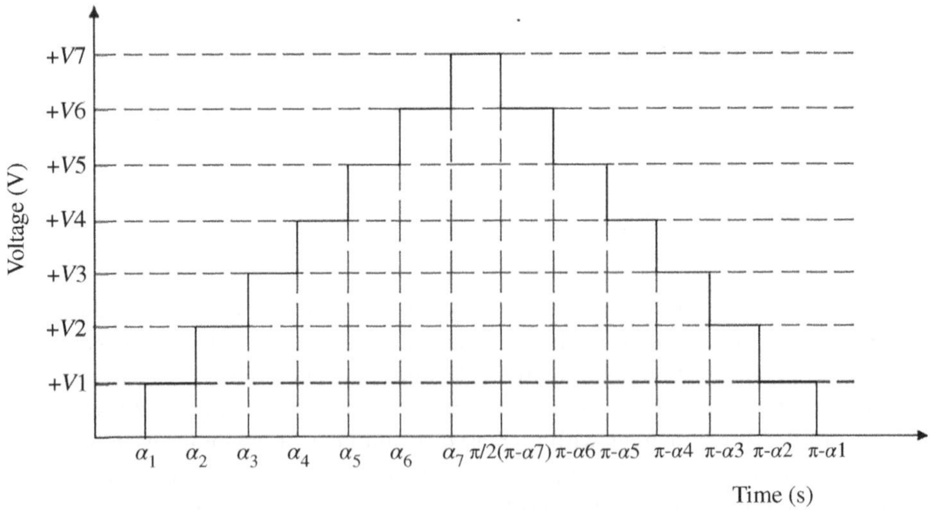

Figure 6.2 Quarter-wave symmetry waveform for 15-level inverters.

voltage's amplitude moving from positive to negative, the identical procedure will be repeated in the following half cycle.

First quarter: Consider the switching angles between α_1 and α_7 between 0 and 2.

The Second quarter:

$$\alpha_8 = \pi - \alpha_7$$
$$\ldots\ldots$$
$$\ldots\ldots$$
$$\alpha_{13} = \pi - \alpha_2$$
$$\alpha_{14} = \pi - \alpha_1$$

Third quarter:

$$\alpha_{15} = \pi + \alpha_1$$
$$\ldots\ldots$$
$$\ldots\ldots$$
$$\alpha_{20} = \pi + \alpha_6$$
$$\alpha_{21} = \pi + \alpha_7$$

Fourth quarter:

$$\alpha_{21} = 2 \cdot \pi - \alpha_7$$
$$\ldots\ldots$$
$$\ldots\ldots$$
$$\alpha_{27} = 2 \cdot \pi - \alpha_7$$
$$\alpha_{28} = 2 \cdot \pi + \alpha_1$$

The amplitude of all even-order harmonics and DC components approaches zero when considering the output waveform's harmonic contents. Therefore, it is important to estimate the odd harmonic in the quarter-wave symmetric CMLI. Equations (6.3)–(6.9) provide the set of nonlinear equations needed to determine the ideal switching angles.

$$V_{\mathrm{PV1}} \cos(\alpha_1) + V_{\mathrm{PV2}} \cos(\alpha_2) + \ldots\ldots\ldots + V_{\mathrm{PV7}} \cos(\alpha_7) = \frac{M\pi}{4} \tag{6.3}$$

$$V_{\mathrm{PV1}} \cos(3\alpha_1) + V_{\mathrm{PV2}} \cos(3\alpha_2) + \ldots\ldots\ldots + V_{\mathrm{PV7}} \cos(3\alpha_7) = 0 \tag{6.4}$$

$$V_{\mathrm{PV1}} \cos(5\alpha_1) + V_{\mathrm{PV2}} \cos(5\alpha_2) + \ldots\ldots\ldots + V_{\mathrm{PV7}} \cos(5\alpha_7) = 0 \tag{6.5}$$

$$V_{\mathrm{PV1}} \cos(7\alpha_1) + V_{\mathrm{PV2}} \cos(7\alpha_2) + \ldots\ldots\ldots + V_{\mathrm{PV7}} \cos(7\alpha_7) = 0 \tag{6.6}$$

$$V_{\mathrm{PV1}} \cos(7\alpha_1) + V_{\mathrm{PV2}} \cos(7\alpha_2) + \ldots\ldots\ldots + V_{\mathrm{PV7}} \cos(7\alpha_7) = 0 \tag{6.7}$$

$$V_{\mathrm{PV1}} \cos(11\alpha_1) + V_{\mathrm{PV2}} \cos(11\alpha_2) + \ldots\ldots\ldots + V_{\mathrm{PV7}} \cos(11\alpha_7) = 0 \tag{6.8}$$

$$V_{\mathrm{PV1}} \cos(13\alpha_1) + V_{\mathrm{PV2}} \cos(13\alpha_2) + \ldots\ldots\ldots + V_{\mathrm{PV7}} \cos(13\alpha_7) = 0 \tag{6.9}$$

where $M = \frac{h_1}{V_{\mathrm{PV(1-7)}}}$

The basic component's amplitude is denoted by h_1. Since the proposed technique employs variable input sources, the V_{PV} is taken at the left-hand side of the previously specified expressions. Equation (6.3) states that the modulation index (M) can be changed to adjust the basic component's amplitude. The extra nonlinear equations that can be eliminated and set to zero are the undesirable harmonic components in Equations (6.4)–(6.9).

The NR approach, which may be implemented in a computer program, uses a consecutive approximation procedure to solve nonlinear equation systems. Equations (6.10)–(6.20) are used in the procedure to get the ideal switching angles.

Step 1:
Equation (6.10), which requires the determination of the unknown the switching angle, is formed by switching angles α_1–α_7 matrix.

$$\alpha^j = \left[\alpha_1^j, \alpha_2^j, \alpha_3^j, \alpha_4^j, \alpha_5^j, \alpha_6^j, \alpha_7^j\right]^T \tag{6.10}$$

Step 2:
Equations (6.11) and (6.12) describe the creation of the nonlinear system matrix.

$$F^j = \begin{bmatrix} \cos\left(\alpha_1^j\right) + \cos\left(\alpha_2^j\right) + \ldots\ldots + \cos\left(\alpha_7^j\right) \\ \cos\left(3\alpha_1^j\right) + \cos\left(3\alpha_2^j\right) + \ldots\ldots + \cos\left(3\alpha_7^j\right) \\ \cos\left(5\alpha_1^j\right) + \cos\left(5\alpha_2^j\right) + \ldots\ldots + \cos\left(5\alpha_7^j\right) \\ \cos\left(7\alpha_1^j\right) + \cos\left(7\alpha_2^j\right) + \ldots\ldots + \cos\left(7\alpha_7^j\right) \\ \cos\left(9\alpha_1^j\right) + \cos\left(9\alpha_2^j\right) + \ldots\ldots + \cos\left(9\alpha_7^j\right) \\ \cos\left(11\alpha_1^j\right) + \cos\left(11\alpha_2^j\right) + \ldots\ldots + \cos\left(11\alpha_7^j\right) \\ \cos\left(13\alpha_1^j\right) + \cos\left(13\alpha_2^j\right) + \ldots\ldots + \cos\left(13\alpha_7^j\right) \end{bmatrix} \tag{6.11}$$

$$\left[\frac{\partial F}{\partial \alpha}\right]^j = \begin{bmatrix} -\sin\left(\alpha_1^j\right), -\sin\left(\alpha_2^j\right), \ldots\ldots, -\sin\left(\alpha_7^j\right) \\ -\sin\left(3\alpha_1^j\right), -\sin\left(3\alpha_2^j\right), \ldots\ldots, -\sin\left(3\alpha_7^j\right) \\ -\sin\left(5\alpha_1^j\right), -\sin\left(5\alpha_2^j\right), \ldots\ldots, -\sin\left(5\alpha_7^j\right) \\ -\sin\left(7\alpha_1^j\right), -\sin\left(7\alpha_2^j\right), \ldots\ldots, -\sin\left(7\alpha_7^j\right) \\ -\sin\left(9\alpha_1^j\right), -\sin\left(9\alpha_2^j\right), \ldots\ldots, -\sin\left(9\alpha_7^j\right) \\ -\sin\left(11\alpha_1^j\right), -\sin\left(11\alpha_2^j\right), \ldots\ldots, -\sin\left(11\alpha_7^j\right) \\ -\sin\left(13\alpha_1^j\right), -\sin\left(13\alpha_2^j\right), \ldots\ldots, -\sin\left(13\alpha_7^j\right) \end{bmatrix}$$
$$\times (f_s(\infty_1, \infty_2, \ldots, \infty_s) = k_s \tag{6.12}$$

Step 3:
Use equation (6.13) to create a harmonic amplitude matrix.

$$T = \left[\frac{M \times 3 \times \pi}{4} 0 \cdot 0 \cdot 0 \cdot 0 \cdot 0 \cdot 0\right]^T \tag{6.13}$$

Equations (6.3) and (6.13) can be rewritten as per Equation (6.14).

$$F(\alpha) = T \tag{6.14}$$

The following is the statement of algorithms for incorporating it into the program using MATLAB software, using the matrices provided in Equations (6.10)–(6.14) and the NR approach.

Step 1:
Estimate the starting values of α using Equation (6.15). The method for determining the initial estimate using the equal area requirement is provided by

$$\alpha^0 = \left[\alpha_1^0,\ \alpha_2^0,\ \alpha_3^0,\ \alpha_4^0,\ \alpha_5^0,\ \alpha_6^0,\ \alpha_7^0\right] \tag{6.15}$$

Step 2:
This step includes Equations (6.16)–(6.20).

$$F(\alpha_0) = F^0 \tag{6.16}$$

Linearizing Equation (6.14) at $\alpha_{0,}$ we get,

$$F^0 + \left[\frac{\partial f}{\partial \alpha}\right]^0 d\alpha^0 = T \tag{6.17}$$

and

$$d\alpha^0 = \left[d\alpha_1^0,\ d\alpha_2^0,\ d\alpha_3^0,\ d\alpha_4^0,\ d\alpha_5^0,\ d\alpha_6^0,\ d\alpha_7^0\right] \tag{6.18}$$

utilizing the inverse of the matrix provided in Equation (6.19) to solve Equation (6.18).

$$d\alpha^0 = \text{Invmatrix}\left[\frac{\partial f}{\partial x}\right]^0 [T - F^0] \tag{6.19}$$

Step 3:
Initial values are updated in accordance with Equation (6.20).

$$\alpha^{j+1} = \alpha^j + d\alpha^j \tag{6.20}$$

The steps (2) and (3) were carried out till is fulfilled to the degree of accuracy, and the condition to be satisfied is $\alpha_1, \alpha_2, \alpha_3, \ldots, \alpha_7 < \pi/2$.

6.4 Artificial Neural Network

To replicate the organization of the human brain, an ANN is composed of fundamental information processing units known as "neurons" that operate in parallel.

By changing the connecting "weights" between the pieces, they are trained to carry out a specific task. The benefit of the ANN technique is that it can automatically adjust the application without requiring an explicit function for control. To make neural networks "experienced" enough to properly categorize unknown patterns, they are trained using sample classification data and weight vector modifications. Depending on the availability of training data – a collection of pairs of input and output vectors – learning can be either supervised or unsupervised. When the net is given an input pattern in the presence of such data, the ANN can calculate the deviation from the intended output values.

The following three entities can be used to summarize the ANN's features.

- **Architecture:** The way neurons are connected, which can be recurrent or feed-forward networks.
- The activation function is a function of the processing unit's net input that converts the net input value to the output signal value.
- Methods of acquiring knowledge. How to choose the right weights for the connections.

Neural networks can be trained both online and offline. Every training pattern – that is, every forward and backward pass – is followed by weight adjustments for the network in online learning. In offline learning, all patterns in the training file have their weight changes collected, and after one complete cycle (epoch) through the training pattern file, the total of all the changes is applied. The first stages before using the ANN-based switching angle determination technique for CMLI are the selection of input–output parameters and data collection.

6.4.1 Dataset Collection

Both online and offline training are possible for neural networks. In online learning, weight adjustments for the network follow each training pattern, that is, each forward and backward pass. In offline learning, the weight changes of every pattern in the training file are gathered, and the total of all the changes is applied following one full cycle (epoch) through the training pattern file. The selection of input-output parameters and data collection are the initial steps prior to applying the ANN-based switching angle determination technique for CMLI.

For a 15-level inverter, the harmonic elimination equations are resolved. The equations in MathCAD can be solved using the following guess values: $\alpha_1 = 11.25°$, $\alpha_2 = 22.5°$, $\alpha_3 = 33.75°$, $\alpha_4 = 45°$, $\alpha_5 = 56.25°$, $\alpha_6 = 67.5°$, and

Table 6.1 ANN training requires a sample dataset.

Input voltage ($V_{PV(1-7)}$)	Switching angles (α_1–α_7 in radians)
[40.4, 52.6, 52.7, 41.6, 53.7, 42.8, 42.5]	[0.112, 0.198, 0.432, 0.587, 0.812, 1.104, 1.53]
[42.4, 52.6, 52.7, 43.6, 53.7, 45.8, 42.5]	[0.117, 0.196, 0.427, 0.584, 0.802, 1.093, 1.526]
[44.4, 52.6, 52.7, 45.6, 53.7, 40.8, 44.5]	[0.136, 0.19, 0.444, 0.594, 0.834, 1.117, 1.533]
[44.4, 40.6, 41.7, 42.6, 52.7, 40.8, 44.5]	[0.111, 0.211, 0.413, 0.559, 0.798, 1.1, 1.535]
[51.4, 43.6, 49.9, 50.8, 42.8, 51.5, 48.5]	[0.162, 0.166, 0.417, 0.598, 0.782, 1.062, 1.523]

$\alpha_7 = 78.75°$. The lower-order harmonic contents, such as the third, fifth, seventh, ninth, eleventh, and thirteenth, are eliminated from the equations by giving the fundamental voltage the desired value and setting the other harmonic components to zero. This keeps the desired fundamental constant. A few of the gathered datasets are listed in Table 6.1. About 1500 datasets are collected in order to obtain the precise set of switching angle (α in radians) values for the appropriate input voltages (V_{PV}) from the solar panels.

This problem may be considered as a combination rather than a permutation in order to reduce the quantity of data collection. Otherwise, 105 different combinations would be produced by a seven-stage converter with ten points evenly spread between 48 and 58 V. Thus, the dataset can be simplified by treating the voltage vector v_1 = [40.4, 52.6, 52.7, 41.6, 53.7, 42.8, 42.5] as the same input as v_2 = [52.6, 40.4, 42.5, 52.7, 42.8, 53.7, 41.6].

6.4.2 Architecture of ANNs

The ANN design under examination is a multilayered feedforward network, as shown in Figure 6.3. A "purelin" activation function with seven neurons in the output layer and a "tansig" activation function with twenty neurons in the first hidden layer and ten neurons in the second are the two hidden layers. The Levenberg–Marquardt backpropagation (LMBPN) technique is used to train the ANN using the input PV voltage and switching angle datasets obtained from the MathCAD software. The ANN's parameters are displayed in Table 6.2.

During the training process, the Levenberg–Marquardt (LM) algorithm alternates between the Gauss–Newton algorithm and steepest descent. LM is the fastest backpropagation (BPN) technique and is highly recommended for supervised learning applications when training moderately sized feedforward neural networks. Three steps make up the traditional BPN algorithm: feedforward of input data, error BPN, and weight updating. In addition to these steps, the LM

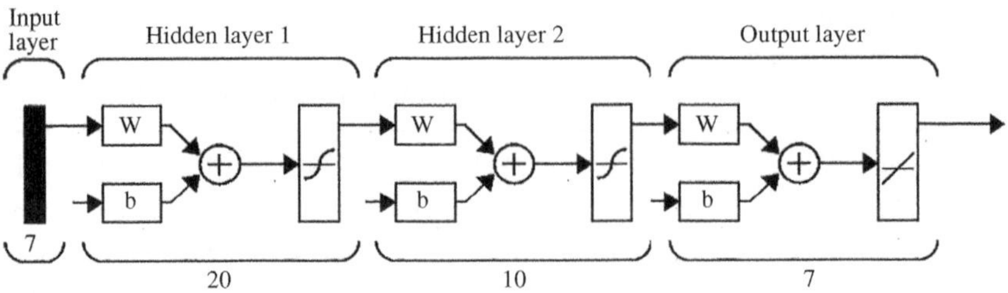

Figure 6.3 ANN architecture.

Table 6.2 ANN requirements.

Parameters	Values
Input layer	7 (input PV voltages)
Hidden layer 1	20
Hidden layer 2	10
Output layer	7 (switching angles)
Activation function (input–hidden 1)	tansig
Activation function (hidden 1–hidden 2)	tansig
Activation function (hidden 2–output)	purelin
Training algorithm	Levenberg–Marquardt
Performance parameter	Mean square error

technique was developed to approach second-order training speed without computing the Hessian matrix. When compared to other algorithms, the LM method has the benefit of faster, steadier convergence and lower memory requirements.

The trained network is then validated for its performance, and a Simulink model of the trained network is generated. For the created model, when provided with input voltage values, the network creates the required switching angles. The triggering circuit converts the switching angles produced by the neural network into the appropriate gating pulses, which are subsequently sent to the gate terminal of the MOSFET switches in the cascaded H-Bridge inverter.

The trained network is then validated for its performance, and a Simulink model of the trained network is generated. For the created model, when provided with input voltage values, the network creates the required switching angles. The triggering circuit converts the switching angles produced by the neural network into the appropriate gating pulses, which are subsequently sent to the gate terminal of the MOSFET switches in the cascaded H-Bridge inverter. The MLI switches' gates receive the generated pulses. The corresponding switching angles are produced based on changes in the inverter's input. It improves the quality of electricity in solar PV by producing the lowest THD and eliminating selective order harmonics.

6.5 Optimization Techniques

The process of achieving the best result possible in any situation is called optimization. In the fields of science and technology, it is frequently used to optimize the desired outcome and reduce the amount of work required. The formulation of an optimization problem begins with identifying the underlying design variables, which are primarily used during the optimization process. The next step is to determine the optimization problem's constraints, which show a functional link between the design variables and other design factors that fulfill physical phenomena and resource limitations.

Constraints fall under two categories: equality and inequality. Equality restrictions are generally more difficult to handle and need to be avoided. Finding the objective function in terms of the design variables and other aspects of the problem is the third step in the formulation approach. Maximization and minimization are the two categories of objective functions. By simply multiplying the objective function by $\neq 1$, the algorithm developed for minimization problems may be applied to maximizing problems based on the duality principle, and vice versa. The final step is to determine the lowest and maximum bounds for each design variable; however, certain optimization methods do not require this information.

6.5.1 Formulation of Problems

Finding the optimal switching angles (α_1, α_2, α_3, α_4, α_5, α_6, and α_7) to lower the THD for a solar-powered 15-level inverter is the goal. Fourier analysis yields the Equation (6.21) for the RMS value of the fundamental and harmonic components of the phase voltage from Figure 6.2.

$$V_n = \left\{ \begin{array}{c} \frac{2\sqrt{2}}{n\pi} V_{PV(1-7)}((\cos n\alpha_1) + (\cos n\alpha_2) + (\cos n\alpha_3) \\ + \ldots + (\cos n\alpha_7)), \text{odd } 0, \text{even} \end{array} \right\} \tag{6.21}$$

Equations (6.22) and (6.23) provide the greatest value of the fundamental component when taking into account the quarter-wave symmetry waveform by making all switching angles zero.

(i.e. $\alpha 1$ to $\alpha 7 = 0$).

$$V_{1\ \max} = \frac{2\sqrt{2}}{\pi} \cdot (V_{PV(1-7)} \times 7) \tag{6.22}$$

$$V_{1\ \max} = \frac{14\sqrt{2}}{\pi} V_{PV(1-7)} \tag{6.23}$$

Equation (6.24) explains the RMS value of the phase voltage, whereas Equation (6.25) shows how the fundamental component is normalized according to its maximum value. Equation (6.26) provides the formula for the produced VRMS.

$$V_1(\text{p.u.}) = \frac{V_1}{V_{1(\max)}} = \frac{2\sqrt{2}V_{PV(1-7)}}{n\pi} \times \frac{\pi}{14\sqrt{2}V_{PV(1-7)}}$$

$$V_1(\text{p.u.}) = \frac{V_1}{V_{1(\max)}} = \frac{1}{7}\{(\cos\alpha_1) + (\cos\alpha_2) + (\cos\alpha_3) + \ldots + (\cos\alpha_7)\} \tag{6.24}$$

$$V_{RMS} = \sqrt{\frac{2}{\pi}\int_0^{\pi/2} V^2 d\omega t} \tag{6.25}$$

$$V_{RMS} = V_{PV(1-7)}\sqrt{\begin{array}{l}\frac{2}{\pi}(\alpha_2 - \alpha_1) + 4(\alpha_3 - \alpha_2) + 9(\alpha_4 - \alpha_3) + 16(\alpha_5 - \alpha_4)\\ +25(\alpha_6 - \alpha_5) + 36\left(\frac{\pi}{2} - \alpha_7\right)\end{array}} \tag{6.26}$$

In per instance, by dividing Formula (6.26) by (6.23), the resulting average voltage RMS is given in Equation (6.27).

$$V_{rms}^{p.u.} = \frac{V_{RMS}}{V_{1(\max)}}$$

$$V_{rms}^{p.u.} = V_{PV(1-7)}\sqrt{\frac{\pi}{196}\begin{bmatrix}(\alpha_2 - \alpha_1) + 4(\alpha_3 - \alpha_2) + 9(\alpha_4 - \alpha_3) +\\ 16(\alpha_5 - \alpha_4) + 25(\alpha_6 - \alpha_5) + 36\left(\frac{\pi}{2} - \alpha_7\right)\end{bmatrix}} \tag{6.27}$$

THD can be defined as the ratio of all fundamental components of the RMS value given in Expression (6.24) to the fundamental element of RMS values given in Equation (6.27). Equation (6.28) expresses it.

$$\text{THD} = \frac{\sqrt{\sum_{n=2}^{\infty} V_n^2}}{V_1}\text{THD}$$

$$= \sqrt{\frac{\pi}{4}\frac{\begin{bmatrix}(\alpha_2 - \alpha_1) + 4(\alpha_3 - \alpha_2) + 9(\alpha_4 - \alpha_3) + 16(\alpha_5 - \alpha_4)\\ + 25(\alpha_6 - \alpha_5) + 36\left(\frac{\pi}{2} - \alpha_7\right)\end{bmatrix}}{\begin{array}{c}(\cos\alpha_1) + (\cos\alpha_2) + (\cos\alpha_3) + (\cos\alpha_4) +\\ (\cos\alpha_5) + (\cos\alpha_6) + (\cos\alpha_7)\end{array}} - 1} \tag{6.28}$$

Equations (6.29) and (6.30) provide the objective function and constraints for a 15-level inverter, respectively.

$$\text{Min Z} = \{\text{W(k)} \times (\text{per unit equivalent} - \text{Equation (6.24)}) + \text{Equation (6.28)}\} \quad (6.29)$$

$$\text{Constraints } 0 \leq \alpha_1 \ldots \ldots \alpha_7 \leq \frac{\pi}{2} \quad (6.30)$$

The symbols α_1, α_2, α_3, α_4, α_5, α_6, and α_7 stand for the switching angles. To make the mistake suitably small and per unit equivalent to the output peak voltage, which has a value between 0 and 1, the weighting factor, W(k), is used. The word (per unit equivalent ÷ Equation (6.24)) in the objective function represents the absolute value of the error in changing the basic component.

6.6 Genetic Algorithm

The symbols α_1, α_2, α_3, α_4, α_5, α_6, and α_7 stand for the switching angles. To make the mistake suitably small and per-unit equivalent of the output peak voltage, which has a value between 0 and 1, the weighting factor, W(k), is used. The word (per unit equivalent ÷ Equation (6.24)) in the objective function represents the absolute value of the error in changing the basic component. The GA, a derivative-free stochastic optimization method, can be used to solve both constrained and unconstrained optimization problems based on natural selection, the process that drives biological evolution. It addresses both maximizing and minimizing problems. The GA repeatedly modifies a population of unique solutions. In order to create the progeny for the next generation, GA randomly chooses members of the existing population to be parents. Natural selection, the mechanism that propels biological evolution, is the basis for both constrained and unconstrained optimization problems that can be solved using the GA, a derivative-free stochastic optimization technique. It deals with both minimization and maximization issues. A population of distinct solutions is repeatedly altered by the GA. At each stage, GA selects individuals at random from the current population to be parents and uses them to generate the offspring for the following generation. The population "evolves" toward an ideal solution over the course of several generations. The population "evolves" toward a perfect answer.

The symbols α_1, α_2, α_3, α_4, α_5, α_6, and α_7 stand for the switching angles. To make the mistake suitably small and per-unit equivalent of the output peak voltage, which has a value between 0 and 1, the weighting factor, W(k), is used. The word (per-unit equivalent ÷ Equation (6.24)) in the objective function represents the absolute value of the error in changing the basic component. GA was first developed in 1975 by John Holland of the University of Michigan. GAs encode each point in a solution space into a binary bit string called a chromosome. For maximizing, each point is associated with a fitness value that is usually equal to the objective function evaluated at the point. GAs often keep a population (or gene pool) of points that are continuously improved in order to increase the overall fitness value. Every generation, the GA produces new populations through genetic operators like mutation and crossover. A greater fitness value increases the likelihood that members will survive and engage in mating behaviors. Members of the population with higher fitness values appear after a few generations.

6.6.1 Computation of Switching Angles

Natural evolution is mimicked using a search heuristic known as GA. Using GA in the recommended manner, the optimal switching angles required for a 15-level inverter are determined. GA is divided into six terminations. GA is divided into six major stages. Here are: (i) Initial generation of chromosomes, (ii) Demographic, (iii) Evaluation of fitness, (iv) Going beyond, (v) Mutation, and (vi) Termination.

6.6.1.1 Generation of Initial Chromosomes

Natural evolution is mimicked using a search heuristic known as GA. Using GA in the recommended manner, the optimal switching angles required for a 15-level inverter are determined. GA is divided into six terminations. The creation of beginning chromosomes is the first phase that occurs in GA. The earliest chromosomes are formed between the lowest and maximum values of the complete chromosome. Since a 15-level inverter requires 7 H-Bridges, each chromosome in this application will have seven switching angles, or α_1 to α_7. The number of variables specific to the problem is represented by the number of genes in the chromosome or the number of H bridges in a CMLI. Once the initial chromosomes are produced, the fitness function needs to be computed.

6.6.1.2 Population

Natural evolution is mimicked using a search heuristic known as GA. Using GA in the recommended manner, the optimal switching angles required for a 15-level inverter are determined. GA is divided into six terminations. The GA starts with the population, or group of chromosomes. One crucial metric that must be adjusted is the population size. The complexity of the problem, which is reflected in the size of the search space, must be considered while calculating the population size. Usually, the initial population is created at random. A greater population increases the execution time even though it may accelerate convergence. There are seven switching angles for each of the population's twenty chromosomes. The population is established using random angles between 00 and 900, taking into mind the quarter-wave symmetry of the output voltage waveform.

6.6.1.3 Fitness Function

Fitness function is one of the most important GA methods for selecting the optimal chromosome. Fitness functions are objective functions that are used to assess a certain solution that is represented by a population's chromosomes. There are two types of fitness functions. The first cannot be changed because it is fixed. The fitness function can be changed in the second scenario. The most important consideration when choosing a fitness function is that it should directly relate to the problem and be easy to compute quickly.

The cost function is the most important factor for the GA to assess each chromosome's fitness. The cost function needs to be linked to certain harmonics since the proposed strategy aims to minimize these harmonics. The objective of this study is to reduce the output of a 15-level inverter's 3rd, 5th, 7th, 9th, 11th, and 13th harmonic orders. The cost function (f) can therefore be selected as the sum of these two harmonics normalized to the fundamental given in Equation (6.31).

$$f(\alpha_1, \alpha_2, \alpha_3, \alpha_4, \alpha_5, \alpha_6, \alpha_7) = 100 \cdot \frac{|V_3| + |V_5| + |V_7| + |V_9| + |V_{11}| + |V_{13}|}{|V_1|} \tag{6.31}$$

The switching angles in each chromosome are used to build a multilayer output voltage waveform, and FFT techniques are used to determine the necessary harmonic magnitudes. The optimal answer for the initial iteration is the switching angle configuration that produces the highest fitness value (low THD).

6.6.1.4 Crossover Function

To create a new chromosome, two chromosomes are crossed over. Based on the crossover rate, the crossover operation is carried out. The genes are chosen, and a new kid chromosome is created based on this crossover. The fitness function is then applied to the newly created child chromosome.

6.6.1.5 Mutation Operation

The mutation operator introduces a little change at a random spot on a chromosome. How the mutation process proceeds is determined by the mutation rate, which is frequently kept very low. The genes are randomly altered to produce the mutation based on the predetermined mutation rate.

6.6.1.6 Termination

During the termination stage, the best chromosome is selected based on the fitness function. Until the maximum number of iterations is reached, the aforementioned procedure is repeated. The optimal switching angles are found once the termination process is complete.

6.7 Particle Swarm Optimization

PSO is a population-based, stochastic optimization technique developed in 1995 by Drs. James Kennedy and Russell Eberhart. PSO is inspired by the social behavior of flocks of birds or schools of fish. A possible solution is represented by each particle in the swarm of individuals that the PSO algorithm maintains. Potential solutions, referred to as particles in PSO, follow the present optimum particles as they fly through the issue space. Particles behave in a simple way: they imitate the achievements of their neighbors and themselves.

The following functional parameters are involved in PSO modeling. A particle “j” in the swarm has three characteristics: (i) Position in search space X_j, (ii) Velocity Y_{jd}, and (iii) My own finest Pbestj: met by the particle thus far. Another important aspect of the swarm is its global best position, Gbest, which is the ideal location of every particle in the population. The position and velocity are updated using Equations (6.32) and (6.33), respectively.

$$y_{ij}(k+1) = w(k)y_{ij}(k) + C_1 r_1(\text{Pbest}_{ij}(k) - X_{ij}(k)) + C_2 r_2(\text{Gbest}_i(k) - X_{ij}(k)) \tag{6.32}$$

$$X_{ij}(k+1) = X_{ij}(k) + Y_{ij}(k+1) \tag{6.33}$$

where k is the number of iterations, W(k) is the inertia constant or weighting function, r_1 and r_2 are the uniformly distributed random values among [0, 1], and C_1 and C_2 are the weighting factors or acceleration coefficients.

Three characteristics, including an inertial component, a cognitive component, and a social component, make up Equation (6.32). The phrase “inertial component” describes the inertial behavior of the bird that propels it in the previous direction. The bird’s memory of its prior position is contained in the cognitive component, while the social component suggests the bird’s recall of the optimal position among the particles and their cooperative behavior to arrive at the global optimal solution. Equation (6.33) states that the new velocity is determined by adding the previous position to the present velocity.

The SHE problem for a 15-level solar-fed inverter is solved using the following algorithmic techniques:

Step 1:
Start with random position (X_i) and velocity (V_i) vectors. Every particle in the population is initialized at random from 00 to 900. Similarly, the velocity vector of every particle must be generated randomly between $-V_{max}$ and V_{max}.

Step 2:
Evaluate the particles using the fitness function of the harmonic minimization problem. The switching angles α_1 to α_7 can be calculated so that the output phase

voltage of the CMLI is free of the selected harmonic orders in order to minimize the cost function. Equation (6.34) generates the fitness function based on the 11-level inverter expression.

$$f(\alpha_1, \ldots, \alpha_7) = 100 \times \left[\left| M - \frac{|V_1|}{V_{PV(1-7)}} \right| + \left(\frac{|V_3| + |V_5| + |V_7| + |V_9| + |V_{11}| + |V_{13}|}{|V_{PV(1-7)}|} \right) \right] \quad (6.34)$$

where M stands for the modulation index. Pbest and Gbest can be obtained by analyzing the fitness function.

Stage 3:
Modify the Pbest (P_i) position of the particles. Replace P_i with the current location X_i if the ith particle's current position is superior to its prior personal best position. In addition, if the best position of the personal bests of the particles is better than the position of the global best, replace Gbest (Pg) with the best position of the personal bests.

Step 4:
Revise the position and velocity vectors. Equations (6.32) and (6.33) provide the velocity and position update rules that are used to update every particle in the population.

Step 5:
The termination criterion is the maximum number of iterations. If the iteration counter reaches its maximum, stop; if not, increase it and return to step 2.

6.8 Bees Optimization

Pham created the Bee algorithm, a derivative-free optimization technique, in 2007. The foraging habits of honey bees served as the model for this algorithm. Bees are well-known for being sociable insects with orderly colonies in the natural world. Researchers have used their activities – such as foraging, mating, and nest site location – to tackle a number of difficult combinatorial optimization and functional optimization issues. For a variety of challenging issues, the bee algorithm has demonstrated an additional dependable performance compared to other clever optimization strategies.

6.8.1 Natural World of Bees

A honeybee colony can fly on itself in multiple directions at once due to the abundance of food sources. Generally speaking, more bees should visit flower patches that are easier to collect and have a lot of nectar or pollen, whereas fewer bees should visit patches that have less of these resources.

A colony consists of three different kinds of bees: employed bees, observer bees, and scout bees. Employed bees carry information about the location and amount of nectar in a particular food source. They use a dance in the hive to convey the information to other bees. The length of the dance determines the nectar content of a food source. This dance incorporates three details about a flower patch: its quality rating (or fitness), its distance from the hive, and the direction in which it will be found. Colony communication depends on this dance, and the information it provides allows the colony to precisely direct its bees to flower patches without the need for maps or guides.

According to the information provided by the dance, the colony is able to determine the relative worth of different patches based on the energy needed to harvest them and the quality of the food they offer. An observer's choice is influenced by the nectar content of a food source. An attractive food supply attracts more onlooker bees.

In their search area, scout bees explore for fresh food sources. While employed, onlooker bees perform an exploitative role, and scout bees oversee the exploring procedure. This algorithm considers food sources as potential solutions. The food supply is a D-dimensional vector, where D is the number of optimization variables. Kavousi et al. (2012) state that the amount of nectar in a food source determines the value of fitness.

6.8.2 Determination of Switching Angles

A colony can flourish by directing its foragers to places that are fertile. The suggested approach uses the BO algorithm to find the ideal switching angles needed for a 15-level solar-fed inverter. The steps of the Bees algorithm are as follows:

Step 1:
Basic food sources are initialized at random. There are half of the fundamental food supplies needed by the bee colony.

Step 2:
Employed bees are sent to the food sources to measure the amount of nectar and evaluate its fitness. For every food source, a single bee is employed. There are as many engaged bees as there are food sources. The employed bees modify the solutions that are stored in memory by scanning the area around their food supply. The employed bees save the new response if it is a better fit than the old one. Employed bees share their solutions with the other bees when they return to the hive.

Step 3:
Another half of the colony, onlooker bees, use a probability-based selection method to choose the best food sources. Bees are drawn to food sources that have more nectar. Observer bees target the selected food sources.

The spectator bees evaluate the chosen ideas and make improvements based on their fitness. The observation is that bees save a new answer if it is superior to an old one, just like hired bees do.

Step 4:
Food sources are dropped if they are not improved after multiple rounds. In order to find new food sources, the employed bee is dispatched as a scout bee. The new food source takes the place of the abandoned one.

Step 5:
The best food source is committed to memory. The maximum number of iterations is verified as a termination condition at the end of each iteration. The algorithm goes back to step 2 for the subsequent iteration if it is not satisfied.

6.9 Simulation Results

The solar PV panel is used as the input in MATLAB/Simulink simulations for all three topologies. Consideration is given to a 7-stage inverter with 15 levels, which can minimize six harmonic orders in the output voltage and lower the THD.

6.10 Optimal Harmonic Stepped Waveform

Equations (6.15)–(6.20) are shown via a program written in the MATLAB-M file editor. The program's execution yields the optimal switching angles based on the initial guess value. To give the signals to the inverter switches, the program first produces radians, which are then converted to degrees and time equivalents.

A programmable pulse generator (PPG) that would instantly supply the optimal switching signals to the switches is proposed once the successive solutions based on this technology have been identified. Figure 6.4 shows the PPG where the resulting switching angles stored in the array are directed toward the inverter gates after the conversion segments.

The different sets of switching angles found for the modulation index $M = 0.36$ and $M = 1$ are shown in Table 6.3. Instead of satisfying the criteria $\alpha_1 \leq \alpha_2 \leq \alpha_3 \leq \ldots \leq \alpha_m \leq$ for the previous SHE problems, the benefit of the suggested technique must satisfy the criterion α_1, α_2, α_3......... α_7<. A 15-level inverter's output voltage waveform is displayed in Figure 6.5, and its FFT analysis,

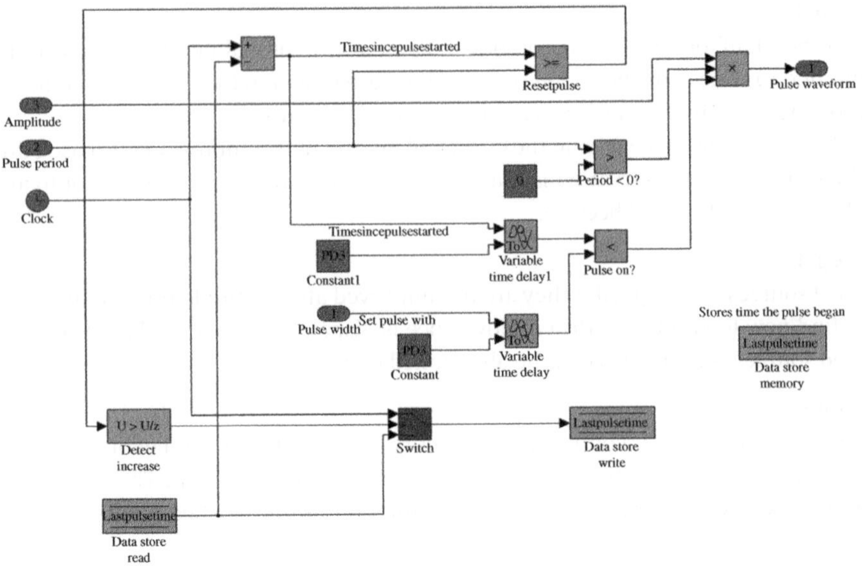

Figure 6.4 Programmable pulse generator.

Table 6.3 Determination of switching angles.

α	1st Quarter	2nd Quarter	3rd Quarter	4th Quarter
		***M* = 0.36**		
α_1	8.857	171.1143	188.8857	351.1143
α_2	14.8027	165.1973	194.8027	345.1973
α_3	27.9257	152.0743	207.9257	332.0743
α_4	47.2589	132.7411	227.2589	312.7411
α_5	57.27771	122.7229	237.2771	302.7229
α_6	72.8232	107.1768	252.8232	287.1768
α_7	87.8539	92.1461	267.8539	272.1461
		***M* = 1**		
α_1	5.0126	174.9874	185.0126	354.9874
α_2	14.2451	165.7549	194.2451	345.7549
α_3	25.9953	154.0047	205.9953	334.0047
α_4	37.6040	142.3960	217.6040	322.3960
α_5	52.0406	127.9594	232.0406	307.9594
α_6	65.0413	114.9587	245.0413	294.9587
α_7	86.2279	93.7721	266.2279	273.7721

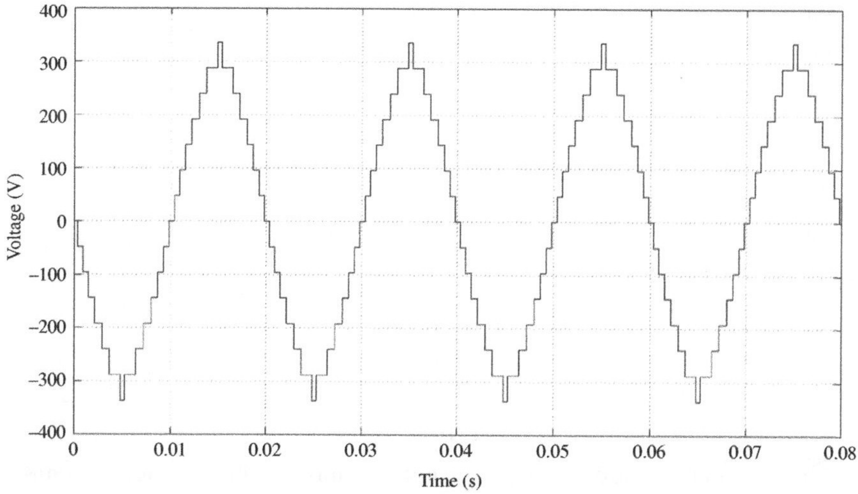

Figure 6.5 15-Level output voltage waveform with OHSW technique.

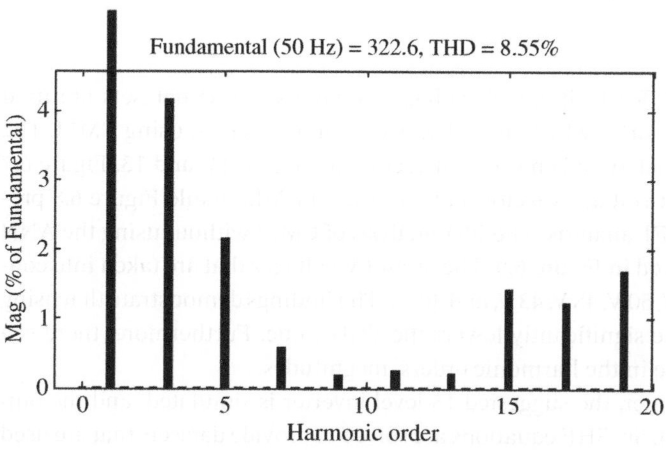

Figure 6.6 FFT analysis of the output waveform with OHSW.

which displays the THD value of 8.55%, is shown in Figure 6.6. The THD acquired for a seven-level inverter using the OHSW approach by Sirisukprasert (1999) is around 32.78%, while the THD obtained for a 15-level solar-fed inverter is 8.55%.

6.10.1 Artificial Neural Networks

ANN-based solar-fed CMLI is simulated and analyzed for 11 and 15 levels, respectively. The dataset for an 11-level inverter is acquired by solving four harmonic

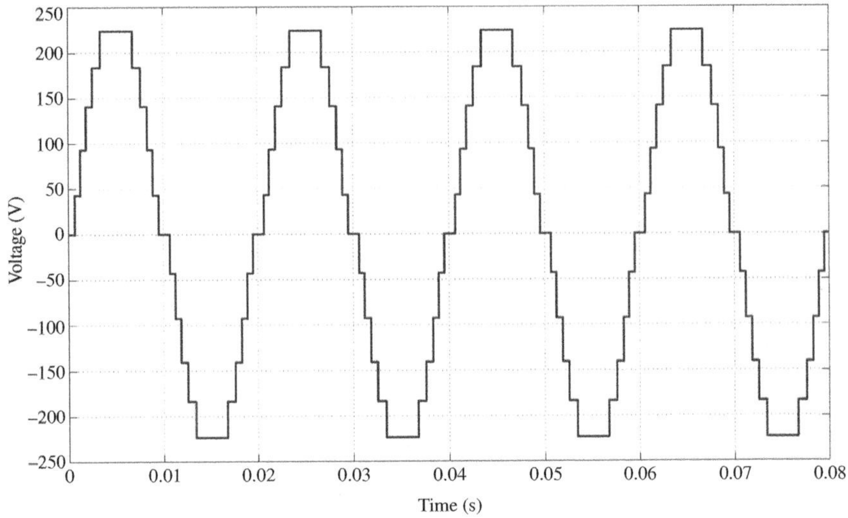

Figure 6.7 Output voltage waveform of an 11-level inverter with ANN.

equations under different input PV voltage situations. These datasets are used to train neural networks, which are subsequently implemented using CMLI. The harmonic orders that are taken into consideration are 5, 7, 11, and 13. Figure 6.7 displays the output voltage waveform of an 11-level CMLI, while Figure 6.8 provides the related FFT analysis. The FFT analysis of CMLI without using the ANN approach is displayed in Figure 6.9. The input PV voltages that are taken into consideration are {43 V, 50 V, 48 V, 43 V, and 40 V}. The findings demonstrate that using the ANN technique significantly lowers the THD value. Furthermore, there is a significant decrease in the harmonic orders' magnitudes.

In a similar manner, the suggested 15-level inverter is simulated, and the outcomes are achieved. Six SHE equations are solved to provide datasets that are used to train the neural network. The solar-fed CMLI is subsequently used to implement the trained network.

Figures 6.10 and 6.11 display the output voltage and current waveforms for the RL load of $R = 100\ \Omega$ and $L = 10\,\text{mH}$, respectively. The analysis is carried out using the input PV voltage values of {49.9 V, 50.7 V, 52.6 V, 54.9 V, 44.7 V, 51.4 V, and 43.5 V}. In this case, floating-point voltage data is also taken into account, as is typically the case in real-world applications. Figures 6.12 and 6.13 show the FFT analysis of the output voltage and current from solar-fed CMLI with ANN. Figure 6.14 shows a comparison of the harmonic order for solar-fed 11-level and 15-level inverters using ANN.

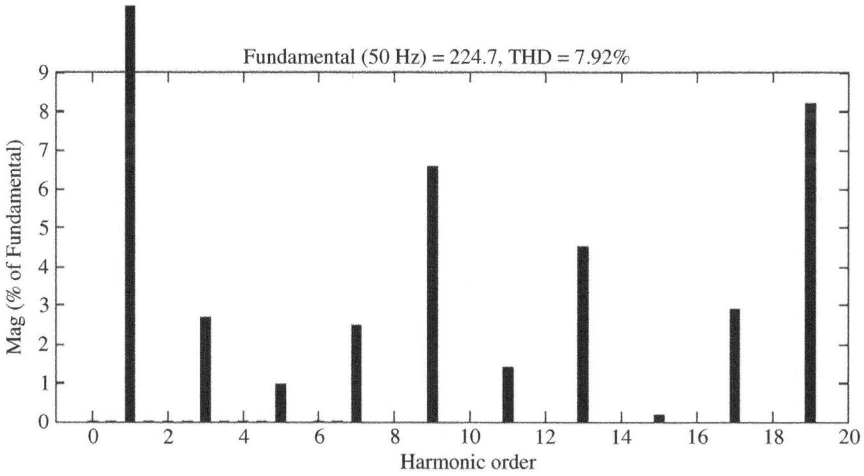

Figure 6.8 FFT analysis of an 11-level inverter with ANN.

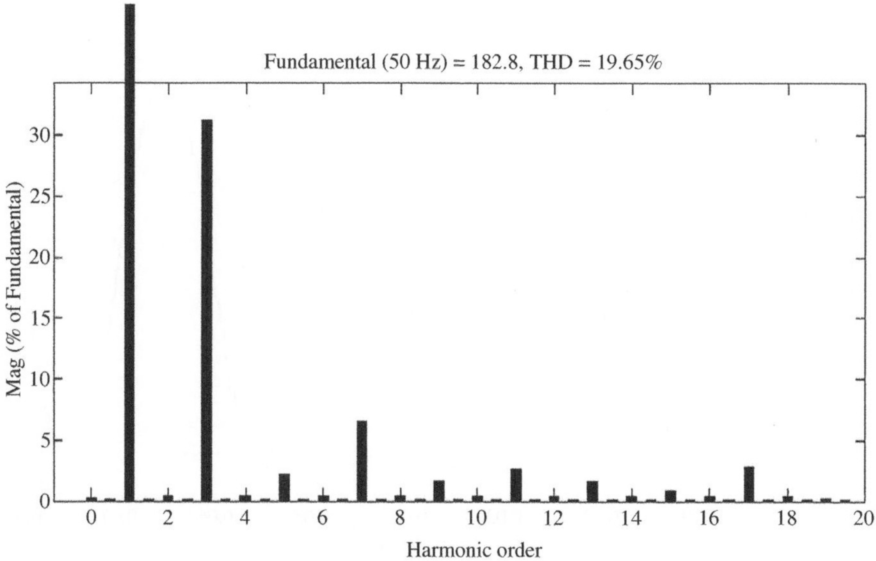

Figure 6.9 FFT analysis of an 11-level inverter without an ANN.

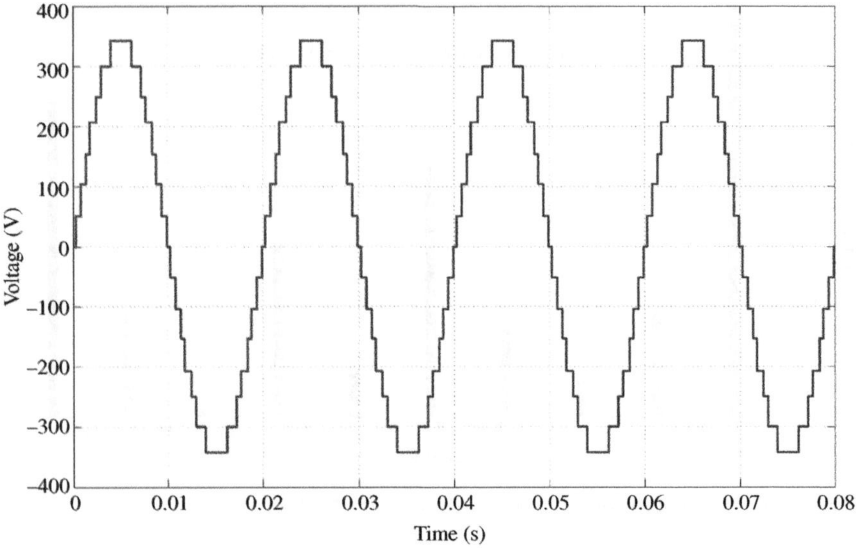

Figure 6.10 Output voltage waveform of a 15-level inverter with ANN.

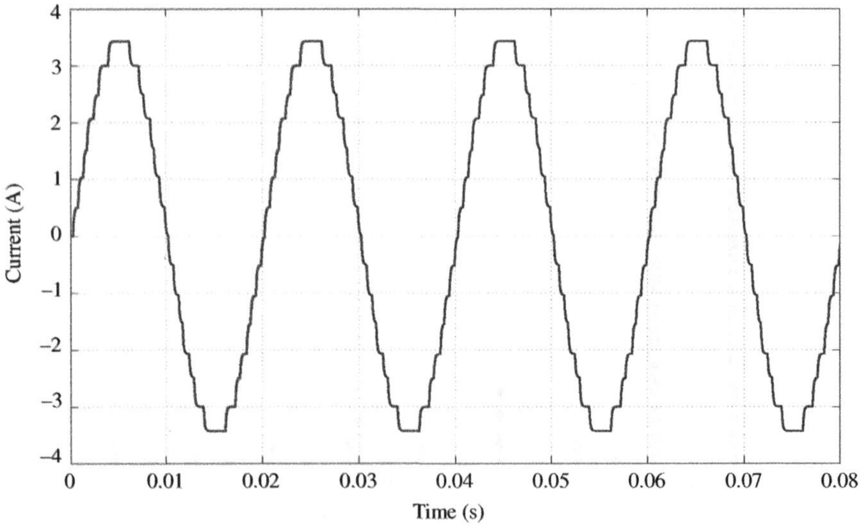

Figure 6.11 Output current waveform of a 15-AQlevel inverter with ANN.

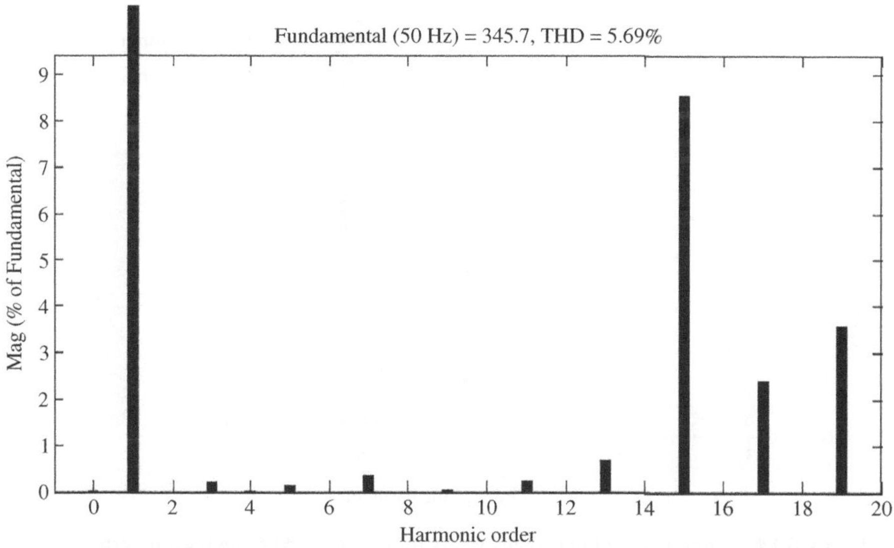

Figure 6.12 ANN-based FFT analysis of the output voltage of a 15-level inverter.

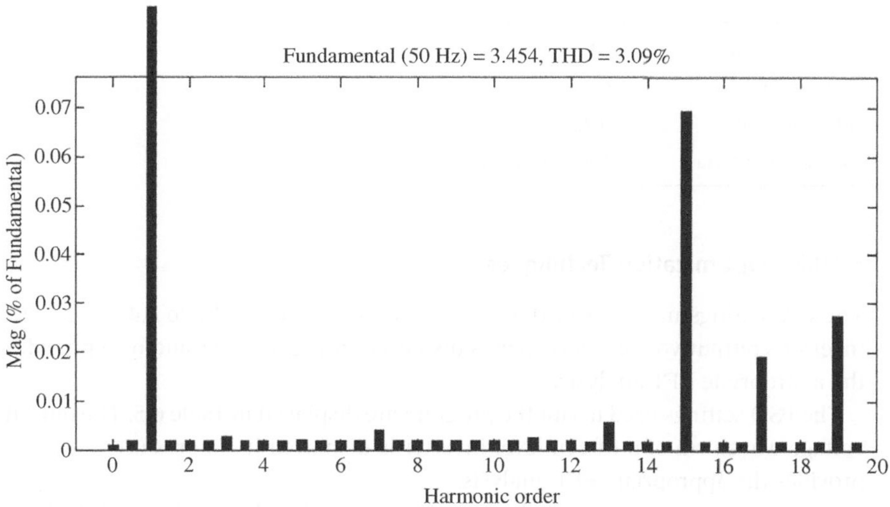

Figure 6.13 FFT analysis of a 15-level inverter output current with ANN.

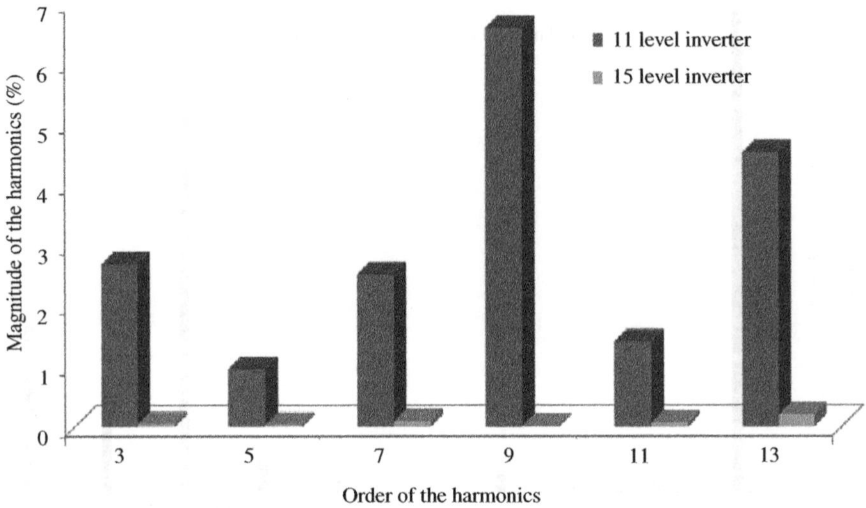

Figure 6.14 Comparison of harmonic orders for eleven and fifteen level inverters.

Table 6.4 GA specifications.

Parameter	Value
Population size	24
Crossover rate	0.5
Mutation rate	0.234
Stopping criteria	100 iterations

6.10.2 Optimization Techniques

The GA settings used to run the program are displayed in Table 6.4. A 15-level inverter's output voltage waveform is displayed in Figures 6.15 and 6.16 provides the appropriate FFT analysis.

The PSO settings used to run the program are displayed in Table 6.5. The output voltage waveform of a fifteen level inverter is displayed in Figures 6.17 and 6.18 provides the appropriate FFT analysis.

The BO settings used to run the program are displayed in Table 6.6. A 15-level inverter's output voltage waveform is displayed in Figures 6.19 and 6.20 provides the appropriate FFT analysis.

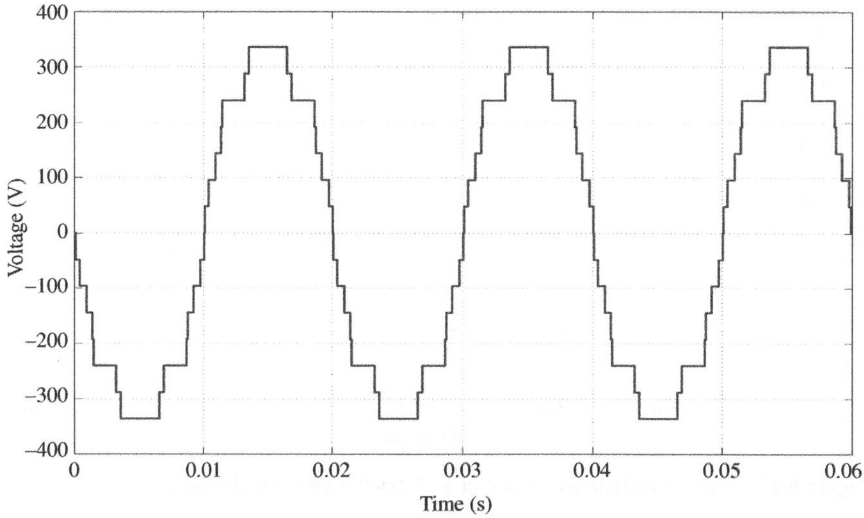

Figure 6.15 Output voltage waveform of a 15-level inverter with GA.

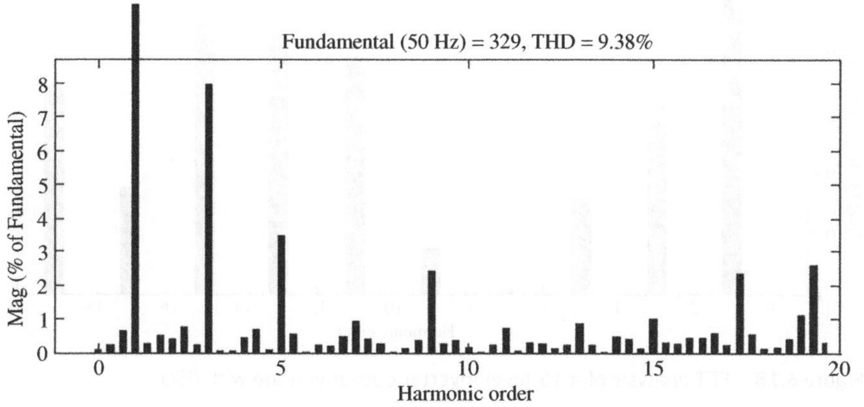

Figure 6.16 FFT analysis of a 15-level inverter output voltage with GA.

Table 6.5 PSO specifications.

Parameter	Value
C_1, C_2	0.3
r_1, r_2	0.9 [0 to 1]
W(k)	0.9
Stopping criteria	100 iterations

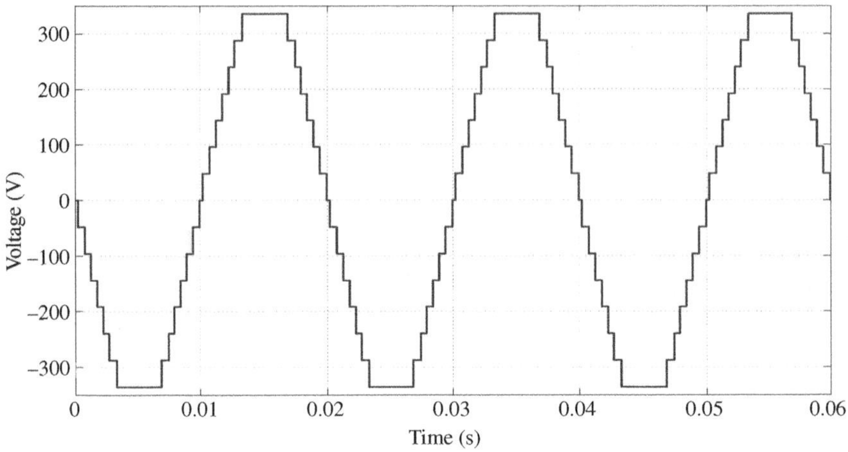

Figure 6.17 Output voltage waveform of a 15-level inverter with PSO.

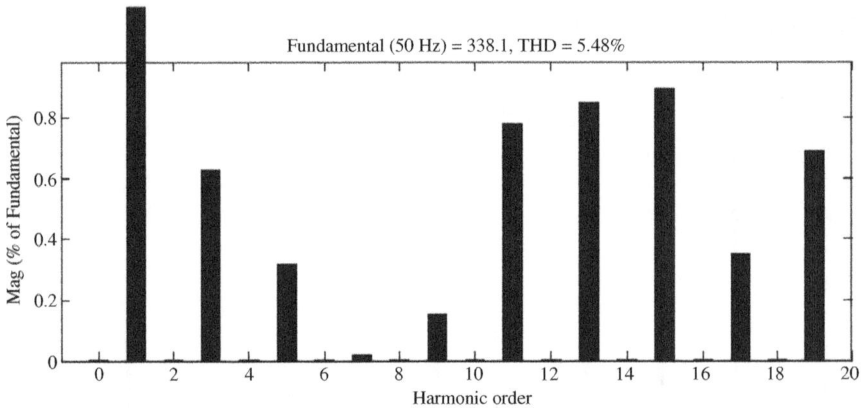

Figure 6.18 FFT analysis of a 15-level inverter output voltage with PSO.

Table 6.6 BO specifications.

Parameter	Value
Number of employed bees	30
Number of scout bees	20
Number of sites selected for neighborhood search	15
Number of elite sites	7
Stopping criteria	100 iterations

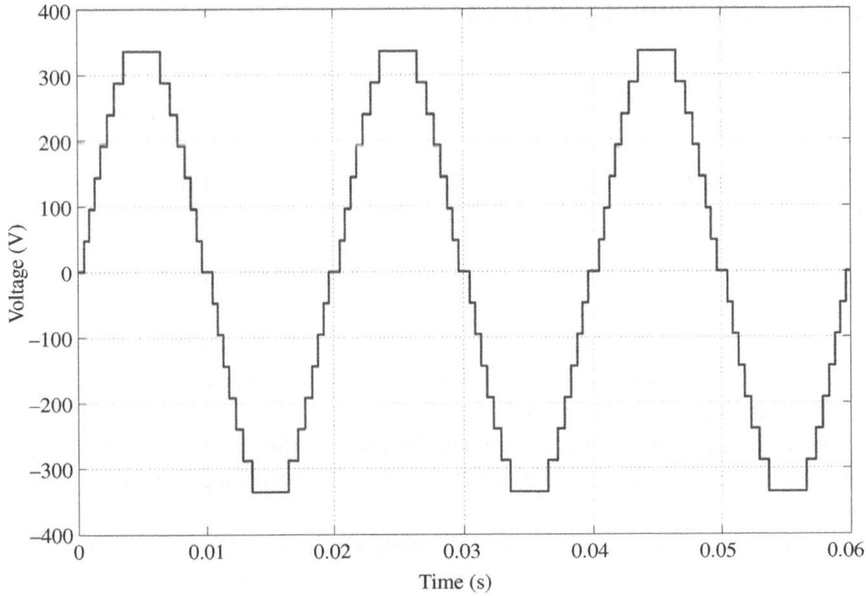

Figure 6.19 Output voltage waveform of a 15-level inverter with BO.

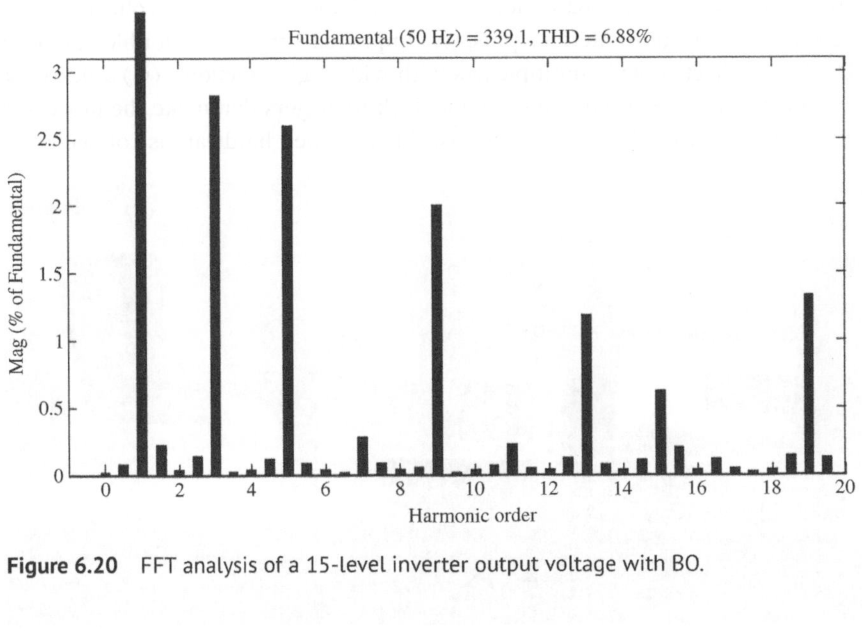

Figure 6.20 FFT analysis of a 15-level inverter output voltage with BO.

6.11 Experimental Results

The hardware implementation uses seven sets of separate PV supply sources, each of which generates $48 \times 7 = 336$ Vp (230 VRMS). The ANN-based system's methodology is expanded for hardware implementations since it offers lower THD. The hardware configuration, which includes seven inverter stages and an FPGA Spartan 3E CPU, is seen in Figure 6.21. The term "inverter" is shown in the figure INV. The FPGA processor receives these inputs after they have been scaled down to the proper range. Each input source to the inverter stages gives a variance of up to $\pm 10\%$.

The switching angles are precalculated and then programmed using an offline approach in the processor. The switching angles are changed at each interval of time due to fluctuations in the solar PV.

The variations of the switching angles observed for the change in input PV voltage are shown in Figures 6.22 and 6.23, which are designated as Cases I and II, respectively.

The CPU determines the appropriate switching angles for each combination of source voltages and sends the information to the driver circuit, which is coupled to a MOSFET to turn it on. As seen in Figure 6.21, seven sets of signal processing components are used to sense voltage. The Xilinx SPARTAN 3E processor's attractive features include: (i) a highly affordable, high-performance logic solution for large-scale, consumer-focused applications; (ii) a proven advanced 90-nanometer manufacturing technology; (iii) multiple voltage, multiple standard selection interface pins; (iv) improved support for double data rate (DDR); (v) effective wide multiplexers with wide logic functions; (vi) quick look ahead carry logic; and (vii) eight digital clock managers that make the processor appropriate for SHE-based applications. The designed hardware is connected to

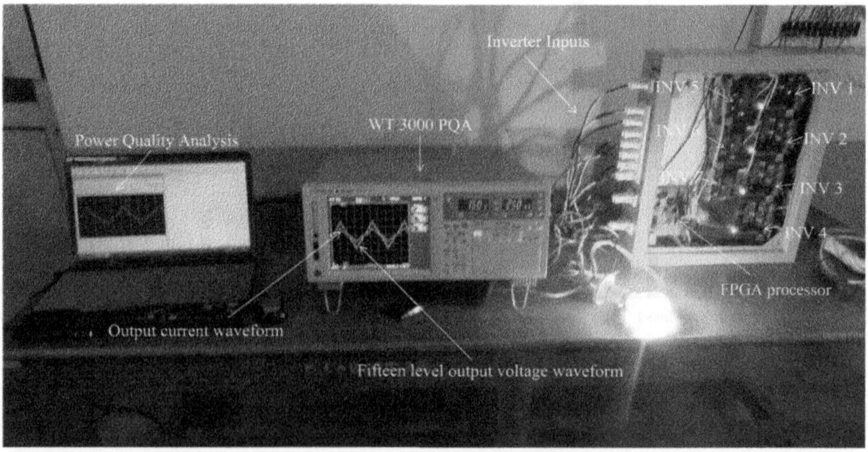

Figure 6.21 Hardware setup for a 15-level inverter with SHE.

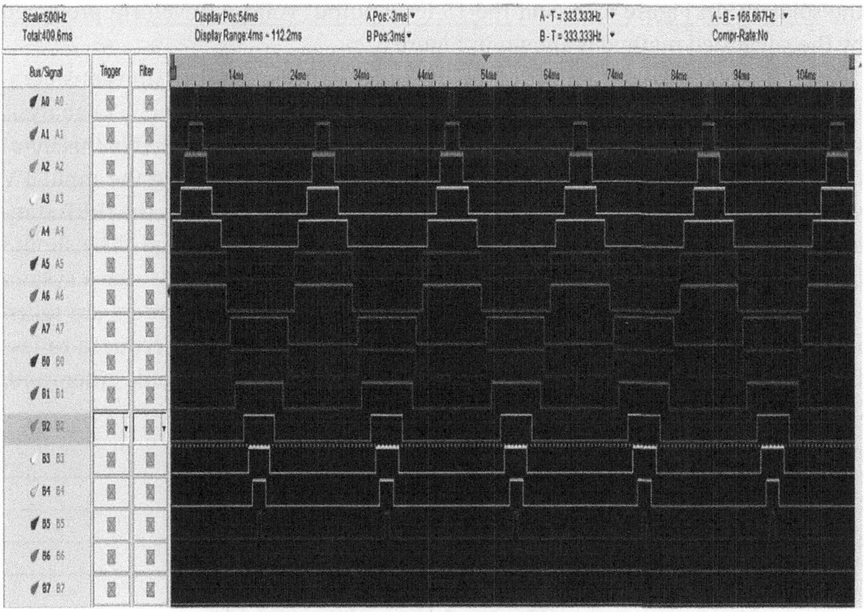

Figure 6.22 Variation of switching angles (case-I).

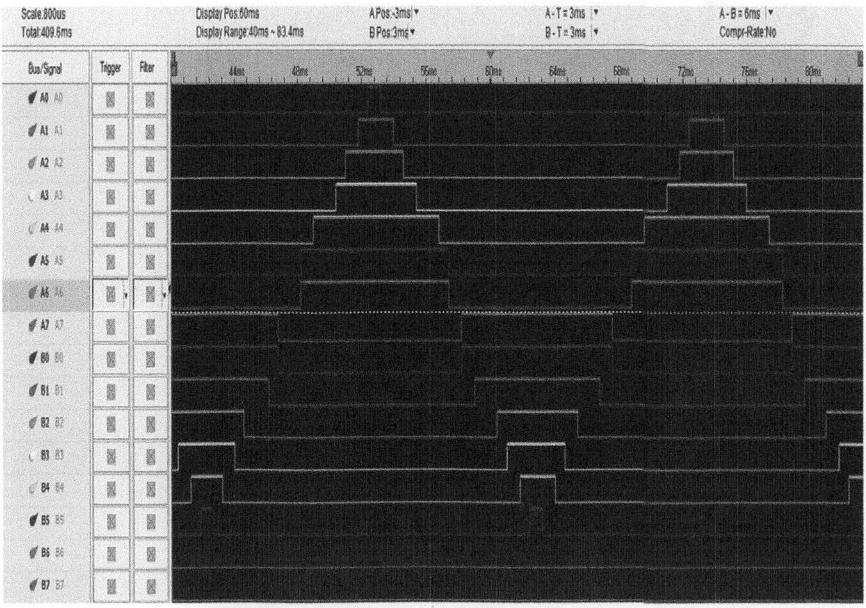

Figure 6.23 Variation of switching angles (case-II).

the solar PV in Figure 6.24, and the 15-level output voltage waveform produced by the suggested method is shown in Figure 6.25.

With the potential minimization of low-order harmonics taken into consideration in this work, Figure 6.26 displays the fast Fourier transform (FFT) analysis using the power quality analyzer (PQA) WT3000. The THD produced in this manner is approximately 5%. SHE equations are solved by measuring the input PV voltage variation at each instant. According to the findings, it meets the IEEE standard for voltage harmonics 519–1992, which states that the voltage THD should be 5% and the individual harmonic order magnitude should be 3% for a system with a bus voltage of less than 69 kV. When the six harmonic orders are taken into account for the given system, the magnitude is significantly reduced to less than 3%. Because of this, the created hardware can lower the harmonic orders and, consequently, the overall harmonic distortions.

Figure 6.24 Hardware setup with solar PV integration.

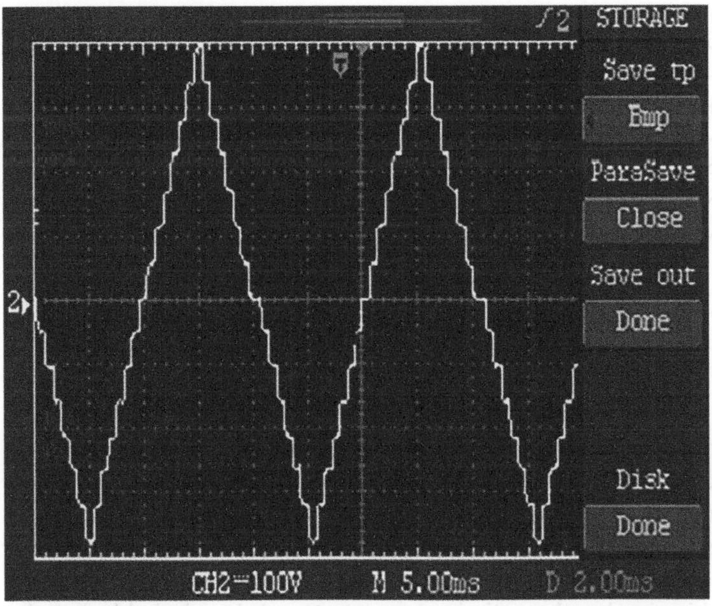

Figure 6.25 15-Level output voltage waveform.

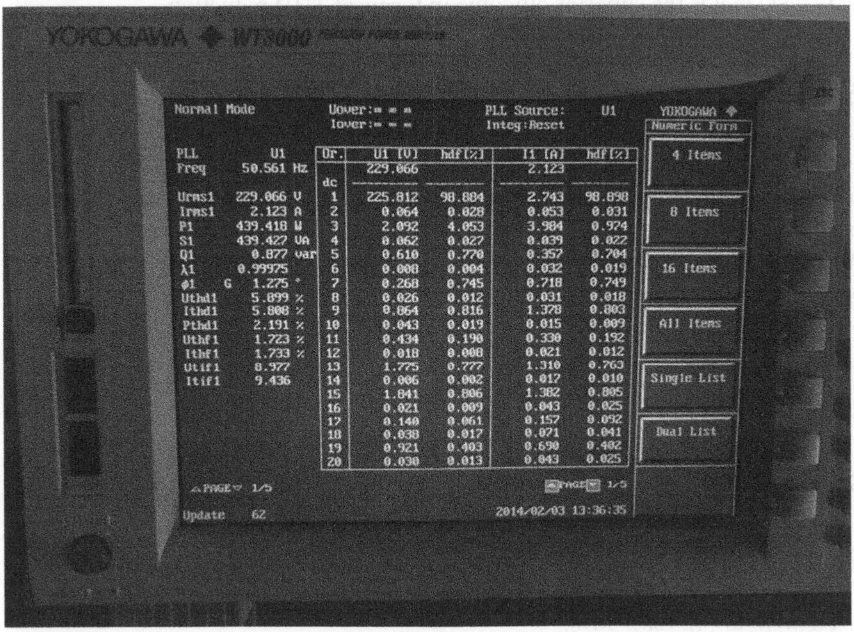

Figure 6.26 Harmonic analysis using WT3000.

Table 6.7 Comparison of results for the proposed methodologies.

S. no.	Method	Level	THD (%)
1	OHSW	15	8.55
2	ANN	11	7.92
4	ANN	15	**5.73**
5	GA	15	9.38
6	PSO	15	**5.48**
7	BO	15	6.88

Table 6.7, which compares the approaches suggested for SHE, demonstrates that the ANN-based approach yields superior outcomes and is generally appropriate for solar PV systems. When constructing the problem, the OHSW approach requires the initial guess values for switching angles. Because the goal function's voltage value is frequently changed to provide a greater degree of freedom in determining the ideal switching angles, the optimization technique is less desirable for solar PV systems. When comparing the three optimization methods – GA, PSO, and BO – PSO yields the superior outcome in terms of THD reduction.

6.12 Summary

For the purpose of improving power quality, a 15-level solar-fed CMLI is developed to eliminate particular harmonic orders. The approach is suggested using optimization techniques (GA, PSO, and BO) and intelligent techniques like ANN. Additionally, a harmonic reduction method based on OHSW has been investigated. In order to reduce six likely harmonic orders, an experimental analysis is conducted for a 3 kWp system with an FPGA processor. The advantages of this approach include a simple computational process and the absence of filters, extensive lookup tables, or output transformers.

References

Kavousi, A., Vahidi, B., Salehi, R., et al. (2012). Application of the bee algorithm for selective harmonic elimination strategy in multilevel inverters. *IEEE Trans. Power Electron.* 27 (4): 1689–1696.

Sirisukprasert, S. (1999). Optimized harmonic stepped waveform for multilevel inverter. M.Sc. thesis, Virginia Polytechnic Institute and State University.

7

Experimental Implementation of PWM Generation

7.1 Introduction

Pulse-width modulation (PWM) signals can be produced on a variety of platforms in experimental research. A thorough introduction to PWM generation is given in this chapter, along with pertinent case studies. PWM can be produced using both standard integrated circuits and sophisticated processors. The PWM generating fundamental functional block diagram is displayed below (Figures 7.1 and 7.2):

7.2 PWM Generation Using IC

The TL494 PWM controller is used to produce both variable and fixed PWM. It consists of two on-chip error amplifiers, a frequency-adjustable oscillator, an output flip-flop with pulse steering control, and an output-controlled circuit with feedback. In a common voltage arrangement, the error amplifiers may compensate voltage between −0.3 and VCC -2 volts. The dead time is managed by the comparator using a fixed offset. The dead-time comparator provides a range of nearly 5%. Additionally, this PWM IC may receive a reference frequency signal from the external oscillator. By attaching RT to the reference output pin, users can avoid an on-chip oscillator. Figure 7.3 depicts the TL494's internal block diagram, while Figure 7.4 shows the pin arrangement. The pin information is provided in Table 7.1 and its specifications in Table 7.2.

7.3 Features of the TL494

Total PWM control: Offers all the functionality needed to control pulse-width modulation. A built-in oscillator that operates in both slave and master modes is included.

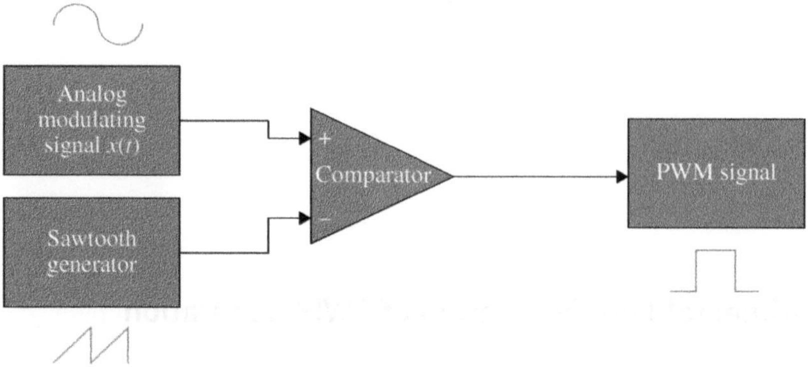

Figure 7.1 Basic functional block diagram for PWM generation.

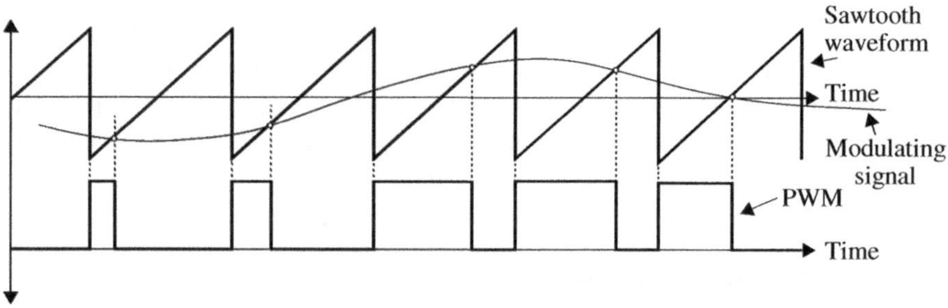

Figure 7.2 Output waveform of PWM generation.

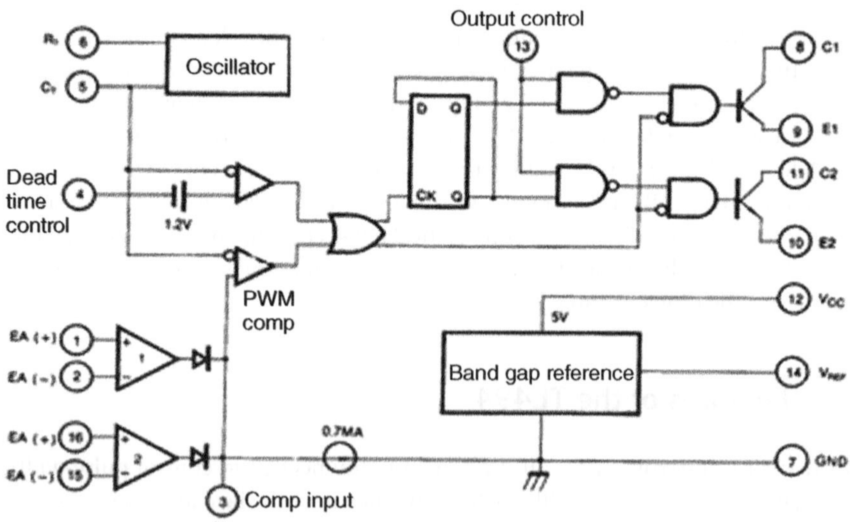

Figure 7.3 Internal block diagram of TL494.

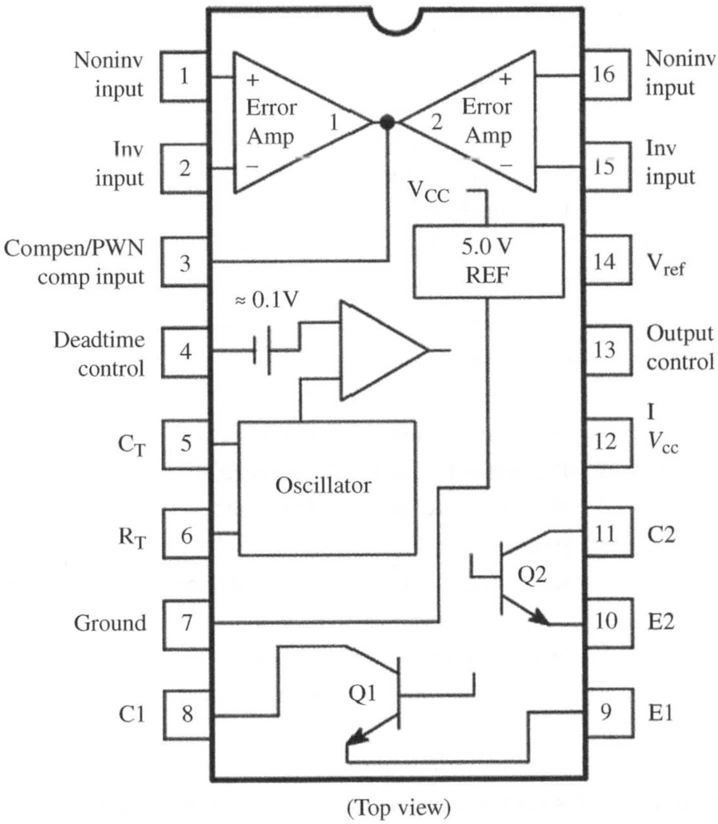

Figure 7.4 Pin diagram of TL494.

Integrated error amplifiers: Provides amplifiers to enhance control and feedback.

5 V internal reference: To maintain stable operation, it has an internal 5 V reference.

You can change the dead time to prevent switching overlap.

Flexible Output Transistors: With a maximum current capacity of 500 mA, output transistors offer versatility for a range of applications.

Output control for modes: Push–pull and single-ended operation are both possible.

Undervoltage lockout: If the voltage is too low for safe operation, the IC cannot function.

Automotive versions: Available for vehicles and other specific applications.

Table 7.1 Pin configuration details.

Pin name	Pin no.	Description
1IN+	1	Noninverting input to error amplifier 1
1IN−	2	Inverting input to error amplifier 1
FEEDBACK	3	Input pin for feedback
DTC	4	Dead-time control comparator input
CT	5	Capacitor terminal is used to set the oscillator frequency
RT	6	Resistor terminal is used to set the oscillator frequency
GND	7	Ground pin
C1	8	The collector terminal of BJT output 1
E1	9	The emitter terminal of BJT output 1
E2	10	The emitter terminal of BJT output 2
C2	11	The collector terminal of BJT output 2
VCC	12	Positive supply
OUTPUT CTRL	13	Selects single-ended/parallel output or push–pull operation
REF	14	The 5-V reference regulator output
2IN-	15	Inverting input to error amplifier 2
2IN+	16	Noninverting input to error amplifier 2

Lead-free options: Provides lead-free packaging for use that is safer and more ecologically friendly. Figure 7.5 shows the internal structure of TL494. Two error amplifiers in the TL494 control output by varying their gain in response to different input conditions. These amplifiers can tolerate a wide input range because they can be powered straight from the source voltage. By supplying power only when needed, they help to fine-tune the PWM output and provide stable current.

Output-control mode: The output-control pin enables the output transistors to be configured in a variety of ways. There are two operating modes available: push-pull mode, in which the outputs alternate, and single-ended mode, in which both outputs function concurrently. This variable can be changed to suit the needs of the application without impacting other components of the TL494, like the oscillator or flip-flop.

Output transistors: Transistors that can switch up to 200 mA of current make up the TL494's output stage. Depending on the requirements of the circuit, these transistors can either source or sink current. The voltage drop across the transistor is less than 1.3 V in a common-emitter arrangement and less than

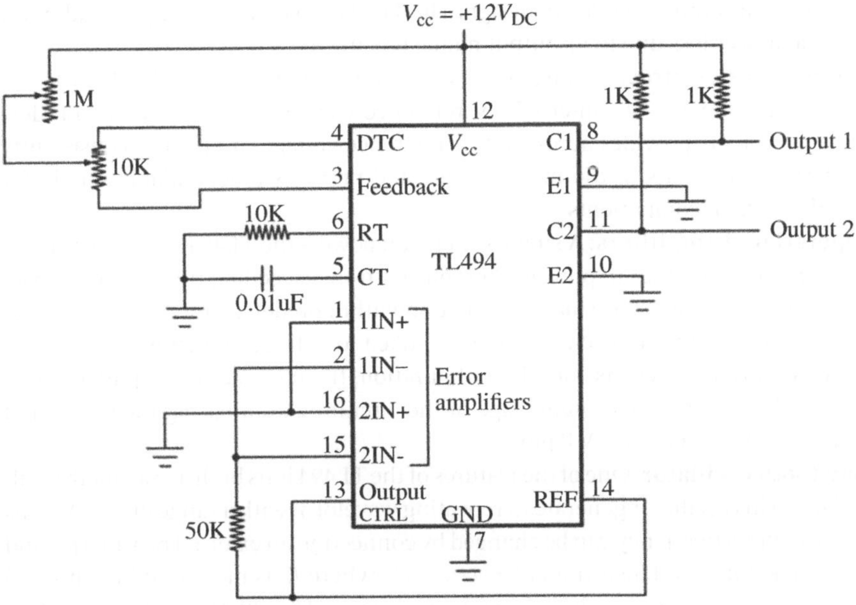

Figure 7.5 Internal structure of TL494.

Table 7.2 Specifications of TL494.

Specifications	Value
Operating voltage range	7–40 V
Number of outputs	2
Switching frequency	300 kHz
Maximum duty cycle	45%
Output voltage	40 V
Output current	200 mA
Maximum output current for both PWMs	250 mA
Temperature range	−65 °C–150 °C
Fall time	40 ns
Rise time	100 ns
Available packages	16-pin PDIP, TSSOP, SOIC, SOP

2.5 V in a common-collector design. The TL494 can drive a variety of loads with little power loss due to its output management.

5 V reference source: As long as the VCC input is more than 7 V (with a 100 mV margin), the TL494's internal 5 V reference voltage stays steady. Pin 14, designated REF, provides access to this reference voltage. It operates consistently despite variations in the input voltage and provides a dependable supply for other circuit components.

Operational amplifiers: A single supply rail powers the TL494's two operational amplifiers. In order to prevent their output from exceeding the system's capacity, these amplifiers are made to function within predetermined voltage restrictions. The output of each amplifier is linked to a diode, which is connected to the COMP pin. Because of this configuration, the more active amplifier is able to influence the subsequent stage of the circuit by dominating the signal that passes through the COMP pin.

Sawtooth oscillator: One of the features of the TL494 is its built-in sawtooth oscillator. This oscillator generates a repeating waveform with a range of 0.3 3 V. This oscillator's frequency can be changed by connecting an external resistor (R_t) and a capacitor (C_t). The formula $f = 1/R_t \times C_t$, where R_t is measured in ohms and C_t in farads, is used to determine the frequency. The PWM timing is based on this adjustable oscillator.

Pulse-width modulation trigger: The interplay between the sawtooth oscillator and the comparator's output's falling edge is what drives the PWM trigger. Depending on the parameters established by the comparator and the sawtooth waveform, the trigger either activates or deactivates one of the output stages when the comparator's output changes.

Comparator functionality: The input signal supplied by the operational amplifiers via the COMP pin is compared to the sawtooth oscillator's waveform by the comparator in the TL494. The comparator output is driven low (0) when the sawtooth voltage is higher than the comparator's input. The output is driven high (1) when the input exceeds the sawtooth voltage.

Dead-time control: A minimum off-time between pulses is specified by pin 4, which is designated dead-time control (DTC). The maximum duty cycle is restricted by this dead time to roughly 45%, or 42% if the DTC pin is grounded. The length of the quiet interval between switching events can be regulated by varying the voltage on this pin, preventing the system from overdriving its components.

By rapidly turning signals on and off, the TL494 aids in power control. It can regulate the amount of power supplied to a gadget thanks to this procedure.

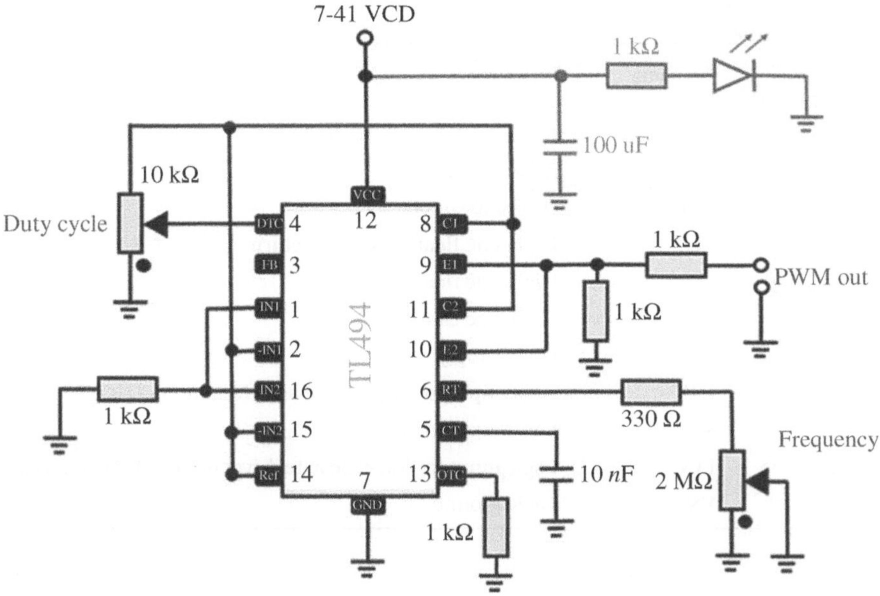

Figure 7.6 Control circuit configuration.

This controller's ability to modify the "duty cycle" – the length of time the signal lasts – while maintaining the same signal speed or frequency is one of its features. Figure 7.6 shows the TL494 control circuit. The TL494 has a switching frequency of up to 300 kHz. The PWM signal's high and low state toggling speed is determined by this frequency. The efficiency and performance of the overall power supply are directly impacted by setting the switching frequency, which entails modifying external or internal timers.

7.4 PWM Driver

A diver IC can be used to direct the PWM OUT from Figure 7.6 to the semiconductor device. An adaptable part of power electronics, the IR2110 is made to efficiently drive MOSFETs in both high-side and low-side topologies. Connect the floating channel to a high-side N-channel power MOSFET or IGBT that can withstand up to 500 volts in order to use the IR2110. Figure 7.7 shows the pin configuration, while Table 7.3 provides a thorough explanation.

Table 7.3 Pin configuration details.

Pin no.	Pin name	Description
1	LO	Low-side gate drive output
2	COM	Low-side return
3	VCC	Low-side supply
5	VS	High-side floating supply return
6	VB	High-side floating supply
7	HO	High-side gate drive output
9	VDD	Logic supply
10	HIN	Logic input for high-side gate driver output (HO), in phase
11	SD	Logic input for shutdown
12	LIN	Logic input for low-side gate driver output (LO), in phase
13	VSS	Logic ground

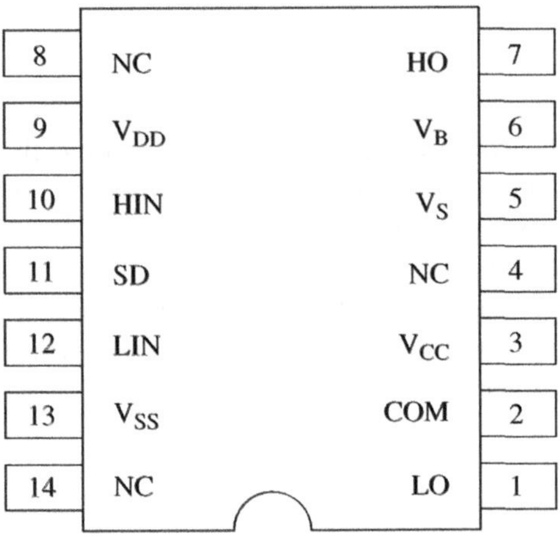

Figure 7.7 Pin diagram of IR2110.

7.5 IR2110 Features

- The floating channel can tolerate negative transient voltages, is completely functional up to +500 V or +600 V, and is suited for bootstrap operation. Additionally, it is unaffected by dV/dt.
- A supply range of 10 V–20 V allows the gate drive to function efficiently.
- To guarantee stability and appropriate operation, both channels have undervoltage lockout.

- The device maintains a ±5 V offset between logic and power grounds and is compatible with 3.3 V logic levels. It can also operate with a separate logic supply that ranges from 3.3 V to 20 V.
- It has a cycle-by-cycle edge-triggered shutdown function for improved efficiency and safety; inputs are CMOS Schmitt-triggered with an inbuilt pull-down resistor.
- To enable synchronization, propagation delays are properly matched between channels.
- Consistent performance is ensured by aligning the output signals in phase with the input signals.

7.6 Case Study

An IR2110 can be connected to a MOSFET by connecting its low-side output (Pin 1) and high-side output (Pin 7) to the corresponding MOSFET gates. The source of the high-side MOSFET is connected to the floating supply rail (VS), whereas the source of the low-side MOSFET is connected to ground (Pin 13/VSS). Both the logic power supply (Pin 9) and the low-side power supply (Pin 3) require decoupling capacitors to their respective grounds. Figure 7.8 illustrates the experimental implementation of TL494-IR2110 for switching the entire bridge inverter.

7.7 Microcontroller-Based PWM Generation

The output voltage waveform for the linear loads at an inverter's output end is displayed in Figure 7.9. In this case, a resistive load of 120 Ω is employed. The result shows that for linear loads, the output voltage waveform is found to be purely sinusoidal. The pulses for the inverter switches are given by the PWM technique. Figure 7.10 depicts the PWM production circuit and its output waveform for the following pulse data: freq = 50 Hz, $V_c = 1.8$ V, and $V_m = 1.2$ V.

Low output voltage distortion is necessary for the nonlinear loads. The circuit simulation and the resulting current and voltage waveform for a nonlinear load (a complete bridge rectifier with a 3200 μF capacitor and a 10 Ω resistor) without the use of a controller are shown in Figure 7.11.

Figure 7.12 shows a block diagram of a UPS system based on a "neural network controller" that generates a smooth sinusoidal waveform even under nonlinear loading conditions. In most UPS systems, the rectifier or charger has two functions. It first provides DC power to the DC bus via a constant output voltage so that the critical load can be served directly or through the inverter. Second, it recharges the system battery after the power outage and maintains it completely charged.

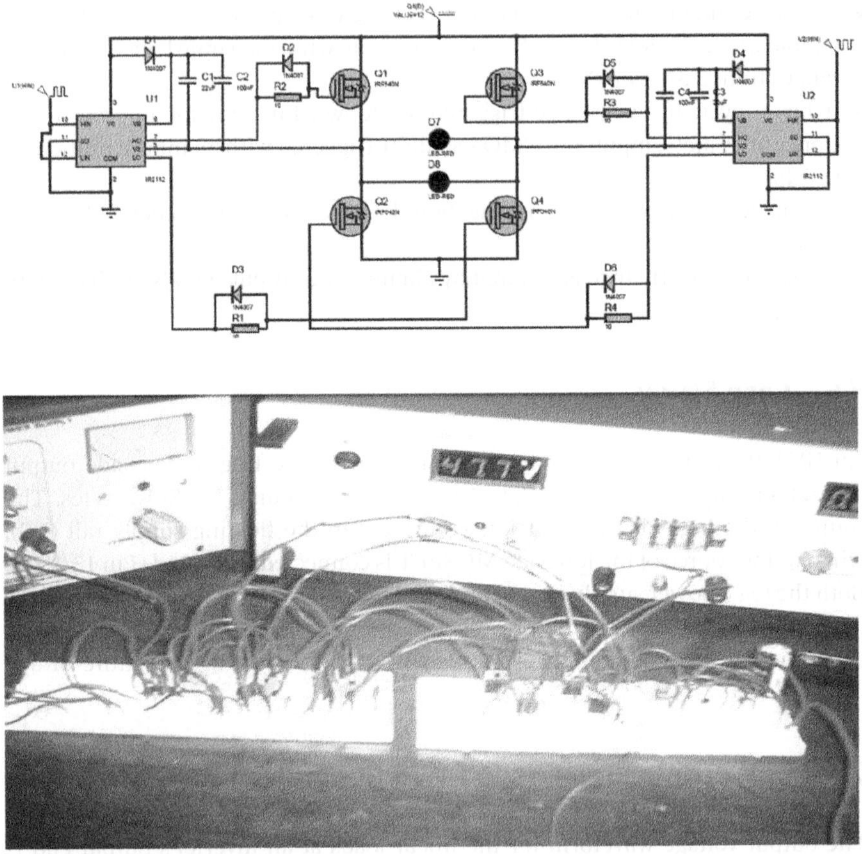

Figure 7.8 Experimental setup.

The sinusoidal pulse-width modulation approach has been used to switch the UPS inverter. Sine and triangle waves are compared to provide the SPWM pulses needed to turn the inverter. The NN controller produces sine and triangle waves and compares them.

The output voltage is entirely sinusoidal for linear loads, but it becomes non-sinusoidal and performs much worse when nonlinear loads are utilized. The simulated controller provides the patterns for the neural network controller.

7.8 Hardware Implementation

Parameters in Table 7.4 serve as the basis for the design of the inverter and filter. Prior to the experimental setup, the following procedures must be completed.

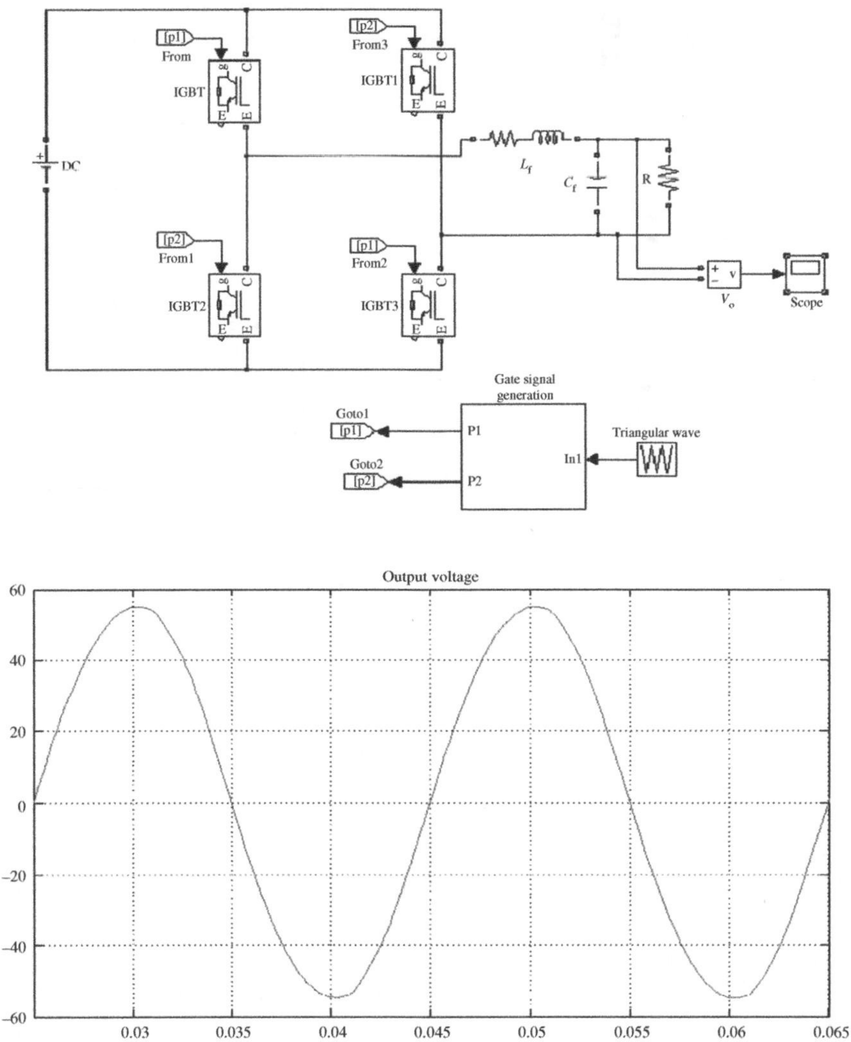

Figure 7.9 Simulation circuit for linear load.

(i) to construct the simulated controller using the inverter's idealized load current reference. (ii) Adjust the controller's parameters to the ideal values for each loading scenario. Next, gather the output voltage, load current, and capacitor current as the neural network's inputs. The compensation signal is the neural network's intended output. In order to train the neural network, these patterns create a pattern database. (iii). To model the simulated controller using the pattern database,

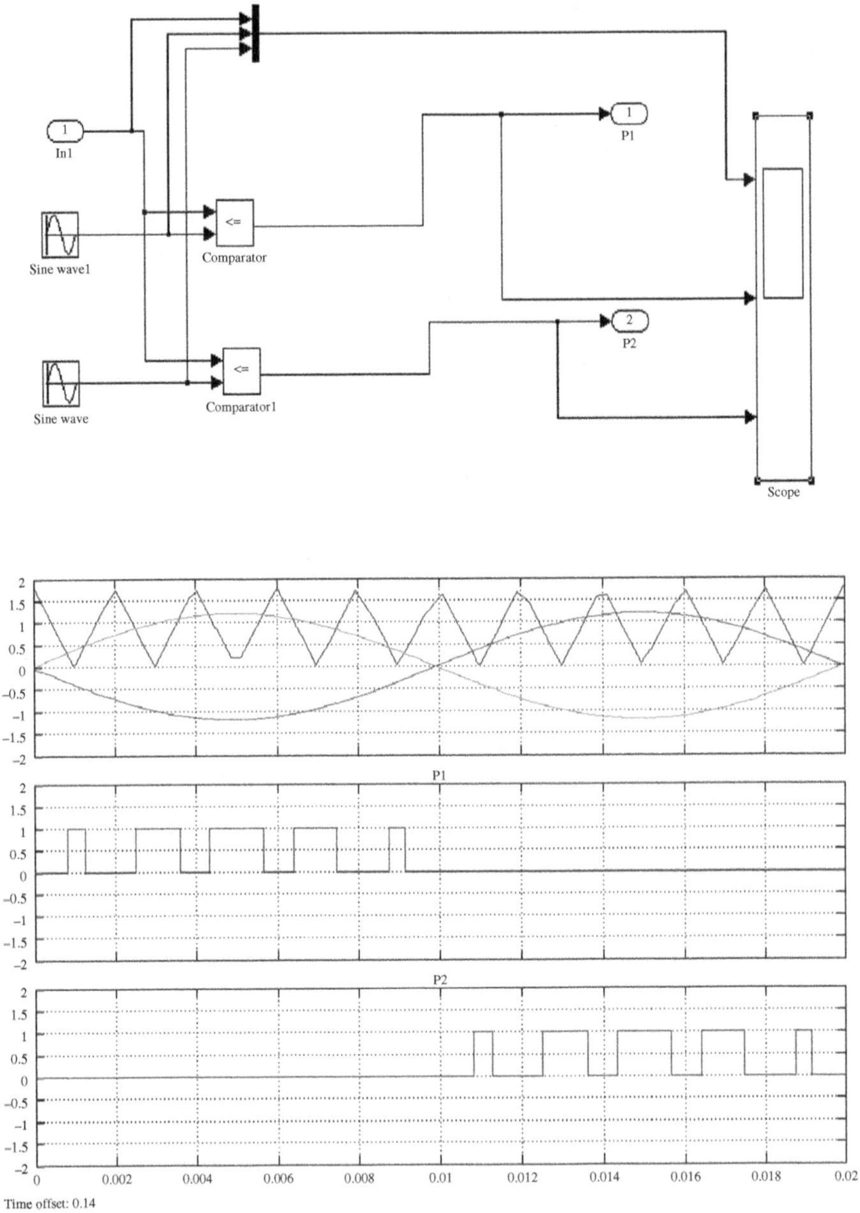

Figure 7.10 Simulation circuit for signal generation.

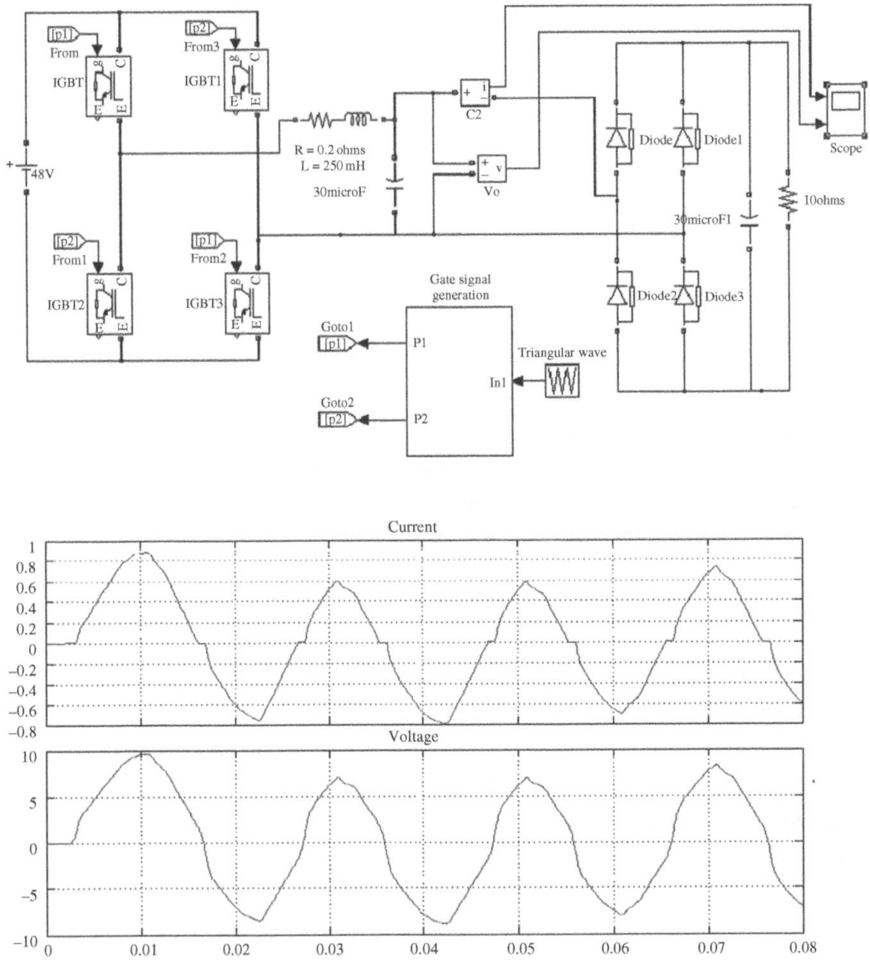

Figure 7.11 Simulation circuit for a nonlinear load.

choose a neural network structure that is both basic and adequate. (iv) Use MATLAB and the Neural Network Toolbox to train the neural network.

7.9 Hardware Circuit

The following circuits make up the hardware implementation: (i) Power supply unit; (ii) Full-bridge inverter; (iii) Nonlinear load; (iv) Microcontroller unit; and (v) Filter circuit. Figure 7.13 depicts the implementation circuit. The nonlinear

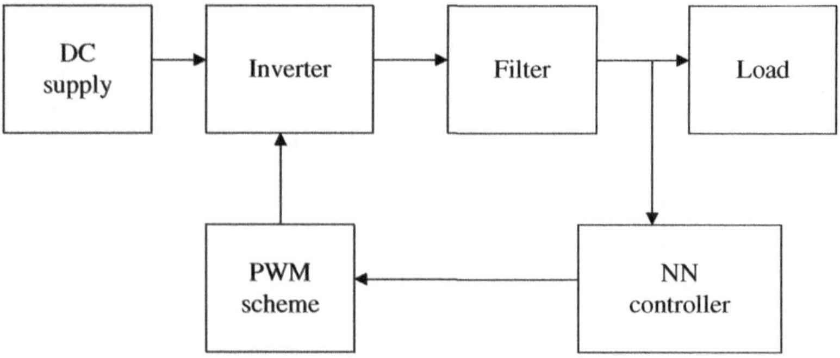

Figure 7.12 Proposed design.

Table 7.4 Design specifications.

Parameter	Value	Unit
Switching frequency, f_s	5	kHz
DC source voltage, V_{dc}	48	V
Rated output voltage	25	V_{rms}
Rated output frequency	50	Hz
Rated output current	5	A_{rms}
Rated output impedance	5	Ω
Filter inductor, L_f	250	mH
Inductor resistance, R_f	0.2	Ω
Filter capacitor, C_f	30	μF

load is connected to a single-phase full-bridge inverter that contains a filter at the output end. The suggested controller receives simultaneous measurements of the voltage and currents. The NN controller is the suggested controller in this case. For this, a microcontroller 16F877 is used. It was programmed to control the inverter switches.

The MOSFET IRF 840 is used to create a single-phase full-bridge inverter. Applications for MOSFETs include motor drivers, relay drivers, switching regulators, switching converters, and drivers for high-power bipolar switching transistors that need low gate drive power and high speed. Integrated circuits can be used to directly operate these kinds. Since the MOSFET cannot be driven by the signal from the pulse generator circuit, an isolation circuit, as illustrated in Figure 7.14, is necessary to provide isolation and enhance the pulses.

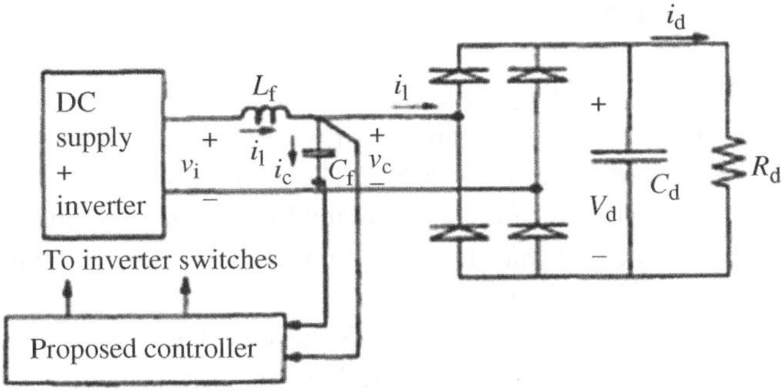

Figure 7.13 Implementation circuit.

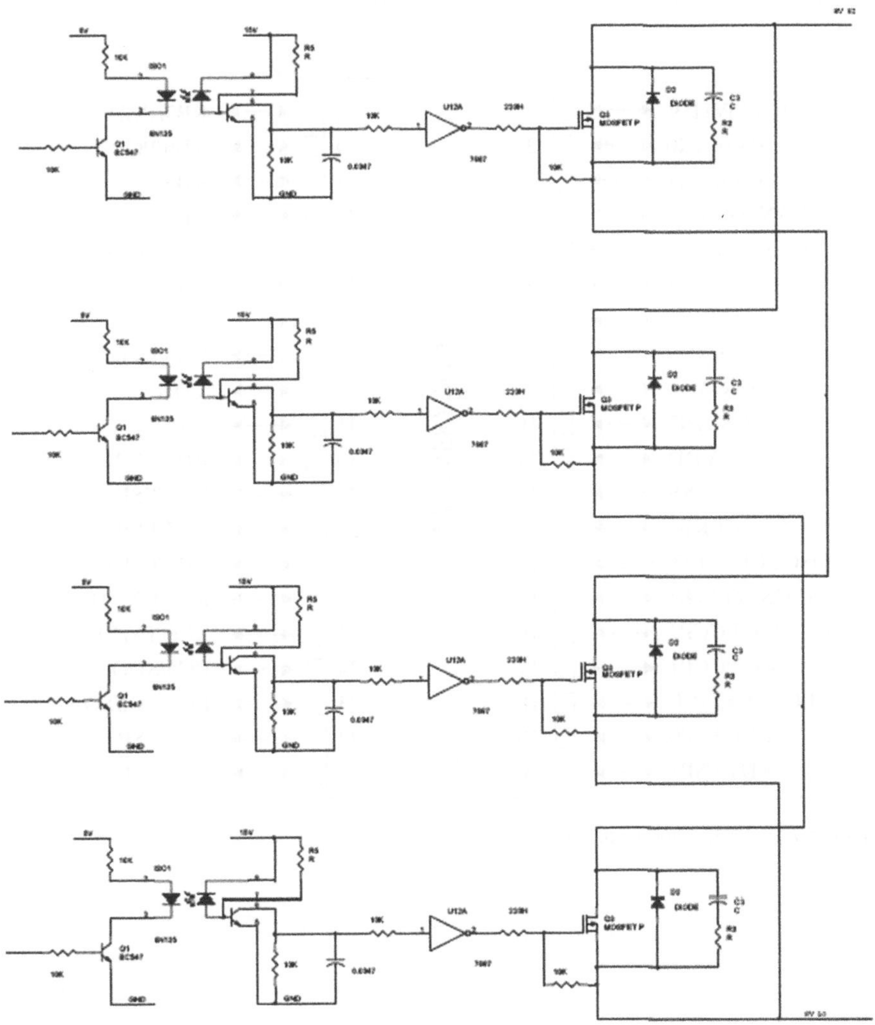

Figure 7.14 Isolation with MOSFET.

7.10 Microcontroller Unit

When a nonlinear load is employed, the microcontroller 16F877 is utilized to generate the switching angles for the inverter switches, resulting in a sinusoidal output voltage. Figure 7.15 shows the pin diagram for 16F877. The program's primary goal is to gather feedback as input so that it can produce switching angles based on the desired value. Figure displays the flowchart for the program that is integrated into the microcontroller (Figure 7.16).

The following figures display the hardware circuit's outputs. The SPWM program is run using the above-described execution mechanism, and the controller generates the SPWM pulses depicted in Figure 7.17. Figure 7.18 displays the output waveform that results from using a linear load.

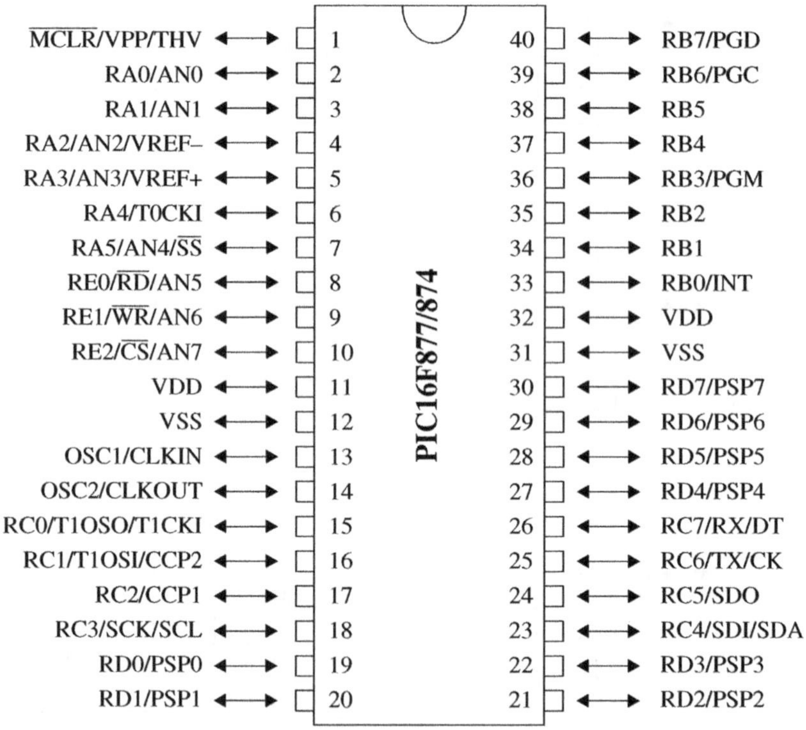

Figure 7.15 Pin diagram of 16F877.

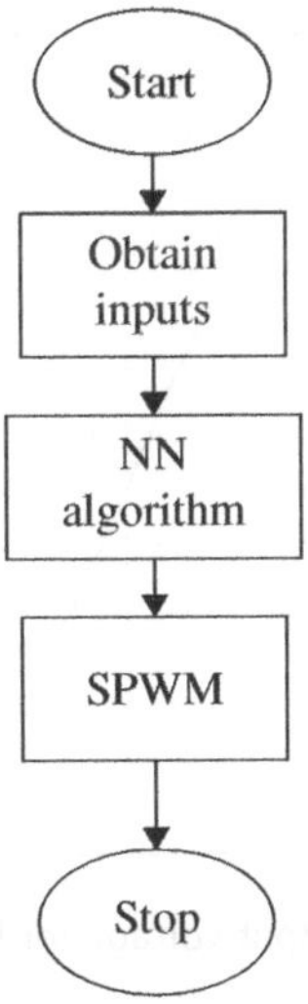

Figure 7.16 Flow chart.

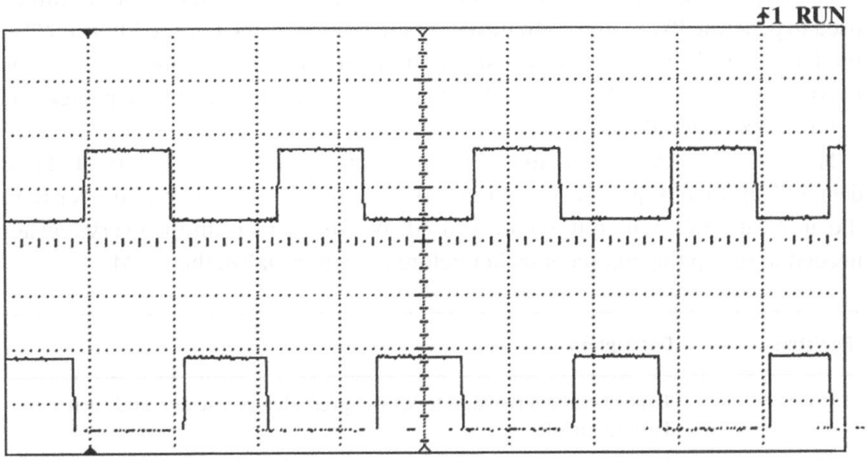

Figure 7.17 Pulses to the inverter switches.

The following waveforms, as seen in Figure 7.19, are formed when an NN controller is utilized with a nonlinear load of a complete bridge rectifier with a capacitor in parallel with a resistor. It demonstrates that even under nonlinear loading conditions, an NN controller generates a sinusoidal output voltage waveform.

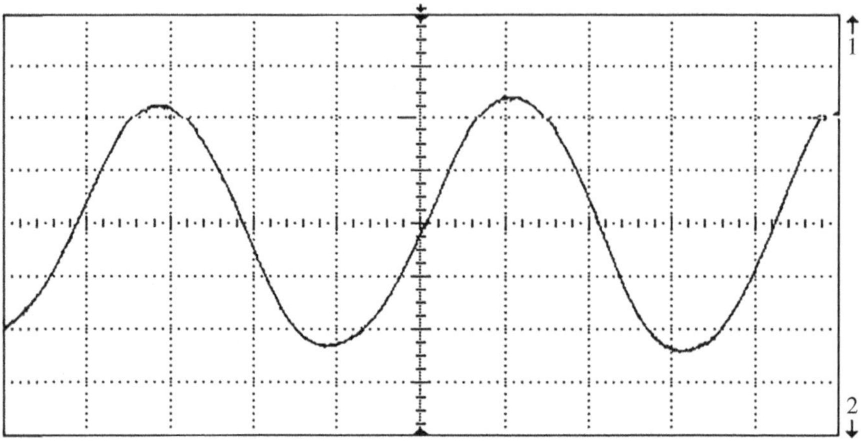

Figure 7.18 Output voltage for linear load.

The Capture, Compare, and PWM (CCP) module with a Timer module can be used to generate PWM signals in nearly all microcontrollers. For the PIC16F877A, the latter is utilized to generate a steady frequency between 1 and 208 kHz. The maximal resolution of the duty cycle will decrease with increasing frequency due to hardware constraints.

The PIC16F877A can maintain a maximum resolution of 10 bits (1024 in decimal) up to a frequency of around 20 kHz, which is significantly higher than the application for which it would be used. We must first examine every register needed to set up this microcontroller before we can initialize the PWM.

Registers	Description
T2CON	Timer2 control register, to set the prescaler, postscaler, and turn Timer2 on/off
PR2	Timer2 Period register, to set the max value of the counter
CCP1CON	This register controls the operations of CCP1
TRISC	This register is used to set PORTC as I/O

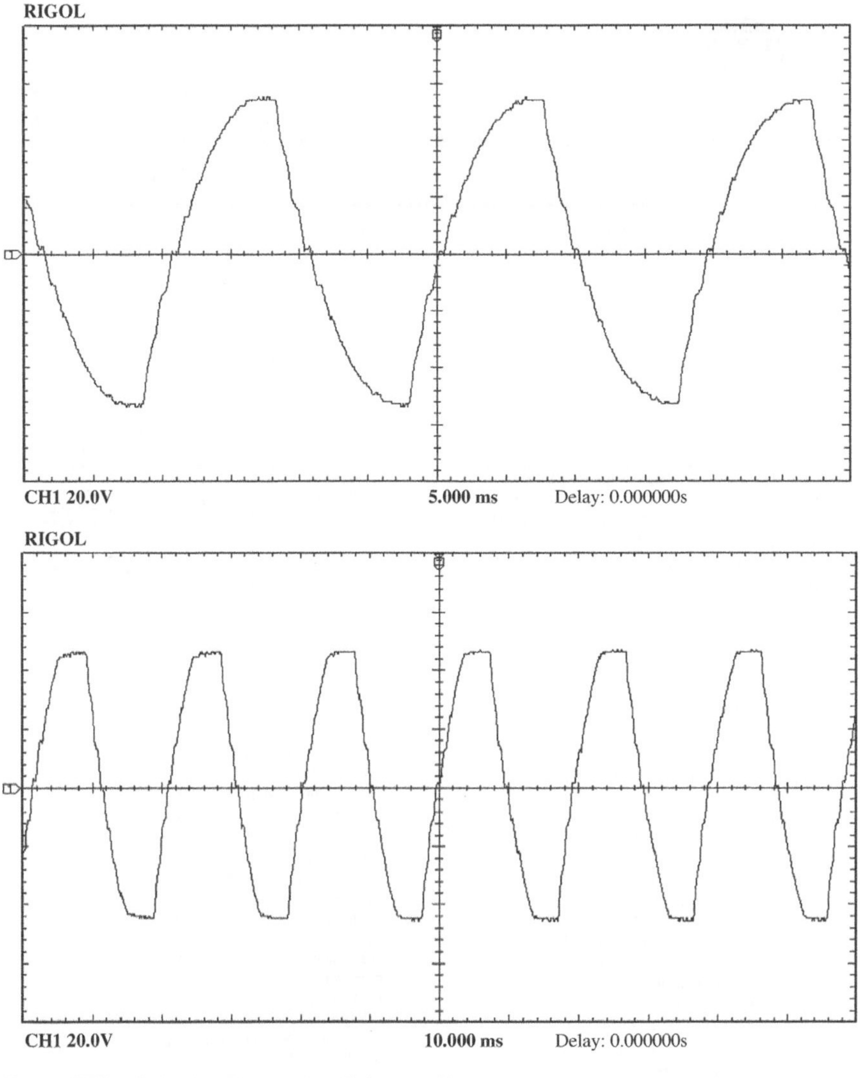

Figure 7.19 Output voltage using NN controller.

MICROCHIP

PIC16F87X

28/40-pin 8-Bit CMOS FLASH Microcontrollers

Devices Included in this Data Sheet:

- PIC16F873
- PIC16F874
- PIC16F876
- PIC16F877

Microcontroller Core Features:

- High-performance RISC CPU
- Only 35 single word instructions to learn
- All single cycle instructions except for program branches which are two cycle
- Operating speed: DC - 20 MHz clock input
  DC - 200 ns instruction cycle
- Up to 8K x 14 words of FLASH Program Memory,
  Up to 368 x 8 bytes of Data Memory (RAM)
  Up to 256 x 8 bytes of EEPROM data memory
- Pinout compatible to the PIC16C73B/74B/76/77
- Interrupt capability (up to 14 sources)
- Eight level deep hardware stack
- Direct, indirect and relative addressing modes
- Power-on Reset (POR)
- Power-up Timer (PWRT) and Oscillator Start-up Timer (OST)
- Watchdog Timer (WDT) with its own on-chip RC oscillator for reliable operation
- Programmable code-protection
- Power saving SLEEP mode
- Selectable oscillator options
- Low-power, high-speed CMOS FLASH/EEPROM technology
- Fully static design
- In-Circuit Serial Programming™ (ICSP) via two pins
- Single 5V In-Circuit Serial Programming capability
- In-Circuit Debugging via two pins
- Processor read/write access to program memory
- Wide operating voltage range: 2.0V to 5.5V
- High Sink/Source Current: 25 mA
- Commercial and Industrial temperature ranges
- Low-power consumption:
  - < 2 mA typical @ 5V, 4 MHz
  - 20 μA typical @ 3V, 32 kHz
  - < 1 μA typical standby current

Pin Diagram

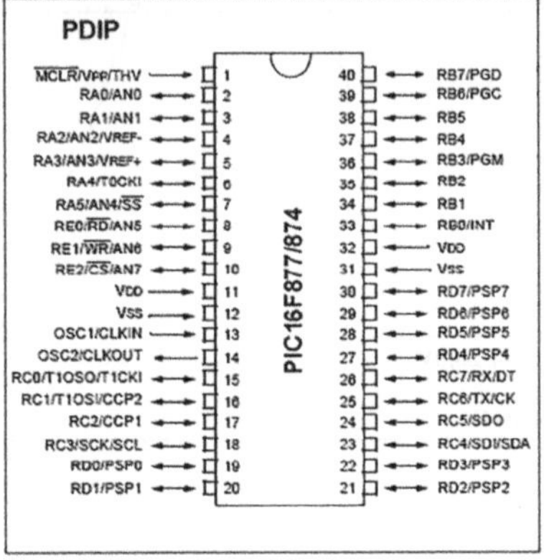

Peripheral Features:

- Timer0: 8-bit timer/counter with 8-bit prescaler
- Timer1: 16-bit timer/counter with prescaler, can be incremented during sleep via external crystal/clock
- Timer2: 8-bit timer/counter with 8-bit period register, prescaler and postscaler
- Two Capture, Compare, PWM modules
  - Capture is 16-bit, max. resolution is 12.5 ns
  - Compare is 16-bit, max. resolution is 200 ns
  - PWM max. resolution is 10-bit
- 10-bit multi-channel Analog-to-Digital converter
- Synchronous Serial Port (SSP) with SPI™ (Master Mode) and I²C™ (Master/Slave)
- Universal Synchronous Asynchronous Receiver Transmitter (USART/SCI) with 9-bit address detection
- Parallel Slave Port (PSP) 8-bits wide, with external RD, WR and CS controls (40/44-pin only)
- Brown-out detection circuitry for Brown-out Reset (BOR)

 DS30292B-page 1

PIC16F87X

Pin Diagrams

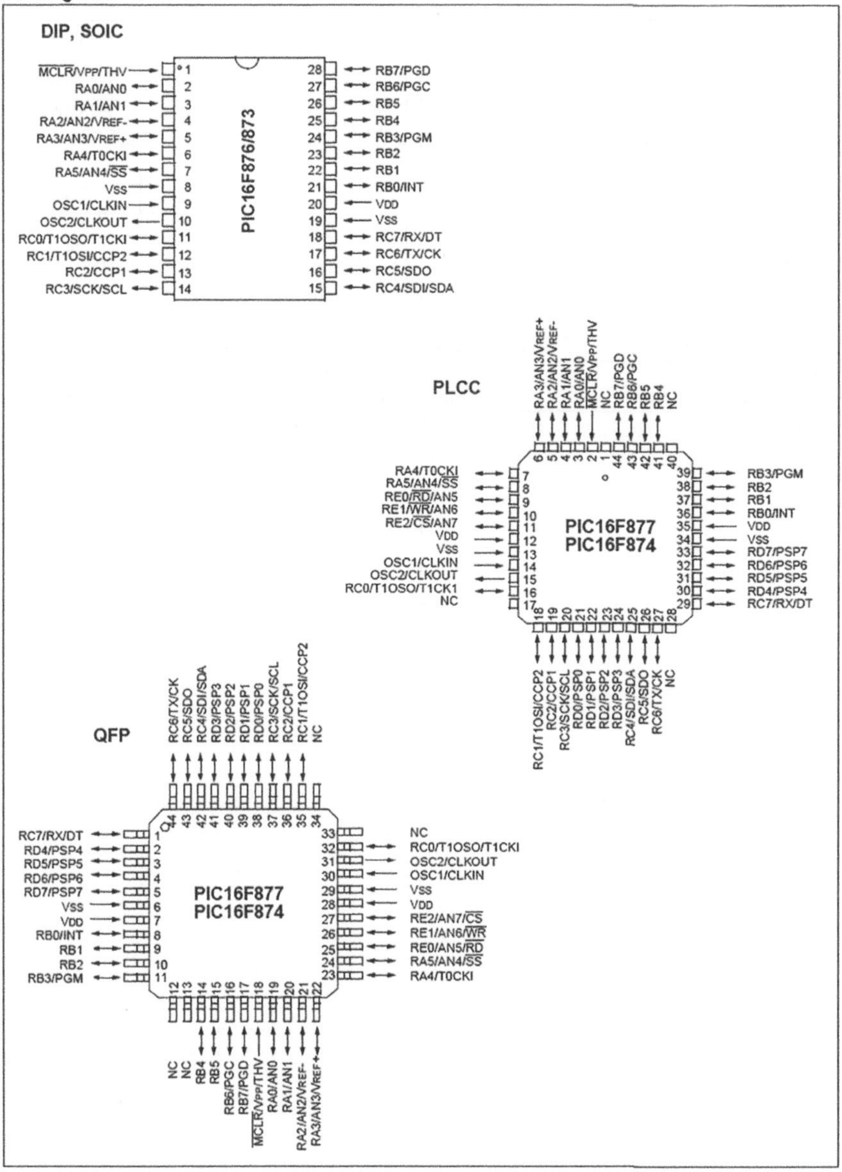

DS30292B-page 2

PIC16F87X

Key Features PICmicro™ Mid-Range Reference Manual (DS33023)	PIC16F873	PIC16F874	PIC16F876	PIC16F877
Operating Frequency	DC - 20 MHz	DC - 20 MHz	DC - 20 MHz	DC - 20 MHz
Resets (and Delays)	POR, BOR (PWRT, OST)	POR, BOR (PWRT, OST)	POR, BOR (PWRT, OST)	POR, BOR (PWRT, OST)
FLASH Program Memory (14-bit words)	4K	4K	8K	8K
Data Memory (bytes)	192	192	368	368
EEPROM Data Memory	128	128	256	256
Interrupts	13	14	13	14
I/O Ports	Ports A,B,C	Ports A,B,C,D,E	Ports A,B,C	Ports A,B,C,D,E
Timers	3	3	3	3
Capture/Compare/PWM modules	2	2	2	2
Serial Communications	MSSP, USART	MSSP, USART	MSSP, USART	MSSP, USART
Parallel Communications	—	PSP	—	PSP
10-bit Analog-to-Digital Module	5 input channels	8 input channels	5 input channels	8 input channels
Instruction Set	35 Instructions	35 Instructions	35 Instructions	35 Instructions

T2CON: TIMER2 CONTROL REGISTER (ADDRESS 12h)

U-0	R/W-0	R/W-0	R/W-0	R/W-0	R/W-0	R/W-0	R/W-0
—	TOUTPS3	TOUTPS2	TOUTPS1	TOUTPS0	TMR2ON	T2CKPS1	T2CKPS0
bit 7							bit 0

TOUTPS3:TOUTPS0: N/A for this application
TMR2ON: Timer2 On bit
1=Timer2 is on
0=Timer2 is off
T2CKPS1:T2CKPS0: Timer2 Clock Prescale Select bits
00 = Prescaler is 1
01 = Prescaler is 4
1× = Prescaler is 16

Sample Code:

```
unsigned int i;
// Lcd module connections
sbit LCD_RS at RB0_bit;
sbit LCD_EN at RB1_bit;
sbit LCD_D4 at RB2_bit;
sbit LCD_D5 at RB3_bit;
sbit LCD_D6 at RB4_bit;
sbit LCD_D7 at RB5_bit;

sbit LCD_RS_Direction at TRISB0_bit;
sbit LCD_EN_Direction at TRISB1_bit;
sbit LCD_D4_Direction at TRISB2_bit;
sbit LCD_D5_Direction at TRISB3_bit;
sbit LCD_D6_Direction at TRISB4_bit;
sbit LCD_D7_Direction at TRISB5_bit;
// End Lcd module connections
void main() {
   Lcd_Init();
   PWM1_Init(1000);
   PWM1_Start();
   Lcd_Cmd(_LCD_CLEAR);
   Lcd_Cmd(_LCD_CURSOR_OFF);
   Lcd_Out(1,1,"DEVELOPED BY");
   Lcd_Out(2,1,"MINA TECHNOLOGY");
   delay_ms(1000);
   Lcd_Cmd(_LCD_CLEAR);
```

```
Lcd_Out(1,7,"PWM");
Lcd_Out(2,1,"Duty Cycle:");
Lcd_Out(2,15,"%");
while(1){
    //duty cycle: duty ratio calculate (percent * 255) / 100
    // 0% duty cycle
    PWM1_Set_Duty(0);
    Lcd_Out(2,12," 0");
    delay_ms(2000);

    // 10% duty cycle
    PWM1_Set_Duty(25.5);
    Lcd_Out(2,12," 10");
    delay_ms(2000);

    // 20% duty cycle
    PWM1_Set_Duty(51);
    Lcd_Out(2,12," 20");
    delay_ms(2000);

    // 30% duty cycle
    PWM1_Set_Duty(76.5);
    Lcd_Out(2,12," 30");
    delay_ms(2000);

    // 40% duty cycle
    PWM1_Set_Duty(102);
    Lcd_Out(2,12," 40");
    delay_ms(2000);

    // 50% duty cycle
    PWM1_Set_Duty(127);
    Lcd_Out(2,12," 50");
    delay_ms(2000);

    // 60% duty cycle
    PWM1_Set_Duty(153);
    Lcd_Out(2,12," 60");
    delay_ms(2000);
```

```
// 70% duty cycle
PWM1_Set_Duty(178.5);
Lcd_Out(2,12," 70");
delay_ms(2000);

// 80% duty cycle
PWM1_Set_Duty(204);
Lcd_Out(2,12," 80");
delay_ms(2000);

// 90% duty cycle
PWM1_Set_Duty(229.5);
Lcd_Out(2,12," 90");
delay_ms(2000);

// 100% duty cycle
PWM1_Set_Duty(255);
Lcd_Out(2,12,"100");
delay_ms(2000);

// 90% duty cycle
PWM1_Set_Duty(229.5);
Lcd_Out(2,12," 90");
delay_ms(2000);

// 80% duty cycle
PWM1_Set_Duty(204);
Lcd_Out(2,12," 80");
delay_ms(2000);

// 70% duty cycle
PWM1_Set_Duty(178.5);
Lcd_Out(2,12," 70");
delay_ms(2000);

// 60% duty cycle
PWM1_Set_Duty(153);
Lcd_Out(2,12," 60");
delay_ms(2000);
```

```
        // 50% duty cycle
        PWM1_Set_Duty(127);
        Lcd_Out(2,12," 50");
        delay_ms(2000);

        // 40% duty cycle
        PWM1_Set_Duty(102);
        Lcd_Out(2,12," 40");
        delay_ms(2000);

        // 30% duty cycle
        PWM1_Set_Duty(76.5);
        Lcd_Out(2,12," 30");
        delay_ms(2000);

        // 20% duty cycle
        PWM1_Set_Duty(51);
        Lcd_Out(2,12," 20");
        delay_ms(2000);

        // 10% duty cycle
        PWM1_Set_Duty(25.5);
        Lcd_Out(2,12," 10");
        delay_ms(2000);
        }
    }
```

7.11 FPGA-Based Implementation of PWM

The simulation circuit and output voltage waveform (+10 V, 0 V, and −10 V) for a single-stage three-level CMLI are shown in Figures 7.20 and 7.21, respectively.

Similarly, the output voltage waveform (+20 V, +10 V, 0V, −10 V, and −20 V) for a two-stage five-level CMLI is shown in Figure 7.22. Figure 7.23 shows the output voltage waveform (+30 V, +20 V, +10 V, 0 V, −10 V, −20 V, and −30 V) for a three-stage seven-level CMLI. Figure 7.8 shows the output voltage waveform (+40 V, +30 V, +20 V, +10 V, 0 V, −10 V, −20 V, −30 V, and −40 V) for a four-stage nine-level CMLI. Figure 7.24 shows the output voltage waveform (+50 V, +40 V, +30 V, +20 V, +10 V, 0 V, −10 V, −20 V, −30 V, −40 V, and −50 V) for a 5-stage 11-level CMLI (Figure 7.25).

Following the simulation's successful operation, the hardware for the two-stage five-level CMLI was put into place. Two stages and five tiers of the hardware are

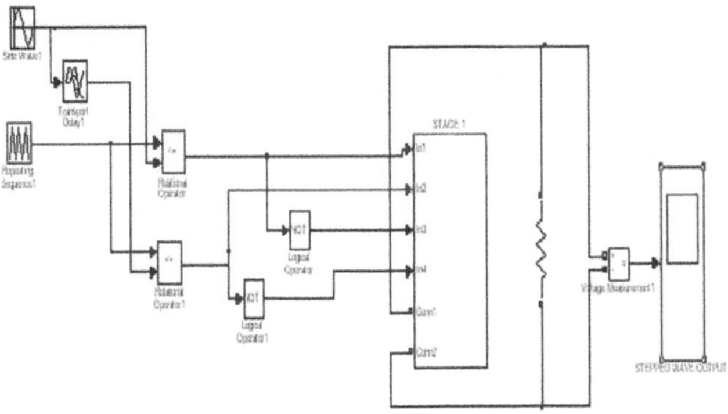

Figure 7.20 Power circuit for single-level CMLI.

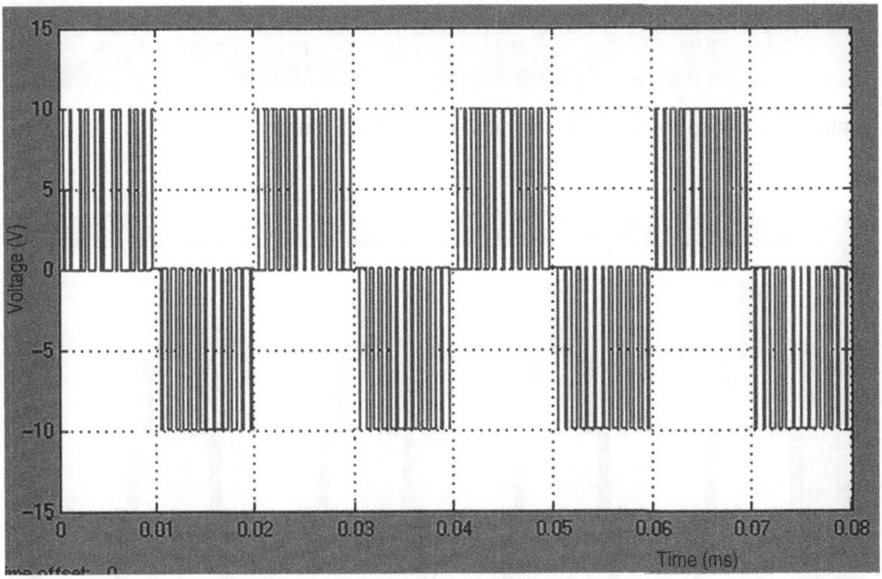

Figure 7.21 Output voltage waveform of a three-stage inverter.

implemented. The pulses for the CMLI are produced in hardware using an FPGA Spartan 3E.

Unlike the other two types of multilevel inverters, the cascaded multilevel inverter's power circuit does not use any unique design or distinct kinds of connections. Four switches are used in a bridge style in the power circuit. The power switches were IRF840 MOSFETs. The rectifier unit and transformer supply the DC voltage of 18 V. The transformers have a 230 V/18 V rating. Here, AC is

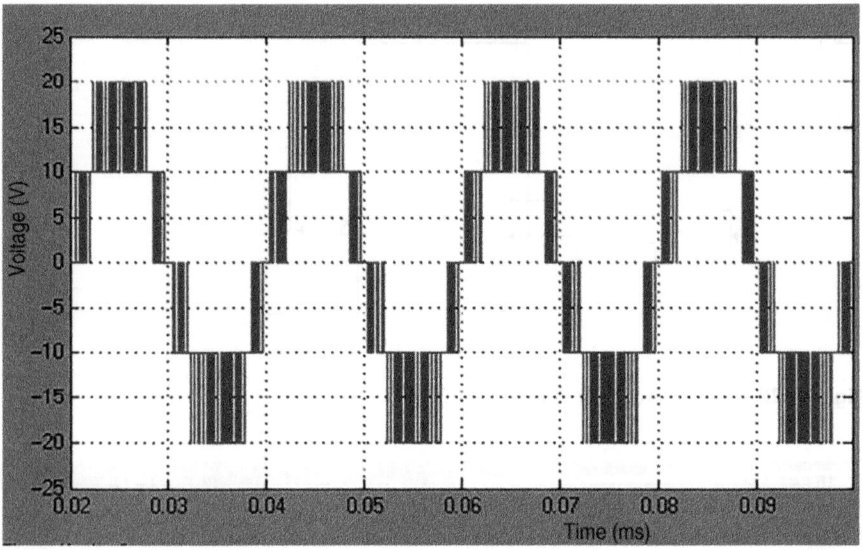

Figure 7.22 Output voltage waveform of a five-level inverter.

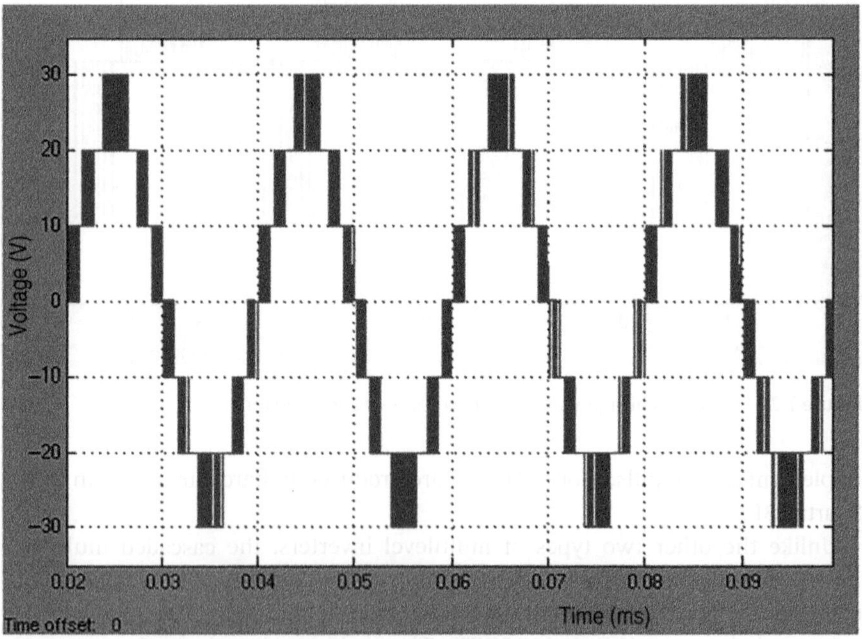

Figure 7.23 Output voltage waveform of a seven-level inverter.

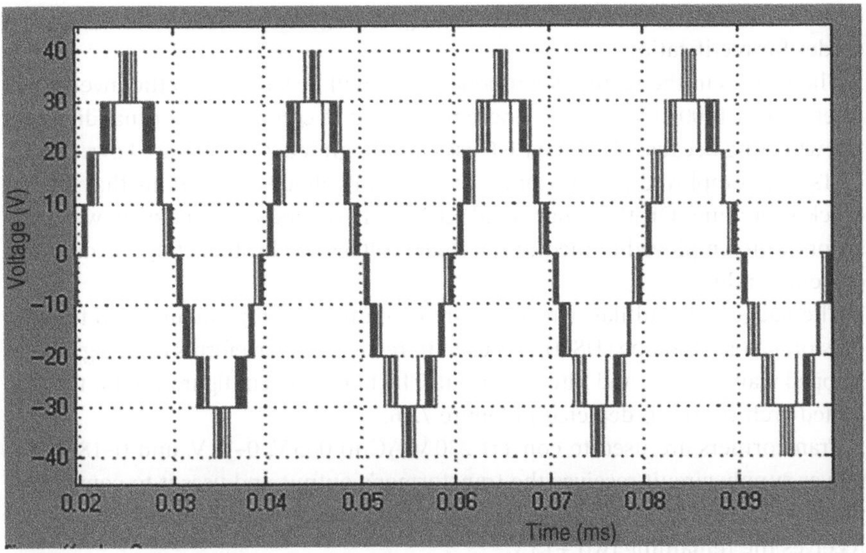

Figure 7.24 Output voltage waveform of a nine-level inverter.

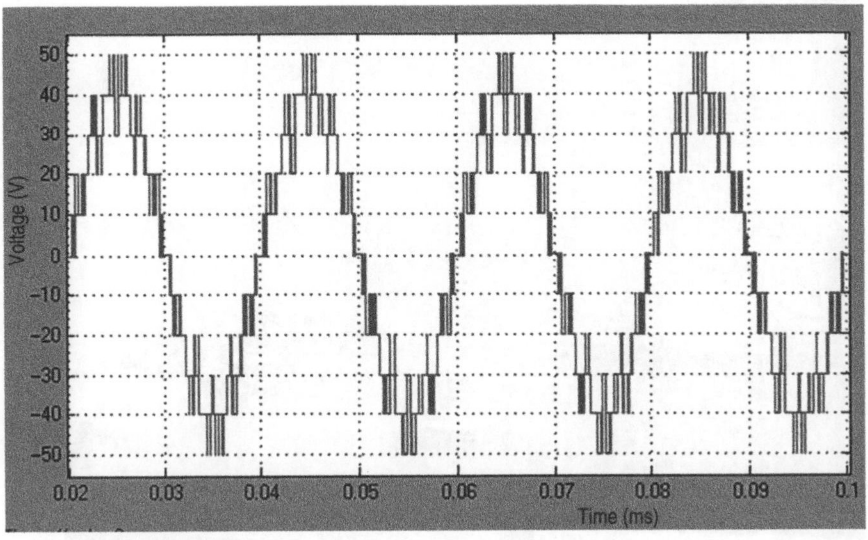

Figure 7.25 Output voltage waveform of an 11-level inverter.

converted into a unidirectional wave using a bridge IC BR W10 rather than four diodes for rectification.

The ripples in the rectified component that will be supplied to the inverter are filtered by a 1000 μF capacitor. There is no need to connect an external diode for freewheeling because the MOSFET uses an integrated Zener diode. Power MOSFETs were employed in low-voltage or medium-voltage applications that needed quick switching. On the other hand, GTOs, thyristors, and bipolar power transistors were employed in medium- to high-voltage applications that needed high current conduction.

The suggested multilayer inverter architecture was implemented, and a digital storage oscilloscope (DSO) was used to record the waveforms of firing pulses, stepped wave output, and filtered output. The hardware configuration for the suggested architecture is depicted in Figure 7.26.

Transformers are used to convert 230 V AC to 0–5 V, 0–18 V, and 0–18 V. The power supply circuit receives the transformer's output and uses it to convert AC to DC +5 V, +15 V, and +15 V. The FPGA receives this +5 V, and each inverter receives the remaining two +15 V.

The FPGA kit drives the firing pulse for MOSFETs used in two inverter circuits. Inverters 1 and 2 get their DC input from a DC power supply circuit or a photovoltaic cell. The circuit of the inverter produced a stepped output voltage wave

Figure 7.26 Hardware setup for a five-level inverter.

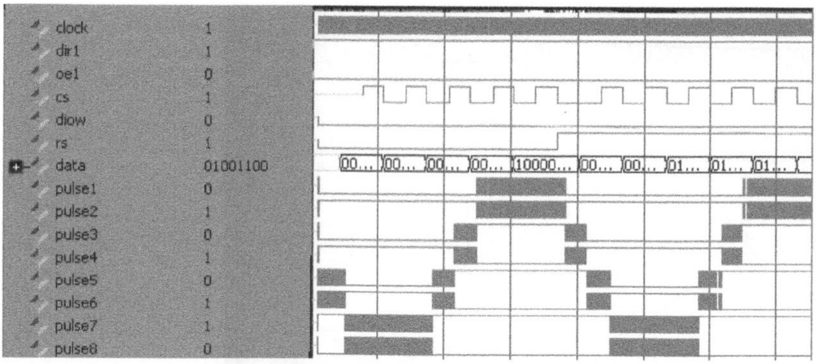

Figure 7.27 Gating pulses generated by the FPGA.

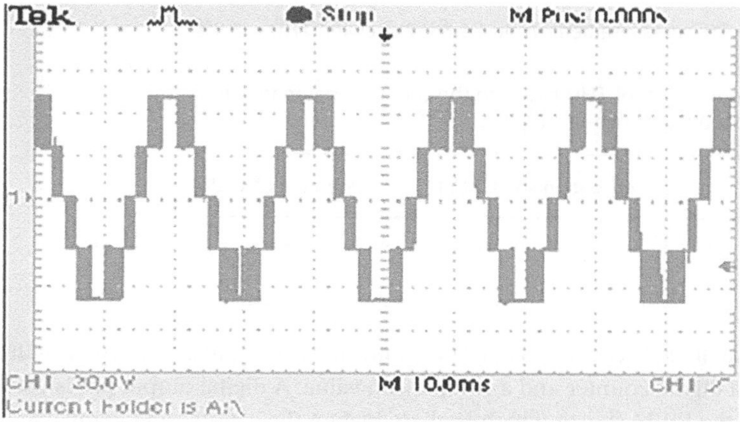

Figure 7.28 Output voltage waveform of a five-level CMLI.

shape. The DSO records this stepped wave. A power quality analyzer (C.A. 8332) is used to measure the THD for hardware setups. The discovered THD level is 3.9% (Figures 7.27, 7.28 and 7.29).

Table 7.5 shows the comparison between the conventional and proposed systems for a five-level cascaded multilevel inverter. PMW is able to cope with the variation of the load that is experienced by the power generation system.

7.12 DSPACE-Based PWM Generation

DSPACE-based PWM generation is the process of creating and controlling PWM signals for applications such as motor control using Simulink and dSPACE

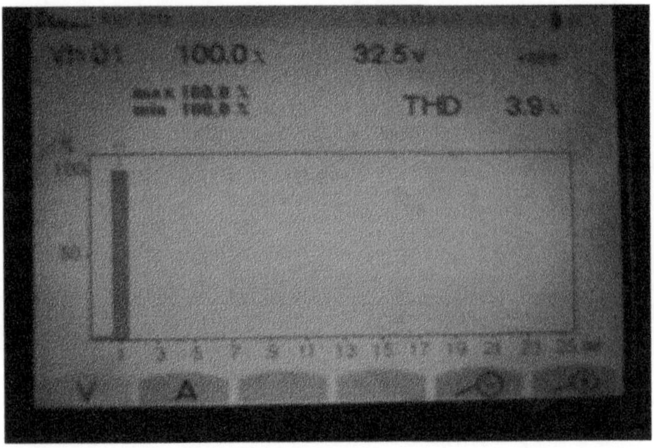

Figure 7.29 THD measurement using CA 8332 showing THD = 3.9%.

Table 7.5 Comparison of THD between the conventional model and the proposed model for five-level CMLI.

THD	Conventional CMLI (%)	Proposed CMLI (%)
With filter	1.6	0.13
Without filter	29.26	27.44

hardware, such as FPGA boards or DSP controllers. The pulse width is usually defined by a digital counter and a comparison value. A digital output pin is used to generate the PWM signal, which is altered when the counter and comparison value coincide.

7.13 MicroLab Box and dSPACE

The equipment MicroLab box is an excellent product for real-time output using MATLAB. The software component of this package, called dSPACE, aids in connecting the hardware section (MicroLab Box) to the user and interfaces it based on user input.

7.13.1 About MicroLab Box

- Compact all-in-one development system for laboratory purposes
- Dual-core real-time processor at 2 GHz

- User-programmable FPGA
- More than 100 channels of high-performance I/O
- Dedicated electric motor control features
- Ethernet and CAN bus interfaces
- Easy I/O access via integrated connector panel

7.13.2 Application Areas

The MicroLab Box is a laboratory development device that combines great performance and diversity with small size and affordability. MicroLab Box facilitates the rapid and simple setup of control, test, or measurement applications and aids in the realization of novel control concepts. MicroLab Box is an adaptable system that can be utilized in mechatronic research and development fields such as robotics, medical engineering, electric drive control, renewable energy, vehicle engineering, or aerospace. It has more than 100 distinct types of I/O channels.

7.13.3 Key Benefits

Excellent real-time performance is achieved by combining high processing power with extremely low I/O latencies. A programmable FPGA offers a great deal of versatility and enables the execution of even very quick control loops, which are necessary in applications like active noise and vibration cancellation or electric motor control. A complete dSPACE software package, such as real-time interface (RTI) for Simulink for model-based I/O integration and the experiment software Control Desk®, which offers graphical instrument access to the real-time application during run time, supports MicroLab Box (Figure 7.30).

7.14 Real-Time Interface (RTI) Using MicroLab box

RTI lets us concentrate fully on the actual design process and carry out fast design iterations. It extends the C code generator Simulink Coder™ (formerly Real-Time Workshop®) for the seamless, automatic implementation of your Simulink and Stateflow models on real-time hardware.

7.15 Working with RTI

Simply drag the I/O module from the RTI block library onto the model and attach it to the Simulink blocks to connect it to a dSPACE I/O board. By selecting the relevant blocks, you can access all of the parameters, including parameterization.

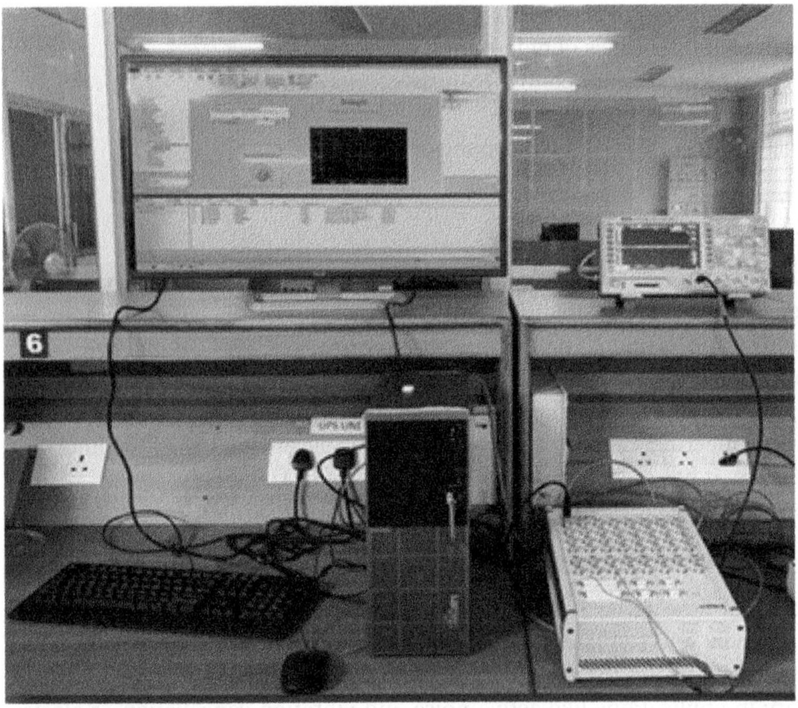

Figure 7.30 MicroLab box.

RTI offers blocks that implement the I/O capabilities of dSPACE systems in Simulink models, preparing the model for the real-time application, while Simulink Coder™ (previously Real-Time Workshop®) creates the model code. Without writing a single line of code, the real-time hardware automatically compiles, downloads, and launches your real-time model. RTI facilitates the configuration process. Potential mistakes can be found and fixed either before or during the construction process thanks to RTI's consistency checks.

7.16 Summary

This chapter presented a comprehensive study of Pulse Width Modulation (PWM) generation techniques and their experimental implementation using various hardware and software platforms. The operation and features of the TL494 PWM controller and IR2110 gate driver were discussed in detail, demonstrating their effectiveness in generating and driving PWM signals for power electronic applications. PWM generation using microcontrollers, FPGA platforms, DSP systems, dSPACE, and MicroLab Box was also explored, highlighting the flexibility and scalability of modern control techniques.

Simulation and hardware case studies showed that PWM-based control significantly improves inverter performance by producing high-quality output waveforms with reduced harmonic distortion. The implementation of neural network-based control demonstrated superior performance under nonlinear loading conditions by maintaining a nearly sinusoidal output voltage. FPGA-based multilevel inverter implementations further validated the effectiveness of advanced PWM strategies in achieving lower Total Harmonic Distortion (THD) and improved power quality.

Overall, the experimental investigations confirm that PWM is a powerful and versatile technique for controlling power converters and inverters. The integration of intelligent controllers, programmable hardware, and real-time development platforms enables the design of efficient, reliable, and high-performance power electronic systems suitable for a wide range of industrial, renewable energy, motor drive, and UPS applications.

Index

h

i

l

Printed and bound by CPI Group (UK) Ltd, Croydon, CR0 4YY

04/08/2026

14935740-0004